Conference Proceedings of the Society for Experimental Mechanics Series

For other titles published in this series, go to
www.springer.com/series/8922

Tom Proulx
Editor

Dynamics of Civil Structures, Volume 4

Proceedings of the 28th IMAC, A Conference on Structural Dynamics, 2010

Springer

Editor
Tom Proulx
Society for Experimental Mechanics, Inc.
7 School Street
Bethel, CT 06801-1405
USA
tom@sem1.com

ISSN 2191-5644 e-ISSN 2191-5652
ISBN 978-1-4419-9830-9 e-ISBN 978-1-4419-9831-6
DOI 10.1007/978-1-4419-9831-6
Springer New York Dordrecht Heidelberg London

Library of Congress Control Number: 2011926995

Printed on acid-free paper

Springer is part of Springer Science+Business Media (www.springer.com)

Preface

Dynamics of Civil Structures represents one of five clusters of technical papers presented at the 28th IMAC, A Conference and Exposition on Structural Dynamics, 2010 organized by the Society for Experimental Mechanics, and held at Jacksonville, Florida, February 1-4, 2010. The full proceedings also include volumes on Structural Dynamics and Renewable Energy, Nonlinear Modeling and Applications, Dynamics of Bridges, and Structural Dynamics.

Each collection presents early findings from experimental and computational investigations on an important area within Structural Dynamics. The current volume on *Dynamics of Civil Structures* includes studies on Modeling and Estimation of Loads on Civil Structures, Model Correlation and Updating (Civil Structures), Modal Parameter Identification (Civil Structures), Damage Detection and Modeling (Civil Structures), Control of Civil Structures/Stadium Dynamics, and Experimental Techniques (Civil Engineering).

Understanding the dynamic response of large civil structures improves design and safety, extends life, and reduces maintenance. IMAC has become a principle technical venue for dissemination of the latest techniques devoted to field testing of civil structural systems and components, processing of response data and identification of dynamic structural properties, calibration and validation of numerical structural models, and assessment of structural condition based on dynamic properties.

The organizers would like to thank the authors, presenters, session organizers and session chairs for their participation in this track.

Bethel, Connecticut

Dr. Thomas Proulx
Society for Experimental Mechanics, Inc

Contents

Proceedings of the IMAC-XXVIII
February 1–4, 2010, Jacksonville, Florida USA

Characterization of a Strongly Nonlinear Laboratory Benchmark System

D. Michael McFarland
Department of Aerospace Engineering
University of Illinois at Urbana-Champaign
Urbana, IL 61801
dmmcf@illinois.edu

Alexander F. Vakakis
Department of Mechanical Science and Engineering
University of Illinois at Urbana-Champaign
Urbana, IL 61801
avakakis@illinois.edu

Lawrence A. Bergman
Department of Aerospace Engineering
University of Illinois at Urbana-Champaign
Urbana, IL 61801
lbergman@illinois.edu

Timothy J. Copeland
m+p international, inc.
5375 Old Jacksonville Road
Springfield, IL 62711
tim.copeland@mpina.com

NOMENCLATURE

A	cross-sectional area
E	Young's modulus
F	static restoring force
L	wire half-span
T_0	wire initial tension
T	wire tension due to deformation
c	viscous damping coefficient
f	dynamic restoring force
k_1, k_2	linear spring stiffness
k_{lin}	coefficient of wire linear stiffness
k_{nl}	coefficient of wire nonlinear stiffness
m_1, m_2	masses of 2-DOF rig
x	wire static displacement
u	displacement of isolated mass
α	exponent of fitted nonlinear stiffness term
ϵ	strain in wire due to transverse displacement
$(\dot{\ })$	time differentiation

T. Proulx (ed.), *Dynamics of Civil Structures, Volume 4*, Conference Proceedings of the Society for Experimental Mechanics Series 13,
DOI 10.1007/978-1-4419-9831-6_1, © The Society for Experimental Mechanics, Inc. 2011

ABSTRACT

A two-degree-of-freedom test structure originally developed to exhibit essentially nonlinear (nonlinearizable) stiffness in one of its connections has been modified to display a variable linear term in parallel with the existing nonlinear spring. This structure will be used in generating data for input to new algorithms for nonlinear system identification. In this paper, we report results from experiments in controlling the linear term introduced by applying an initial tension (preload) to the nonlinear spring, including the estimation of the parameters of a nonlinear force-displacement relation. It is shown that a sufficiently large initial tension results in predominantly linear behavior, meaning the same test setup can be used to produce dynamic behavior ranging from nearly linear to essentially nonlinear.

1 Introduction

While the identification of models of linear structures from experimental data has become routine, the identification of nonlinear systems remains a much more challenging problem. To support recent theoretical and applied work reported in Refs. [1, 2, 3, 4, 5, 6, 7], we have developed a simple test structure in which the effects of nonlinear stiffness can be adjusted. This is achieved by varying one of the linear stiffness coefficients, from zero to a value large enough to reduce the influence of the nonlinear stiffness to a parasitic role.

The model structure taken as a starting point here was originally designed as a linear two-degree-of-freedom (2-DOF) model with an added essentially nonlinear spring (that is, a spring with no linear term in its force-displacement relation) between one of the structural masses and ground. That essentially nonlinear spring was formed from transversely deflected piano wire, and the essential nonlinearity was achieved only if the initial tension in the wire was zero. Several experiments were reported using this configuration and a similar arrangement in which the wire was connected between moving masses rather than between a mass and ground [4, 5, 6, 7]. In the present work, we exploit the linear stiffness terms introduced when a significant initial tension (preload) is established in the wire in its undeformed state.

2 Modeling

The geometry of the piano-wire spring is show in Fig. 1. The wire, of cross-sectional area A, Young's modulus E and total length $2L$, is deflected a distance x at its center in response to a load F. The resulting length of half of the wire is $\sqrt{L^2+x^2}$, and assuming linear material response we can compute the resulting increase in tension of the wire due to this stretching as

$$T = EA\epsilon \tag{1}$$

where

$$\epsilon = \frac{\sqrt{L^2+x^2}-L}{L} \tag{2}$$

is the (engineering) strain. Denoting by T_0 the initial tension in the wire (before deformation), we can express the equilibrium of the wire as

$$F = 2(T_0+T)\sin\theta, \tag{3}$$

where θ is the angle formed by the wire with its initial position and thus

$$\sin\theta = \frac{x}{\sqrt{L^2+x^2}}. \tag{4}$$

Substituting, we obtain an expression for the force F in terms of the transverse displacement x,

$$F = 2\left(T_0 + EA\frac{\sqrt{L^2+x^2}-L}{L}\right)\frac{x}{\sqrt{L^2+x^2}}. \tag{5}$$

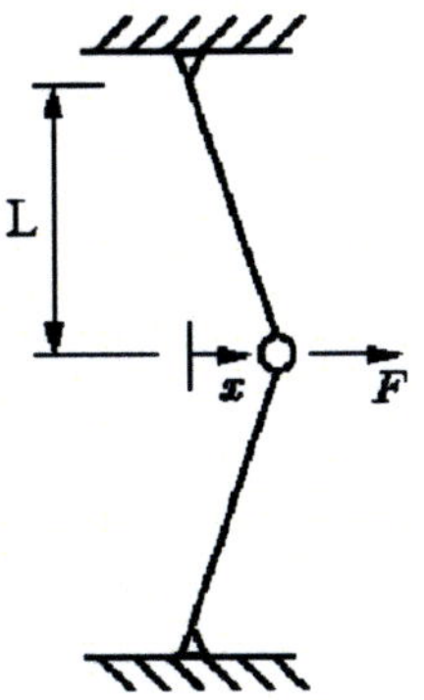

Fig. 1. Geometry of the deflected piano wire providing the nonlinear and variable linear stiffness between the second (NES) mass and ground.

This expression, while exact, is not convenient to work with. If x is small (compared to the half-length L), we can expand $F(x)$ in a Taylor series about $x = 0$, with the result

$$F(x) \approx \frac{2T_0}{L}x - \frac{T_0 - EA}{L^3}x^3 + \frac{3(T_0 - EA}{4L^5}x^5 + O(x^7). \tag{6}$$

In practice, it is sufficient to retain only the leading terms of this expression to represent the force-displacement relation of a real spring,

$$F(x) = \frac{2T_0}{L}x - \frac{T_0 - EA}{L^3}x^3, \tag{7}$$

from which form it is easy to see that the linear term will exist only when there is a preload in the wire. When $T_0 > 0$, we obtain the simple expression

$$F(x) = \frac{EA}{L^3}x^3, \tag{8}$$

which has been used successfully to represent this type of essentially nonlinear spring in several experiments.

3 Experiments and Data Reduction

The physical system incorporating two linear degrees of freedom along with the piano-wire spring of the previous section is show schematically in Fig. 3, where the wire connects the second mass to ground. Both masses are made of aluminum angle stock and run on a straight air track, connected by linear leaf springs. The laboratory model, set up atop an optical table and instrumented with force transducers and accelerometers, is shown in the photograph of Fig. 2. The wire was in this case 0.020 in. in diameter and of total length $2L = 20.0$in.

The parameters of the linear system resulting when the wire spring was disconnected were determined by standard techniques of linear modal analysis, and are summarized in Table 1. This structure is very similar to that described in Ref. [3], where is was configured with the wire installed with zero initial tension and thus providing an essentially nonlinear restoring force. The focus of that work was targeted energy transfer ("energy pumping") from the primary mass to the secondary (the nonlinear energy sink, or NES), a phenomenon that depends on the essential nonlinearity.

m_1, kg	m_2, kg	c_1, N/m/s	c_2, N/m/s	k_1, N/m	k_2, N/m
1.288	518	0.112	0.079	1228	680

Table 1. Identified values of the parameters of the linear 2-DOF structure (with the piano-wire spring removed).

The nonlinear spring was identified here using the restoring force surface method, in which the ring-down response of (a portion of) the structure is fitted with a model equation including viscous linear damping as well as

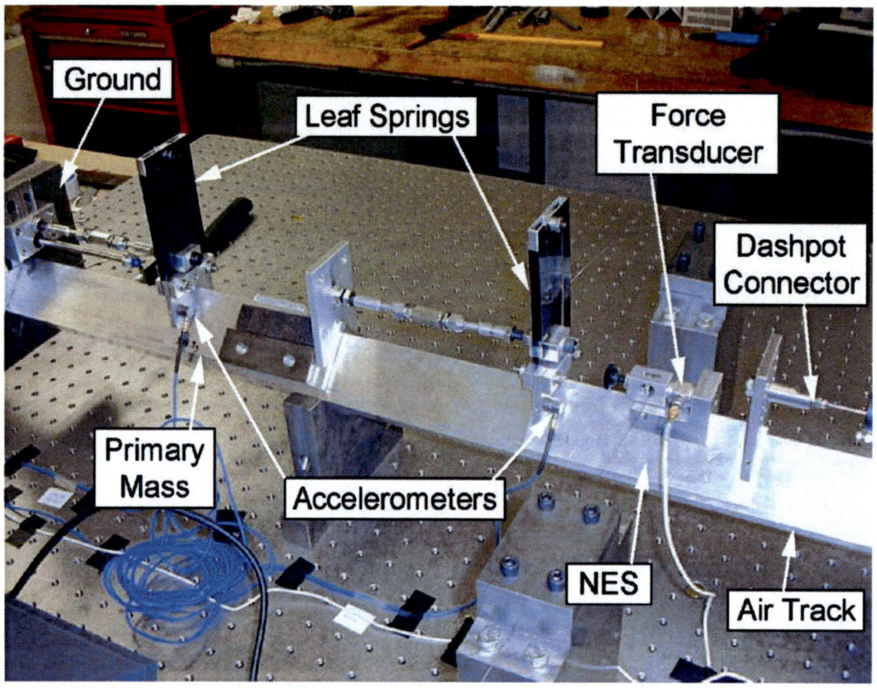

Fig. 2. The 2-DOF system of Fig. 3 realized as two cars running on a rectilinear air track. The car labeled "NES" here is the secondary mass, connected to ground by both nonlinear and (with preload on the piano wire) linear springs.

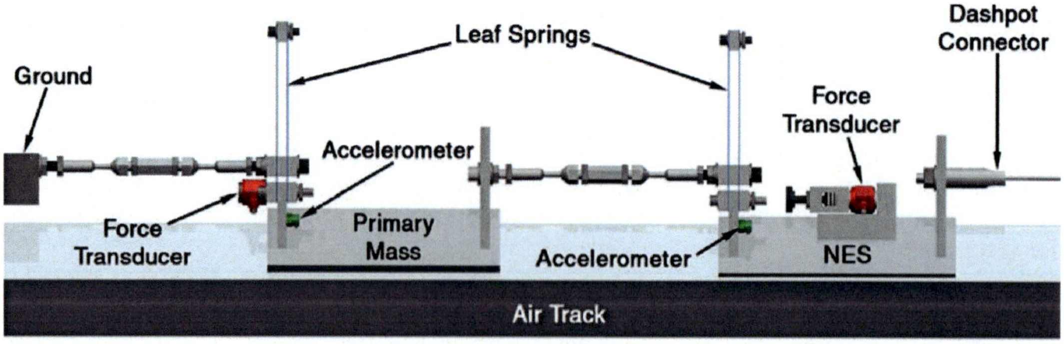

Fig. 3. Schematic of 2-DOF system with variable linear stiffness in parallel with cubic-hardening spring between second (right-hand) mass and ground. The nonlinear, piano-wire spring is normal to the page in this view.

linear and nonlinear (cubic) terms. Specifically, we assumed the restoring force was of the form

$$f = c\dot{u} + k_{\text{lin}}u + k_{\text{nl}}|u|^{\alpha} \operatorname{sgn}\ u, \tag{9}$$

where u is the displacement of the mass attached to the wire spring, $\dot{u}$ is its velocity, c is the damping coefficient, k_{lin} and k_{nl} are the coefficients of the linear and nonlinear stiffness terms, and α is the exponent of the nonlinear term (nominally equal to 3 for a purely cubic spring). The results obtained with several values of initial tension T_0 are reported in Table 2, where it may be seen that the results of increasing preload are generally in agreement with the predictions of eq. (7). A linear term does appear even when $T_0 = 0$, reflecting the finite bending stiffness of the piano wire, but insofar as our goal here was to add a significant linear stiffness to an essentially nonlinear spring this linear term is, if anything, beneficial.

Preload, lb	c, N/m/s	k_{lin}, N/m	k_{nl}, N/m$^{\alpha}$ $\times 10^3$	α	k_{lin}, N/m
5.0	1.02	245	1138	2.73	—
7.5	1.03	355	891	2.70	—
10.0	1.44	584	50	2.07	—
20.0	3.04	1507	-0.073	0.72	1335

Table 2. Parameters for the nonlinear model of eq. (7), estimated with various preloads applied to the piano-wire spring.

4 Conclusion

A structure used in a series of bench-top experiments comprises a linear oscillator of one degree of freedom coupled to a strongly nonlinear, single-degree-of-freedom substructure. The nonlinear stiffness of the attachment, created by the transverse deflection of a piano-wire spring, is essential (lacking a linear part) if the wire is installed with no pretension. If the wire is preloaded with an initial tension, the spring will exhibit a linear term in its force-displacement relation in addition to a strong cubic term and generally negligible higher-order terms.

While in previous work we have sought to minimize the pretension on the wire spring and thus produce a nearly pure, cubic-hardening spring characteristic, here we intentionally apply a significant preload to create a linear restoring force comparable to the nonlinear spring force for displacements of the order observed in tests on the combined system. Taken as a whole, the structure displays strong, controllable, but not essential nonlinearity. We have demonstrated that the linear term produced by the pretension can be varied and can in fact be used to produce essentially nonlinear, strongly nonlinear, or effectively linear dynamics in the same test rig. In future work, this variability will be used to generate experimental data for input to nonlinear system identification algorithms that do not require isolating the nonlinear degree(s) of freedom.

Acknowledgments

This work was supported in part by U.S. National Science Foundation Grant Number 0927995. Mr. Xing Wang and Ms. Mercedes Mane performed the experiments described herein. The nonlinear parameter identification was done by Mr. Sean Hubbard.

References

[1] Jiang, X., D. M. McFarland, L. A. Bergman and A. F. Vakakis, "Steady State Passive Nonlinear Energy Pumping in Coupled Oscillators: Theoretical and Experimental Results," *Nonlinear Dynamics,* 33(1):87102, 2003.

[2] Vakakis, A. F., D. M. McFarland, L. A. Bergman, L. I. Manevitch and O. Gendelman, "Isolated Resonance Captures and Resonance Capture Cascades Leading to Single- or Multi-Mode Passive Energy Pumping in Damped Coupled Oscillators," *ASME Journal of Vibration and Acoustics,* 126(2):235244, April 2004.

[3] McFarland, D. M. , L. A. Bergman and A. F. Vakakis, "Experimental Study of Nonlinear Energy Pumping Occurring at a Single Fast Frequency," *International Journal of Nonlinear Mechanics,* 40(6):891899, July 2005.

[4] McFarland, D. M., G. Kerschen, J. J. Kowtko, L. A. Bergman and A. F. Vakakis, "Experimental Investigation of Targeted Energy Transfers in Strongly and Nonlinearly Coupled Oscillators," *Journal of the Acoustical Society of America,* 118(2):791799, August 2005.

[5] Kerschen, G., D. M. McFarland, J. J. Kowtko, Y. S. Lee, L. A. Bergman and A. F. Vakakis, "Experimental Demonstration of Transient Resonance Capture in a System of Two Coupled Oscillators with Essential Stiffness Nonlinearity," *Journal of Sound and Vibration,* 299(45):822838, 2007.

[6] Kerschen, G., J. J. Kowtko, D. M. McFarland, L. A. Bergman and A. F. Vakakis, "Theoretical and Experimental Study of Multimodal Targeted Energy Transfer in a System of Coupled Oscillators," *Nonlinear Dynamics,* 47(13):285309, 2007.

[7] Vakakis, A. F., O. Gendelman, L. A. Bergman, D. M. McFarland, G. Kerschen and Y. S. Lee, *Nonlinear Targeted Energy Transfer in Mechanical and Structural Systems,* Volume 156 of the series Solid Mechanics and Its Applications, Springer-Verlag, New York, 2008.

[8] Kerschen, G., A. F. Vakakis, Y. S. Lee, D. M. McFarland and L. A. Bergman, "Toward a Fundamental Understanding of the Hilbert-Huang Transform in Nonlinear Structural Dynamics," *Journal of Vibration and Control,* 14(12):77105, 2008.

[9] Lee, Y. S., A. F. Vakakis, L. A. Bergman, D. M. McFarland, G. Kerschen, F. Nucera, S. Tsakirtzis and P. N. Panagopoulos, "Passive Nonlinear Targeted Energy Transfer (TET) and Its Applications to Vibration Absorption: A Review," Proceedings of the Institution of Mechanical Engineers, Part K: *Journal of Multi-body Dynamics,* 222(2):77134, 2008.

[10] Y. S. Lee, A. F. Vakakis, D. M. McFarland and L. A. Bergman, "Time-Domain Nonlinear System Identification of the Dynamics of Aeroelastic Instability Suppression Based on Targeted Energy Transfers ," *The Aeronautical Journal,* in press.

Proceedings of the IMAC-XXVIII
February 1–4, 2010, Jacksonville, Florida USA

A non-destructive technique for the health monitoring of tie-rods in ancient buildings

R. Garziera, L. Collini

Dipartimento di Ingegneria Industriale, Università di Parma
Viale G.P. Usberti 181/A, Parma, 43100 Italy. rinaldo.garziera@unipr.it

ABSTRACT. A technique is developed to identify in-situ the tensile force in tie-rods which are used in ancient monumental masonry buildings to eliminate the lateral load exercised by vaults and arcs. The technique is a frequency-based identification method that allows to minimize the measurement error and that is of simple execution. In particular, the first natural frequencies of the tie-rod are experimentally identified by measuring the FRFs with instrumented hammer excitation. Then, a numerical model, based on the Rayleigh-Ritz method, is developed for the axially-loaded tie-rod by using the Timoshenko beam theory retaining shear deformation and rotary inertia. Non-uniform section of the rod is considered since this is often the case for hand-made tie-rods in old buildings. The part of the tie-rod inserted inside the masonry wall is also modeled and a simple support is assumed at the extremities inside the walls. The constraints given to the part of the tie-rod inserted inside the masonry structure are assumed to be elastic foundations. The tensile force and the stiffness of the foundation are the unknown. In some cases, the length of the rod inside the masonry wall can be also assumed as unknown. The numerical model is used to calculate the natural frequencies for a given set of unknowns. Then, a weighted difference between the calculated and identified natural frequencies is calculated and this difference is minimized in order to identify the unknowns, and in particular the tensile force. An estimation of the error in the identification of the force is given. The technique has been tested on six tie-rods at the central vault of the famous Duomo of Parma, Italy.

Keywords: Tie-rods, frequency-based identification method, Rayleigh-Ritz method.

INTRODUCTION

Tie-rods were often used in ancient monumental masonry buildings to eliminate the lateral load exercised by the vaults and arcs. They give a fundamental contribution to the structural equilibrium. As a consequence of foundation settlements, the tensile force on tie-rods can surpass the yield strength of the material, which is not particularly high since old-time metallurgy was not able to obtain high-strength rods. Also corrosion can play a role in decreasing the strength of ancient tie-rods. For these reasons, it is important to identify the tensile force in tie-rods of masonry building, especially in case of evident deformation of arcs and vaults, in order to substitute them in dangerous cases.

Unfortunately there is no non-destructive technique for a direct in-situ measurement of the force on the rod. Several techniques have been proposed for an indirect measurement of the force. Blasi and Sorace [1, 2] introduced a technique based on a combination of static and dynamic identification. In particular, they modeled the tie-rod as a simply supported Euler beam with two identical rotational springs of unknown stiffness at the edges. The two unknowns, i.e. the tensile force and the stiffness of the two identical rotational springs, were identified by two equations: (i) a static equation giving the central deflection of the rod under a given load, and (ii) a dynamic equation giving the fundamental natural frequency of the rod. The method was tested in laboratory giving good results. Anyway, it requires two different in-situ experiments (measurement of the central static deflection under static load and of the fundamental natural frequency) and can give results with significant error in case of measurement error since the two unknowns are determined by only two data points.

T. Proulx (ed.), *Dynamics of Civil Structures, Volume 4*, Conference Proceedings of the Society for Experimental Mechanics Series 13,
DOI 10.1007/978-1-4419-9831-6_2, © The Society for Experimental Mechanics, Inc. 2011

Briccoli Bati and Tonietti [3] introduced a single static test to identify the force. It requires the measurement (i) of three vertical displacements under a concentrated static load, and (ii) of the strains variations at three sections of the rod. Also in this case good agreement with laboratory experiments has been found.
Fully dynamical identification of cable tension force has been recently proposed by Kim and Park [4]. It allows to identify the tension force, flexural rigidity and axial rigidity of the cable from measured natural frequencies. Anyway this technique is not immediately applicable to tie-rods since they cannot be modeled as cables and present uncertain constraints due to the portion of the rod inserted into the masonry wall or column.
In the present study, a technique is developed to identify in-situ the tensile force in tie-rods in ancient monumental masonry buildings. The technique is based on a frequency-based identification method that allows to minimize the measurement error and that is of simple execution. In particular, the first natural frequencies of the tie-rod are experimentally identified by measuring the FRFs with instrumented hammer excitation; four to six natural frequencies can be easily identified with a simple test. Then, a numerical model, based on the Rayleigh-Ritz method [5], is developed for the axially-loaded tie-rod by using the Timoshenko beam theory retaining shear deformation and rotary inertia. Non-uniform section of the rod is considered since this is often the case for hand-made tie-rods in old buildings. The part of the tie-rod inserted inside the masonry wall is also modeled and a simple support is assumed at the extremities inside the walls. The constraints given to the part of the tie-rod inserted inside the masonry structure are assumed to be elastic foundations. The tensile force and the stiffness of the foundation are the unknown parameters. In some cases, the length of the rod inside the masonry wall can be also assumed as unknown. The numerical model is used to calculate the natural frequencies for a given set of unknowns. Then, a weighted difference between the calculated and identified natural frequencies is calculated and this difference is minimized in order to identify the unknowns, and in particular the tensile force. An estimation of the error in the identification of the force is given. The technique has been tested on the six tie-rods at the ancient Dome of Parma, Italy, famous for the frescos painted by the master Francesco Mazzolla, called *Il Parmigianino*, in 1500.

2. ANALYTICAL MODEL

The tie-rod is modeled as a simply supported Timoshenko beam; the supports are assumed at the beam edges inside the masonry wall and the portion of the beam inside the wall is subjected to an elastic Winkler foundation simulating the interaction between the beam and the wall, as shown in Figure 1. An isotropic beam of length L and non-uniform prismatic section A is considered. A Cartesian coordinate system (x, y, z) is assumed on the beam where the x-axis is coincident with the centroidal axis and the y and z axes are coincident with the principal axes of the root cross-section. It is assumed that the centroidal axis is coincident with the elastic axis so that bending-torsion coupling is negligible [6]. The analysis is here limited to the x-y plane, and the kinematic displacements u, v, along the x and y axes respectively, are given by

$$u(x,y,t) = -y\;\theta(x,t)\,, \qquad v(x,y,t) = v(x,t)\,, \tag{1a,b}$$

where v is the transverse displacement and of the centroidal axis and θ is rotation of the cross-section about the positive z-axis. By using equation (1), the nonzero strain components of the beam are

$$\varepsilon_{xx} = -y\frac{\partial\theta}{\partial x}\,, \qquad \gamma_{xy} = \frac{\partial v}{\partial x} - \theta\,. \tag{2a,b}$$

The flexural displacement of the simply supported beam is expanded by using the following series of admissible functions

$$v(x) = \sum_{i=1}^{N} a_i \sin\left(\frac{i\pi x}{L}\right). \tag{3}$$

The rotation of the cross-section is expanded as

$$\theta(x) = \sum_{i=1}^{N} b_i\, i\cos\left(\frac{i\pi x}{L}\right) \tag{4}$$

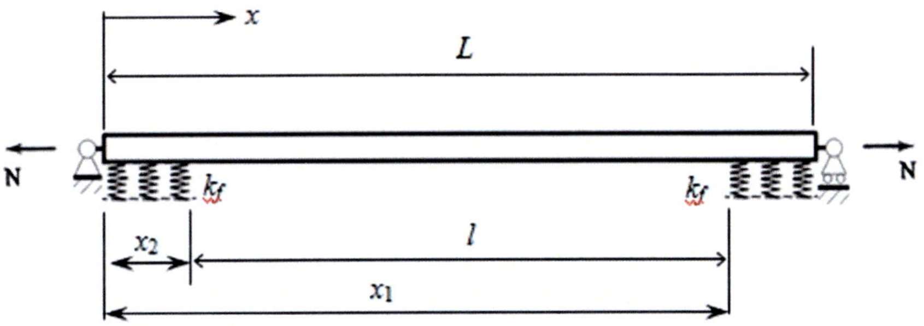

Figure 1. Analytical model of a tie-rod.

In equations (3) and (4) the same number of terms N has been used for the expansions of v and θ, in general, this number of terms can be different. The potential strain energy of the beam is given by [6]

$$V_B = \frac{1}{2}\int_0^L \int_A \boldsymbol{\sigma}^{\mathrm{T}} \boldsymbol{\varepsilon} \, \mathrm{d}A \, \mathrm{d}x \; , \tag{5}$$

where $\boldsymbol{\varepsilon}$ is the strain vector and $\boldsymbol{\sigma}$ is the stress vector. By using Hooke's stress-strains relationships, equation (5) can be rewritten as [6, 7]

$$V_B = \frac{1}{2}\int_0^L \left[EJ\left(\frac{\partial \theta}{\partial x}\right)^2 + kGA\left(\frac{\partial v}{\partial x} - \theta\right)^2 \right] \mathrm{d}x \; , \tag{6}$$

where EJ and kGA are the flexural rigidity and the shear rigidity, respectively, with E being the Young modulus, J the second moment of inertia of the beam cross-section about the y-axis, k the shear coefficient [8] and G is the shear modulus. In particular, J, k and A are functions of x. The potential energy associated with the axial load F (positive is for traction) is expressed by

$$V_F = \frac{1}{2} F \int_0^L \frac{\partial^2 v}{\partial x^2} \mathrm{d}x = \frac{F\pi^2}{4L} \sum_{i=1}^{N} i^2 a_i^2 \; . \tag{7}$$

The potential energy associated with the elastic foundation is given by

$$V_W = \frac{1}{2}\int_{x_1}^{x_2} k_f v^2 dx = \frac{1}{2}\sum_{i=1}^{N}\sum_{j=1}^{N} a_i \, a_j \int_{x_1}^{x_2} k_f \sin\left(\frac{i\pi x}{L}\right) \sin\left(\frac{j\pi x}{L}\right) \mathrm{d}x \; , \tag{8}$$

where k_f is the stiffness of the foundation, assumed uniform for simplicity, and the portion of the beam inserted in the walls is comprised between x_1 and x_2. In particular, $x_1 = 0$ for the left edge; for the right edge, $x_2 = L$. Both energies associated to elastic foundations at the left and right ends of the beam must be included. The global potential energy is

$$V = \frac{1}{2}\int_0^L \left[EJ(x)\left(\frac{\partial \theta}{\partial x}\right)^2 + k(x)GA(x)\left(\frac{\partial v}{\partial x} - \theta\right)^2 + F\frac{\partial^2 v}{\partial x^2} \right] \mathrm{d}x + \frac{1}{2} k_f \left[\int_0^{x_2} v^2 \, \mathrm{d}x + \int_{x_1}^{L} v^2 \, \mathrm{d}x \right] . \tag{9}$$

The kinetic energy of the beam is given by

$$T_B = \frac{1}{2}\int_0^L \int_A \rho \left[\left(\frac{\partial u}{\partial t}\right)^2 + \left(\frac{\partial v}{\partial t}\right)^2 \right] \mathrm{d}A \, \mathrm{d}x \, , \tag{10}$$

where ρ is the mass density. By using equations (1a,b) and integration over the cross-section A, equation (10) can be rewritten as [6]

$$T_B = \frac{1}{2}\int_0^L \left[\rho J \left(\frac{\partial \theta}{\partial t} \right)^2 + \rho A \left(\frac{\partial v}{\partial t} \right)^2 \right] \mathrm{d}x \ . \tag{11}$$

The reference kinetic energy of the beam, i.e. the maximum kinetic energy divided by ω^2, is given by

$$T_B^* = \frac{1}{2}\int_0^L \left[\rho J(x) \theta^2 + \rho A(x) v^2 \right] \mathrm{d}x \ . \tag{12}$$

By introducing the following vectorial notation

$$\mathbf{q}^{\mathrm{T}} = \left(a_1, \ldots, a_N, b_1, \ldots, b_N\right)^{\mathrm{T}} \ , \tag{13a}$$

$$V = \mathbf{q}^{\mathrm{T}} \mathbf{K} \mathbf{q} \ , \tag{13b}$$

$$T_B^* = \mathbf{q}^{\mathrm{T}} \mathbf{M} \mathbf{q} \ , \tag{13c}$$

the natural circular frequencies ω of the tie-rod are obtained by solving the following eigenvalue problem

$$\mathbf{K} - \omega^2 \mathbf{M} = 0 \ . \tag{14}$$

The shear coefficient k in equations (6, 9) for rectangular cross-section of Timoshenko beams is given by [8]

$$k = -\frac{2(1+v)}{\left[\dfrac{9}{4a^5 b} C_4 + v\left(1 - \dfrac{b^2}{a^2}\right) \right]} \ , \tag{15a}$$

where

$$C_4 = \frac{4}{45} a^3 b \left(-12a^2 - 15va^2 + 5vb^2\right) + \sum_{n=1}^{\infty} \frac{16 v^2 b^5 \left[n\pi a - b \tanh\left(\dfrac{n\pi a}{b} \right) \right]}{\left(n\pi^5\right)(1+v)} \ , \tag{15b}$$

v is the Poisson's coefficient, $2a$ is the depth of the beam in *y*-direction and $2b$ is the width of the beam in *z*-direction; a and b are functions of x.

3. IDENTIFICATION METHOD

In order to identify the axial force F and the stiffness k_f of the elastic foundation, the weighted difference between the calculated and identified natural frequencies is introduced

$$err_{RMS}\left(F, k_f\right) = \sqrt{\frac{1}{M} \sum_{i=1}^{M} \delta_i^2} \ , \tag{16}$$

where M is the number of the natural modes included in the identification process (5 to 10) and δ_i is the difference between the *i*-th computed and experimentally measured natural frequency. The function given in equation (16) is minimized over the full set of the useful values of the two unknowns (F, k_f), defined in the pre-processing, giving finally the identification of the axial force F. This minimization process requires the calculations of a large set of eigenvalue problems (14). The advantage of a self-developed optimized codes for the calculations of the natural frequency of the beam is fundamental versus a commercial FEM code. In the present case, the self-developed program has been validated by comparison to the ANSYS commercial program, and the results obtained are practically coincident. In particular, the first two mode shapes of one of the tie-rods studied in Section 4 are shown in Figure 2; they have been obtained by the ANSYS program and are coincident to those computed by the self-developed Rayleigh-Ritz program. However, since the numerical program implementing equation (14) uses a largely reduced size of the model with respect to the FEM commercial program, the time saving is huge.

The minimization process of function (16) can be implemented versus three unknowns F, k_f and the length of the elastic foundation, in case that the length of the tie-rod inserted into the masonry wall is unknown.

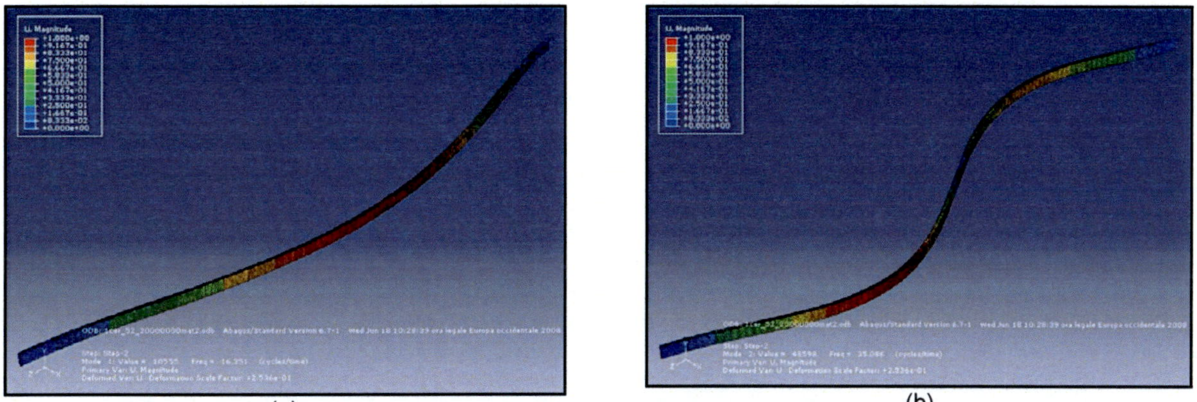

(a) (b)

Figure 2. Numerical simulation of a tie-rod dynamics showing (a) mode I and (b) mode II

4. EXPERIMENTS AND IDENTIFICATION AT THE CASTLE OF FONTANELLATO

Experimental measurement of the natural frequencies of the six tie-rods of the central vault of the ancient Dome of Parma, Italy (see Figures 3 and 4), was performed in September 2008. The used instrumentation is composed by an accelerometer *B&K* 4370, an instrumented hammer *B&K* 8202, two charge amplifiers *B&K* 2635 and a *Pimento* 8 channel system by *LMS* for data acquisition and experimental modal analysis.

The tests have been conducted in two days, and an articulated aerial platform has been used to reach the 6 tie-rods at 21 meters of height. The tie-rods have a circular cross-section of 40 mm in diameter; all tie-rods have been changed in '50 because the rupture of one of them.

Experimental tests, showed in figure 4, have been made putting ne or two accelerometer on the rod, and gently hammering it by hand. One of the six measured Frequency Response Functions (FRFs) is shown in Figure 5 for the tie-rods "3": eight peaks are clearly visible that identify the first eight natural vibration modes of the tie-rod.

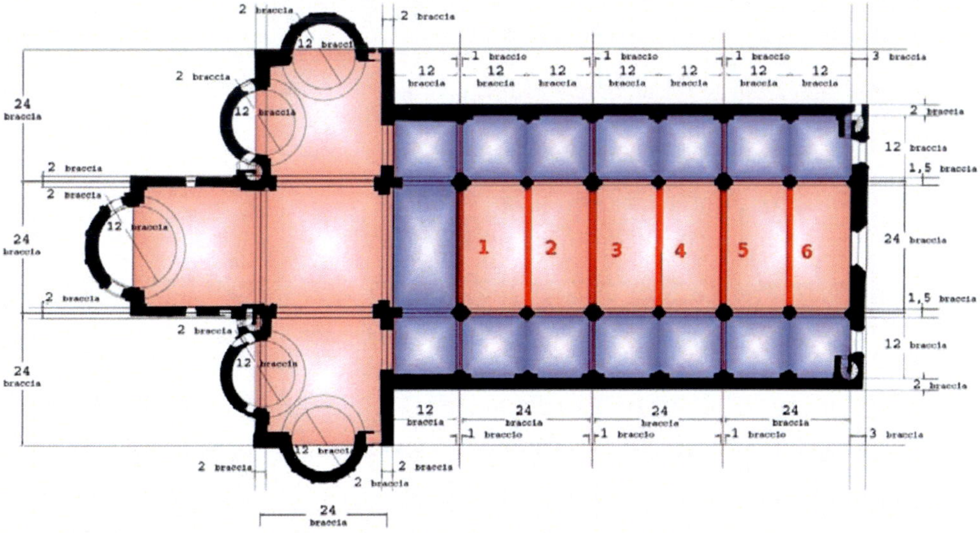

Figure 3. The Dome of Parma and the 6 tie-rods at 21 meters height.

Elaboration of experimental data has been conducted in conjunction with the numerical calculation of the axial load effect on the vibration modes of each tie-rods. The assumed material characteristics are: $E = 193\times10^9$ N/m^2, $G = 74.8\times10^9$ N/m^2, $\rho = 7870$ kg/m^3, $\nu = 0.29$. From $M = 2$ to $M = 6$ natural modes are used in the identification process, and $N = 15$ terms in the expansions (3) and (4). The effect of the axial force F on the natural frequencies is increase the computed natural frequencies in the direction of higher frequency. Otherwise, the effect of the stiffness k_f of the elastic foundation changes the slope of the curve. Therefore, the two unknowns have two different type of effect on the dynamics of the tie-rod, simplifying the identification process.

Figure 4. Experimental testing at the Dome of Parma in 2008.

The results of the identification process are given in Table 1. Results are presented for identification using M = 2 to 6 modes and show a convergence of the natural frequencies both experimentally and numerically determined.

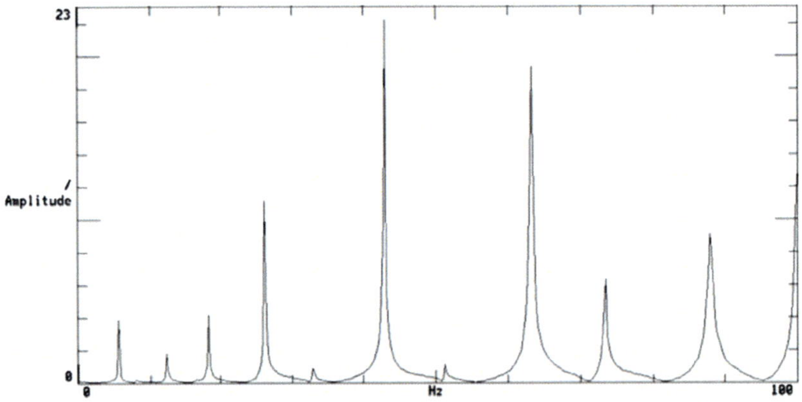

Figure 5. Natural modes of tie-rods "3".

Table 1. Experimental and computationally determined natural frequencies of the six tie-rods of Parma Dome.

Tie-rod	Freq. Nat. 1 [Hz]	Freq. Nat. 2 [Hz]	Freq. Nat. 3 [Hz]	Freq. Nat. 4 [Hz]	Freq. Nat. 5 [Hz]	Freq. Nat. 6 [Hz]
	Measured					
1	5,9	13,0	19,0	27,3	34,4	44,7
2	6,1	13,4	19,8	28,3	35,2	45,4
3	5,6	12,2	18,3	26,1	33,0	43,0
4	6,6	14,4	21,2	29,8	37,4	47,4
5	6,8	15,4	22,0	31,7	39,1	50,3
6	6,6	13,7	20,3	28,6	35,9	45,4
	Simulated					
1	5,9	12,7	18,6	26,9	33,6	43,6
2	6,1	13,1	19,2	27,6	34,6	44,6
3	5,6	12,2	17,9	26,1	32,9	43
4	6,6	14,1	20,6	29,5	36,6	47
5	6,8	14,4	21,2	30,1	37,7	48,3
6	6,6	13,8	20,5	28,8	36,2	46

Verification of the mechanical resistance of the tie-rods is made by the simple formula of axial stress calculation in a uniform cross-section beam:

$$\sigma_A = \frac{F}{A_{res}}, \quad (17)$$

where A_{res} is the design area of the metric M40 bolt in the medium section of the tie-rod. Results are shown in graphical form in figure 6. The stress state of at least 3 of the 6 tie-rods is over the admissible design stress, that for this steel is 160MPa.

5. CONCLUSIONS

The fully dynamic identification method allows to determine with a simple experiment the axial force on tie-rods. Since the experimental apparatus is compact and no fixed and accurate reference is necessary, the technique is particularly suitable for in-situ measurement on monumental buildings. The technique presented here has the advantage of using redundant data with respect to the unknowns in the identification process, minimizing the measurement and modeling errors.

Moreover, it allows to estimate the accuracy in the estimation of the axial force acting on the tie-rod: the experimental testing at the Parma Dome showed a dangerous state of the tie-rods of the central vault.

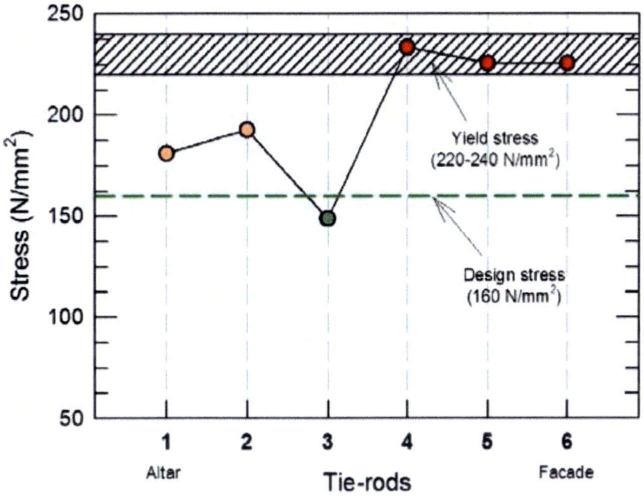

Figure 6. Stress state in the six tie-rods of Parma Dome.

REFERENCES

[1] C. Blasi, S. Sorace, Sulla determinazione del "tiro" nelle catene mediante prove statiche e dinamiche. Atti III Congresso Nazionale ASS.I.R.CO., Catania, Italy, November 1988.

[2] C. Blasi, S. Sorace, Determining the axial force in metallic rods. Structural Engineering International (IABSE) 4 (1994) 241-246.

[3] S. Briccoli Bati, U. Tonietti, Experimental method for estimating in situ tensile force in tie-rods. Journal of Engineering Mechanics 127 (2001) 1275-1283.

[4] B. H. Kim, T. Park, Estimation of cable tension force using the frequency-based system identification method. Journal of Sound and Vibration 304 (2007) 660-676.

[5] M. Amabili, R. Garziera, A technique for the systematic choice of admissible functions in the Rayleigh-Ritz method. Journal of Sound and Vibration 224 (1999) 519-539.

[6] J. B. Kosmatka, An improved two-node finite element for stability and natural frequencies of axial-loaded Timoshenko beams. Computers and Structures 57 (1995) 141-149.

[7] M. O. Kaya, O. Ozdemir Ozgumus, Flexural-torsional-coupled vibration analysis of axially loaded closed-section composite Timoshenko beam by using DTM. Journal of Sound and Vibration 306 (2007) 495-506.

[8] J. R. Hutchison, Shear coefficients for Timoshenko beam theory. Journal of Applied Mechanics 68 (2001) 87-92.

Proceedings of the IMAC-XXVIII
February 1–4, 2010, Jacksonville, Florida USA
©2010 Society for Experimental Mechanics Inc.

Estimating Effective Prestress Force on Grouted Tendon by Impact Responses

*Byeong Hwa Kim, **Soo Jin Kim, **Keum Soo Yeo, ***Jung-Bum Jang and ***Hong-Pyo Lee

*Assistant Professor, Department of Civil Engineering, Kyungnam University, 449 Wolyoung, Masan, Kyungnam, 631-701, Republic of Korea
**Master Student, Department of Civil Engineering, Kyungnam University, 449 Wolyoung, Masan, Kyungnam, 631-701, Republic of Korea
***Researcher, Structure Safety Group, Environment & Structure Laboratory, Korea Electric Power Research Institute, 103-16 Munji-Dong, Yuseong-Gu, Daejeon, 305-380, Republic of Korea.

ABSTRACT: A way of evaluating effective tensile force on a grouted seven-wire strand embedded in the post-tensioned beams has been examined experimentally. The proposed approach makes use of the longitudinal stress wave responses collected at the end of strands induced by a mechanical impact at the other end. The six 8m-long post-tensioned concrete beams have been specially designed and constructed. For each beam, the different tension levels are applied. The impact tests are conducted. The results show that longitudinal frequency, elastic wave velocity, and elastic modulus are nonlinearly increased as the tensile force level increases. It is thus expected that the longitudinal vibration characteristics of the existing bonded tendons can be of promising indicators for the inspection of a tensile force loss.

INTRODUCTION

Primary objective of nuclear containments is to prevent a leakage of a radioactive substance in the event of an extreme loading. In order to prevent such leakage and maintain a high level of integrity for such important structures, a prestressed concrete (referred to as PSC hereafter) has been widely used. However, time dependent loss of a prestressd force on tendons embedded in PSC is inevitable due to both creep and shrinkage of concrete and relaxation of tendons. Thus, most of regulations and guidelines of nuclear related structures have specified a periodic in-service inspection of tendon force.

For unbonded tendons in the containments, average tendon force can be evaluated by measuring the elongation at the end of active tendon induced by an applied jacking force [1]. For bonded tendons, however, direct estimation of tendon force is not feasible because of cement grouting between tendon and duct. Accordingly, alternative approaches have been proposed for the periodic in-service inspection of the containments with bonded tendons. US RG 1.9 [2] has suggested a lift-off measurement for unbonded sample tendons and a deformation measurement for the containments during periodic full scale pressure tests. Pandey [3] has proposed a reliability-based assessment of integrity of bonded prestressed concrete containments. The study has recommended that inspection results of lift-off, flexural and destructive tests for a set of beam specimens can be used in order to update the probability distribution of the prestressing force and the number of degraded tendons on the containments. Sun et al. [4] have also studied strength monitoring of a prestressed concrete containment with grouted tendons. Work done by Sun et al. [4] has utilized a real time monitoring system for overall containment deformation of such structure. However, the above approaches are neither fully satisfactory for strength monitoring of tendons nor promising methods for ensuring the structural integrity of containments and tendons.

A measurement of the tendon force has been achieved in terms of using a low and a high frequency approach. Whereas the former is in the range of vibration frequency level, the latter is in the range of ultrasonic frequency level. For a bridge cable, the existing tensile force on the bridge cable or a hanger can be evaluated by applying

T. Proulx (ed.), *Dynamics of Civil Structures, Volume 4*, Conference Proceedings of the Society for Experimental Mechanics Series 13,
DOI 10.1007/978-1-4419-9831-6_3, © The Society for Experimental Mechanics, Inc. 2011

the frequency-based system identification algorithm to the measured flexural frequencies [5]. For the unbounded PSC beam, tensile force and flexural rigidity can be evaluated by measuring the displacement, strain, and load time histories of the beam [6]. This is based on the results of Saiidi et al. [7] reporting that flexural natural frequencies increase as the applied compressive force on tendon increases. Without experimental verifications, however, Hamed and Frostig [8] have insisted that the flexural natural frequencies of a PSC beam are irrelevant not only to the applied tension force but also to the grouting conditions. Thus, the above two research results conflict each other.

Meanwhile, the high frequency approach is based on the physical phenomenon that the velocity of stress wave on a rod or a strand relies on the applied stress level. Chen and Wissawapaisal [9] have estimated the tensile stress on seven-wire strand by means of detecting the arrival group velocity of stress wave at one end of the strand after injecting longitudinal ultrasonic stress wave (150kHz~350kHz) at the other end of the strand. In addition, using magnetostrictive ultrasonic transducers, Kwun et al. [10] have also reported the existence of notch frequency (75 kHz~110 kHz) that is a disappeared frequency induced by the applied stress level in a seven-wire strand. Using the same concept of the arrival velocity difference on a stress field, Scalea et al. [11] and Chaki and Bourse [12] have conducted a similar pulse-echo technique utilizing magnetostrictive guided ultrasonic waves.

However, it is not feasible to apply such high frequency methods to the case of bonded tendon. This is due to the fact that the cement grouting surrounding steel tendons causes significant energy attenuation. Moreover, for the bonded tendons, it is known that the maximum inspection range of ultrasonic waves is less than 1.5m [13] while the typical height of nuclear containments is over 40m. Subsequently, no attempt has been made to directly evaluate tensile forces on the bonded tendons.

Since the first nuclear power plant was constructed in 1978, 20 unit nuclear power plants have currently been operated and maintained in Korea [14]. Among such nuclear containments, only Wolsong Units 1,2,3,4 and Ulchin Units 1, 2 were designed as prestressed concretes with bonded tendon systems. Since Wolsong Units started a commercial service in 1983, 30 years of design life time has now been approaching. Consequently, Korea Hydro Nuclear Power Company (KHNP) that owns the plant now concerns an important pending problem for continuous operation beyond the design life. Accordingly, the periodic in-service inspection has been arranged for Wolsong Units and a long-term research project has been designed to assess directly the effective prestress force on the bonded tendons embedded in the nuclear containments using nondestructive techniques. For this purpose, the indirect test beam method by Pandey [3] has been employed and thus a number of test beams have been prepared. The present study therefore represents the first feasible study of the research project.

Main objective of this study is to investigate a feasibility for nondestructively evaluating prestress force on bonded tendons embedded in the nuclear containments. Since the previous high frequency pulse approach has a strict limitation on energy attenuation in the case of bonded tendons, the current study focuses primarily on the utilization of low frequency pulse method. Rationale behind using the low frequency pulse method is due to the fact the low frequency propagates a long distance regardless of grouting conditions.

LONGITUDINAL VIBRATION TEST

Figure 1 shows the typical test set-up for the longitudinal vibration test. To excite the longitudinal modes of the specimens, an impact hammer, PCB Piezotronics model of 086C04, has been used. Acceleration responses caused by the impact hammer have been collected by PCB Piezotronics model of 352B10. Use is made of NI model CRIO-9073, four channels of NI 9233 and a laptop computer for data acquisition. In order to avoid an electric noise, two parallel 12 Volt batteries are employed as a power supply of the acquisition facilities. A sampling ratio of 50 kHz is utilized for the collections of the data.

A total six PSC beam specimens with bonded strands have been tested in the present study. Layout of the specimens is shown in Fig. 1, and dimension details and material properties of six specimens are summarized in Table 1. The PSC beams represent mock-up specimens of vertical walls for nuclear containments with unit width.

A B-type of KSD 7002 seven-wire strand is installed at the center of a concrete cross section. Nominal diameter, area and ultimate strength of the strand are 15.2mm, 138.7mm^2 and 260.68 kN, respectively. A VSL three hole-wedges system of P 6-3 anchors is installed. Internal and external diameters of the duct are 51mm and 54mm, respectively. After applying tensile force on the strands, the space between duct and strands is filled with cement paste grouting. Three cylinder tests of grouting materials are carried out and average compressive strength is 71.7MPa. In addition, characteristic tests of concrete materials being used for specimens are conducted, and average values of elastic modulus and Poisson's ratio are 32,644MPa and 0.22, respectively.

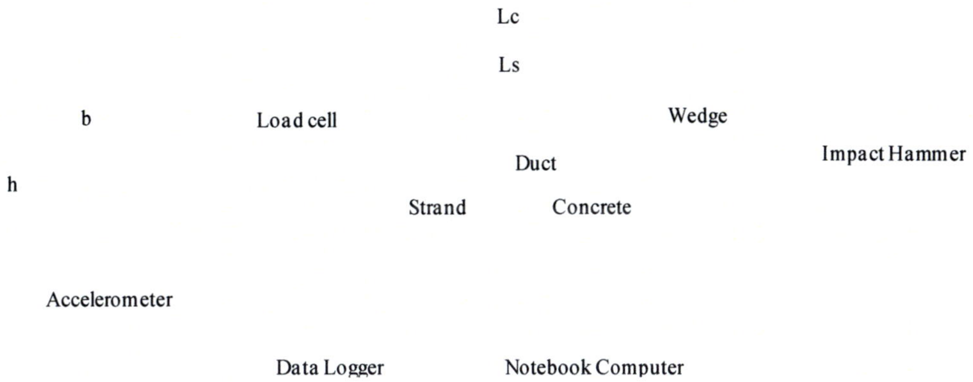

Fig.1 Test Specimen

Table 1. Test Specimens

PSC No.	Concrete				Strand			Remark
	Length Lc(m)	Width b(m)	Height h(m)	Strength (MPa)	Length Ls(m)	Anchor (VSL Type)	Tension T(kN)	
1	7.999	0.302	0.302	37.08	8.278	P 6-3	0.0	0
2	8.000	0.303	0.301	37.08	8.439	P 6-3	145.6	19% Tu
3	7.995	0.300	0.300	37.08	8.444	P 6-3	263.8	34% Tu
4	7.994	0.302	0.301	37.08	8.433	P 6-3	355.8	46% Tu
5	7.998	0.303	0.299	37.08	8.435	P 6-3	465.0	59% Tu
6	7.993	0.303	0.302	37.08	8.433	P 6-3	522.5	67% Tu

All of specimens are assumed to be simply-supported by two roller conditions and a tensile force is introduced at one end of each specimen. A load cell has been installed in each of specimens in order to estimate the applied tensile force introduced on strands. Whereas a load cell of ZIS model, ELC-100S is employed for specimens.

Six tests have been carried out for the specimens. In the tests, three accelerometers are clamped at core wires since the specimens have three strands. An impact is loaded at the opposite side of accelerometers and arbitrary support locations are employed as illustrated in Fig. 2.

(a) (b)

Fig.2 Support Condition and Accelerometer Locations in Test No.4: (a) supports; (b) accelerometers

EXPERIMENTAL RESULTS

Longitudinal impact tests of 160 times are repeatedly conducted for each specimen. A typical impact time history is collected with the sampling rate of 50 kHz as shown in Fig.3a. Averaged power spectrum of acceleration responses is evaluated using Hanning window and is depicted in Fig. 3b. To obtain the spectrum, 2^{13}, 2^{14} and 1.562Hz are employed for a data length, a number of FFT and a frequency solution, respectively. As observed from the spectrum, the impact has been excited up to 6 kHz.

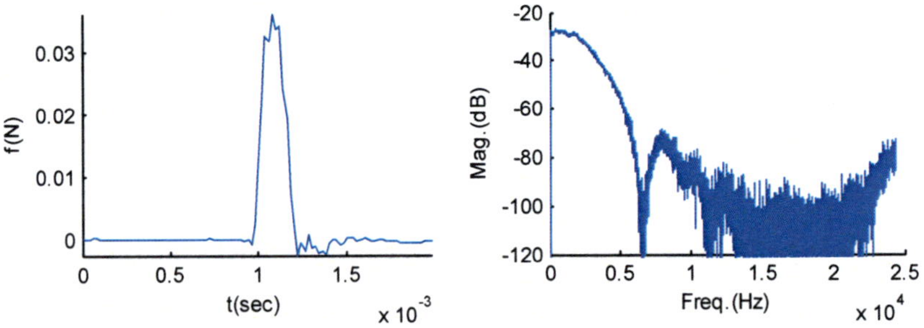

(a) (b)

Fig.3 Typical Impact Signal: (a) time history; (b) frequency spectrum

For each specimen and the resulting transfer functions are illustrated in Fig. 4. It is observed in general that magnitudes of transfer functions in the frequency band of 4 to 5 kHz increase as the applied tensile force level increases. In addition, a number of noticeable peaks in the transfer function of PSC specimen No. 6 is more than 30, while those in the transfer function of PSC specimen No. 1 is at most 8. This can be attributed to the introduced tensile force level on the specimens. As the tensile force is increased, axial force level is also increased. This leads to an increase in both stiffness and strength of the specimens, but a decrease in internal damping.

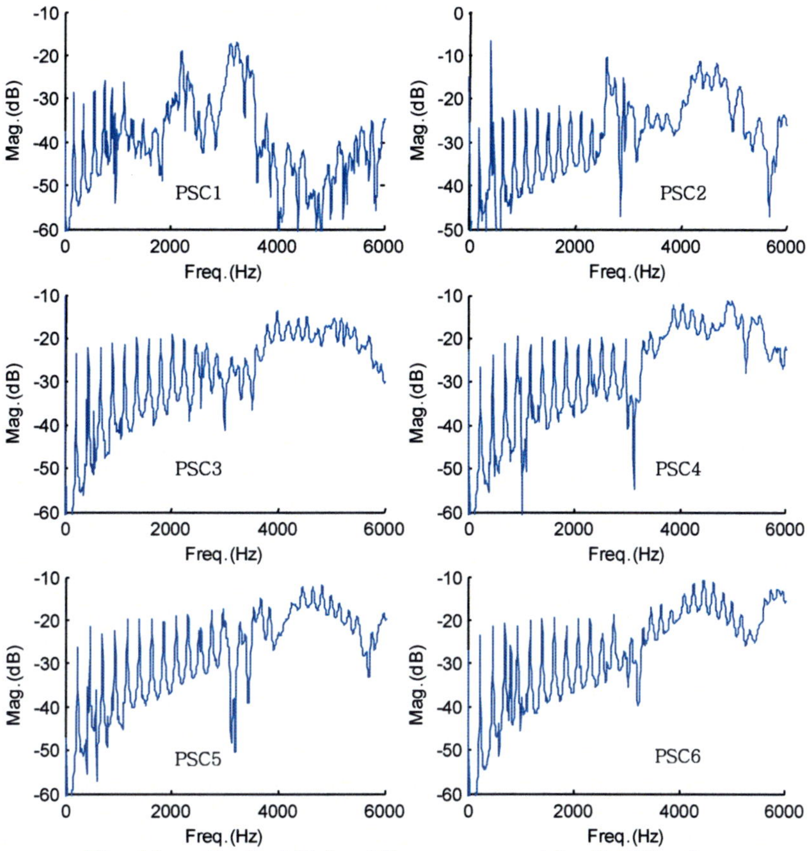

Fig. 4 Longitudinal Natural Frequency and Applied Tension

In order to investigate the longitudinal natural frequency, a simple equation is employed for the calculation of theoretical longitudinal frequency. For a homogeneous axial member with free-free boundary conditions, the *n*th theoretical longitudinal natural frequency in Hz is obtained by.

$$f_n = \frac{n c_0}{2L} \tag{1}$$

$$c_0 = \sqrt{\frac{E}{\rho}} \tag{2}$$

where L, c_o, E, and ρ denote length of a member, elastic wave velocity, elastic modulus, and mass density for the condition of no tension, respectively. To compute theoretical longitudinal frequencies, the wave velocity of PSC specimen No. 1 is firstly identified using the known length given in Table 1 and the first measured natural frequency. This is reasonable since no tension is applied to the PSC specimen No. 1. Calculated wave velocity is 2930.4 m/s. Then, the *n*th higher modes are generated in terms of multiplying the wave velocity by *n*. For the case of no tension, the measured eight frequencies agree very well with the predictions. This implies that the existing linear theory in Eq. (1) is based on a constant wave velocity assumption. Extracted longitudinal natural frequencies of the specimens are shown in Fig. 5. It is clearly observed that the longitudinal natural frequencies are increased when the applied tensile force level increases. This leads to the fact that the longitudinal natural frequency depends on the introduced tensile force level.

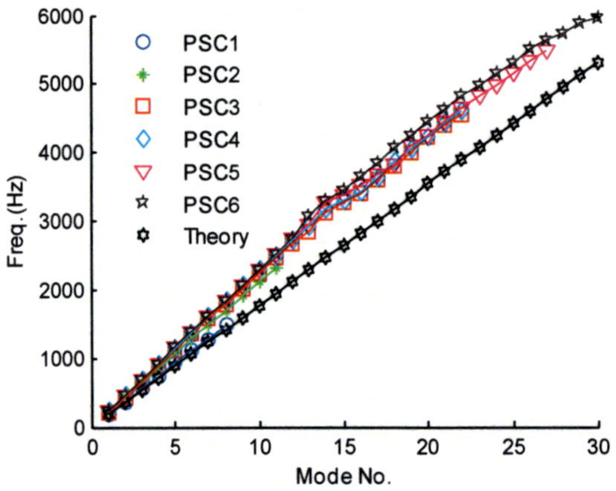

Fig.5 Extracted Natural Frequency

DISCUSSIONS

Frequency and Tension

For all PSC specimens, longitudinal natural frequencies extracted by a peak-peaking technique are summarized in Table 2. It is obvious from the Table 2 that the frequencies are increased as the applied tensile force level increases. This has been graphically represented in Fig. 6 which shows variations of frequencies with reference to the applied tensile force levels for various modes. In general, the frequencies vary linearly with tensile force in relatively low levels of the applied tension.

Table 2. Extracted Natural Frequency

PSC	Frequency (Hz)							
No.	1	2	3	4	5	6	7	8
1	177.0	366.2	567.6	750.7	939.9	1117	1257	1483
2	207.5	427.2	640.9	854.5	1068	1282	1483	1703
3	225.8	451.7	677.5	903.3	1129	1361	1587	1807
4	231.9	463.9	689.7	921.6	1154	1379	1611	1837
5	231.9	463.9	689.7	927.7	1160	1385	1624	1843
6	231.9	463.9	695.8	927.7	1160	1392	1624	1855

However, the frequencies increase asymptotically in high levels of the applied tension. This suggests that the frequencies can be a sensitive feature for PSC specimens with a level of the applied tension. In addition, for higher modes (i.e., from the 3rd mode), frequencies in high levels of the applied tension (approximately 50% of the ultimate strength of a strand) tend to be slightly increased as the mode is increased. This phenomenon is particularly important since typical tension levels of nuclear containments in practice are within a range of 60% to 70% of the ultimate strength of a strand. It is thus expected that a continuous monitoring of the longitudinal frequencies may enable change of tension levels to be identified.

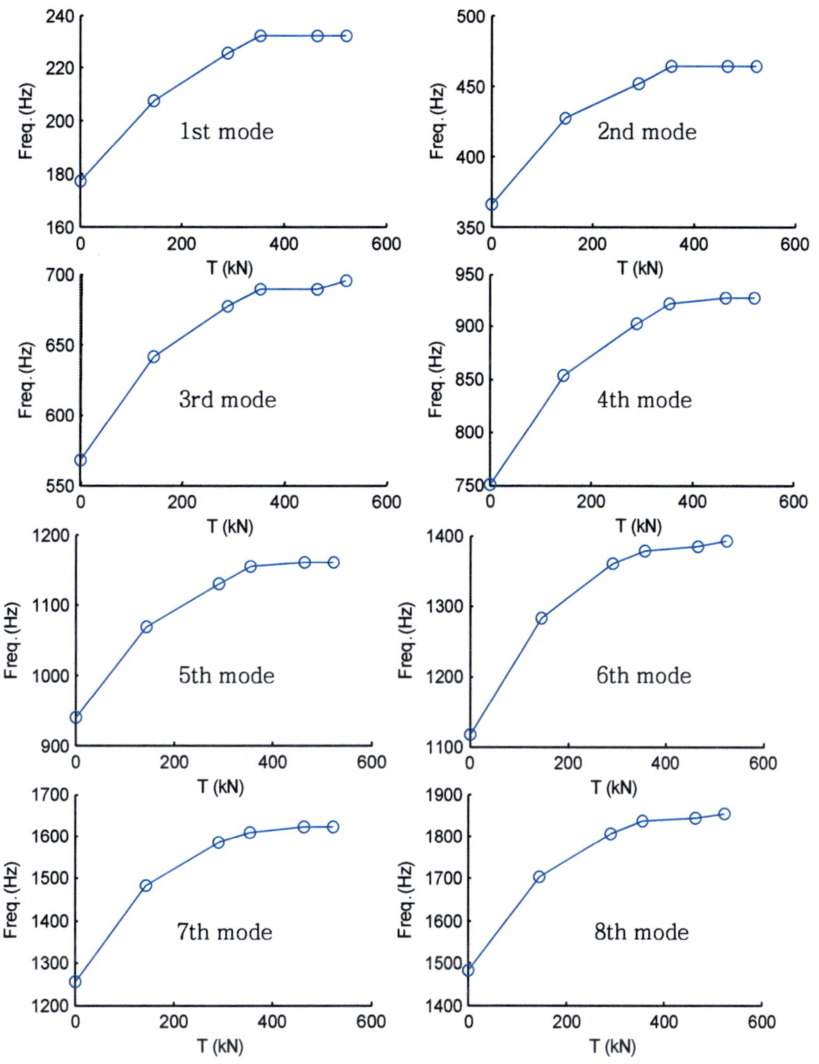

Fig.6 Longitudinal Natural Frequency and Applied Tension

Wave Velocity and Tension

An elastic wave velocity can be calculated for each mode in terms of substituting the measured frequency given in Table 2 and a specimen length into Eq. (1). Shown in Fig. 7 is a comparison of change rates for the estimated elastic wave velocities with respect to the case of no tension, i.e., PSC specimen No. 1. In general, change rates of elastic wave velocities are increased as the applied tensile force level increases. In particular, when the applied tensile force level is corresponding to 67% of the ultimate strength of a strand, the maximum increment rate of 33.5% occurs in the 1st mode and the minimum increment rate of 24.9% takes place in the 3rd mode. Meanwhile, in case that the applied tensile force level is reduced from 67% to 46% of the ultimate strength of a strand, little variation of elastic wave velocity is happened in the 1st mode, and a minor decrease of 1.24% in elastic wave velocity occurs in the 8th mode. This suggests that when the applied tension level is decreased approximately 30%, a variation in the elastic wave velocity is within a margin of a measurement error. Thus, the elastic wave velocity does not seem to be a feasible parameter for the identification of tension level variation.

However, in case that the applied tension level is further reduced from 67% to 19% of the ultimate strength of a strand, estimated elastic wave velocities in the 1st and 8th modes are reduced to 13.96% and 10.38% in comparison with those of no tension specimen, respectively. This indicates that, when an approximately 72% of reduction in applied tension level occurs, a variation rate of estimated elastic wave velocities seems to make the tension levels to be identified. In short, tension levels can be identified in terms of using the variation rate of the elastic wave velocities, only for the case that the applied tensile force level is reduced in 30% or more.

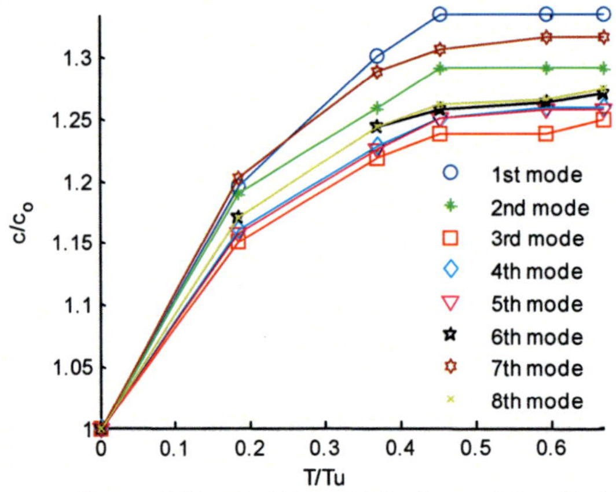

Fig.7 Change Rate of Elastic Wave Velocity and Applied Tension

Elastic Modulus and Tension

In the linear elastic theory of Eq. (1), a basic assumption is that the longitudinal natural frequency and elastic wave velocity do not vary with a change in the applied tension. However, experimental results demonstrate obviously that the natural frequency is nonlinearly increased as the applied tension level increases. Substituting Eq. (2) into Eq. (1), an elastic modulus can be given as follows.

$$E = \rho\left(\frac{2Lf_n}{n}\right)^2 \quad (3)$$

With the assumption that a small variation in the mass density is negligible, the elastic modulus relies on the mode number and longitudinal frequency. Since the frequency has a nonlinear relationship with the applied tension as illustrated in Fig. 6, the elastic modulus can also be relied on both the applied tension level and the mode number. This is supported by the research results of Kim and Park [5]. Work done by Kim and Park [5] confirmed that a flexural rigidity of a bridge cable is proportional to the applied tension on the cable.

Substituting the measured frequency listed in Table 2 and the mode number into Eq. (3), the elastic modulus of a strand can be achieved and summarized as Table 3. Use is made of an electronic scale for the mass density (7729.779 kg/m^3) of the strand. Coefficient of variation (COV) reveals that the variation of the estimated elastic modulus with respect to the modes is negligible, while it is sensitive to the applied tension level. Hence, the elastic modulus can be of more useful indicator for the change of the applied tension levels rather than frequencies and elastic wave velocities. This can be further supported by fact that the elastic modulus has a square sensitivity to the measured frequency as given in Eq. (3), while the elastic wave velocity has a linear sensitivity to the frequency in Eq. (1).

Table 3. Estimated Elastic Modulus (GPa) of Strand

PSC No.	Mode No.								Mean	COV
	1	2	3	4	5	6	7	8		
1	66.38	71.03	75.84	74.63	74.87	73.43	68.32	72.81	72.16	0.0464
2	94.81	100.5	100.5	100.5	100.5	100.5	98.83	99.78	99.48	0.0199
3	112.4	112.5	112.4	112.4	112.4	113.4	113.3	112.5	112.7	0.0039
4	118.2	118.3	116.2	116.7	117.1	116.1	116.5	115.9	116.9	0.0079
5	118.3	118.4	116.3	118.3	118.4	117.2	118.4	116.8	117.8	0.0074
6	118.2	118.3	118.3	118.3	118.3	118.3	118.3	118.2	118.3	0.0004

Assuming that the mass of density of concrete is 2450 kg/m^3, the elastic modulus of concrete has been evaluated and given in Table 4. A similar trend as the above case of a strand is achieved. In general, an estimated elastic modulus of a concrete specimen is insensitive to the mode number, while it is affected by the applied tension levels. Graphical representation is given in Fig. 8 which demonstrates that normalized elastic moduli of both strand and concrete vary with the applied tension levels. In Fig. 8, E_o denotes the estimated elastic modulus of the PSC specimen No. 1 which has no tension. As observed, when the applied tension level of 67% is reduced to 59%, 46%, 34%, and 19% of the ultimate strength of a strand, average elastic modulus of the strand is decreased as much as 0.5%, 1.2%, 4.8%, 15.9%, and 39%, respectively, while that of concrete is diminished as much as 0.4%, 1.2%, 5.0%, 15.9%, and 37%, respectively. Taking into account approximately 5% of a measurement error, a change of the applied tension level can be of measurable when the applied tension level of 67% is reduced to 46% or less of the ultimate strength of a strand.

Table 4. Estimated Elastic Modulus (GPa) of Concrete

PSC No.	Mode No.								Mean	COV
	1	2	3	4	5	6	7	8		
1	19.6	21.0	22.4	22.1	22.2	21.7	20.2	21.5	21.4	0.0464
2	27.0	28.6	28.6	28.6	28.6	28.6	28.2	28.4	28.3	0.0199
3	31.9	32.0	31.9	32.0	31.9	32.2	32.2	32.0	32.0	0.0039
4	33.7	33.7	33.1	33.2	33.4	33.1	33.2	33.0	33.3	0.0079
5	33.7	33.7	33.1	33.7	33.7	33.4	33.7	33.3	33.6	0.0074
6	33.7	33.7	33.7	33.7	33.7	33.7	33.7	33.7	33.7	0.0004

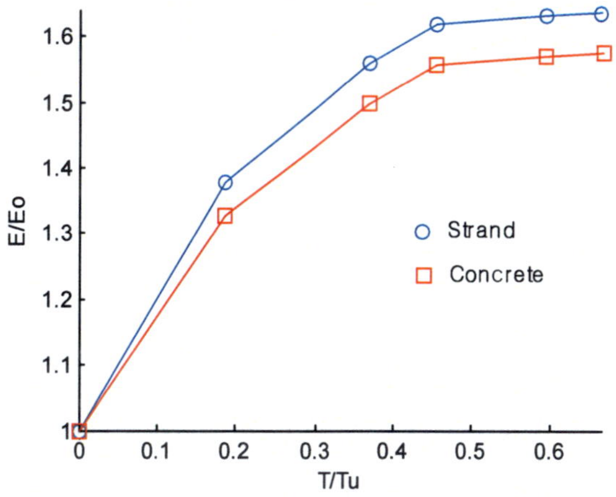

Fig.8 Change Rate of Elastic Modulus and Applied Tension

SUMMARY AND CONCLUSIONS

A feasibility study has been conducted for the evaluation of tensile force levels applied on grouted tendons which are embedded in nuclear containments. For this purpose, an experimental programme has been carried out for PSC specimens subjected to longitudinal vibration of an impact. Based on detailed analyses and evaluation for experimental responses of acceleration time histories and power spectrums, the following conclusions can be made.

In the existing linear theory, longitudinal natural frequency, elastic wave velocity, and elastic modulus are independent of the applied tension. However, this may not be true since present experimental results demonstrate that there is a functional relationship between them. The results show that they are nonlinearly increased as the applied tensile force level increases. In addition, present study proves that variation rates of the applied tension levels are noticeable in elastic modulus rather than in either frequency or elastic wave velocity. Hence, the elastic modulus can be of the most promising parameter for the identification of the applied tensile force level. It is thus expected that monitoring the longitudinal vibration responses enables the applied tension level on the grouted tendon embedded in a nuclear containment to be accurately captured.

However, there is certainly a room for improvements of the present study if experimental study is extensively carried out for further analyses of longitudinal vibration responses.

ACKNOWLEDGEMENTS

This research is sponsored by the Korean Ministry of Science and Technology under the National Mid-term and long-term Atomic Energy R&D Program.

REFERENCES

[1] Anderson, P., Berglund, L.E., 2005. Average force along unbounded tendons: a field study at nuclear reactor containments in Sweden. Nuclear Engineering and Design, Vol.235, pp. 91-100.
[2] USNRC, 1990, Determining Prestressing Forces for Inspection of Prestressed Concrete Containments. RG 1.35.1.
[3] Pandey, M.D., 1997. Reliability-based assessment of integrity of bonded prestressed concrete containment structures, Nuclear Engineering and Design, Vol.176, pp. 247-260.
[4] Sun, Z., Liu, S., Lin, S., Xie, Y., 2002. Strength monitoring of a prestressed concrete containment with grouted tendons. Nuclear Engineering and Design, Vol. 216, pp.213-220.
[5] Kim, B.H. and Park, T. (2007) Estimation of cable tension force using the frequency-based system identification method. *Journal of Sound and Vibration*, Vol. 204, pp. 660-676.
[6] Law, S.S. and Lu, Z.R. (2005) Time domain responses of a prestressed beam and prestress identification. *Journal of Sound and Vibration*, Vol.288, pp. 1011-1025.
[7] Saiidi, M., Douglas, B. and Feng, S. (1994) Prestress force effect on vibration frequency of concrete bridges", *Journal of Structural Engineering*, ASCE, Vol. 120, No. 7, pp. 2233-2241.
[8] Hamed, E. and Frostig, Y. (2006) Natural frequencies of bonded and unbonded prestressed beams-prestress force effects. *Journal of Sound and Vibration*, Vol. 295, pp. 28~39
[9] Chen, H.L., Wissawapaisal, K. (2002) Application of Wigner-Ville Transform to evaluate tensile forces in seven-wire prestressing strands. *Journal of Engineering Mechanics*, ASCE, Vol. 128, No. 11, pp. 1206-1214.
[10] Kwun, H. Bartels, K.A. and Hanley, J.J. (1998) Effects of tensile loading on the properties of elastic-wave propagation in a strand. *Journal of Acoustic Society of America*, Vol.103, No.6, pp. 3370-3375.
[11] Scalea F L, Rizzo P, and Seible F, "Stress measurement and defect detection in steel strands by guided stress waves", Journal of Materials in Civil Engineering, ASCE, 2003 15(3) 219-227.
[12] Chaki S and Bourse G, "Guided ultrasonic waves for non-destructive monitoring of the stress levels in prestressed steel strands", Utrasonics, 2009 49(2) 162-171.
[13] Beard M D, Lowe M J S, and Cawley P, "Ultrasonic guided waves for inspection of grouted tendons and bolts", Journal of materials in civil engineering, ASCE, 2003 15(3) 212-218.
[14] Park, J. Hong, J., 2009. Present status of nuclear containments and ISI in Korea, Progress in Nuclear Energy, doi:10.1016/j.pnucene.2009.05.005, article in press.

Proceedings of the IMAC-XXVIII
February 1–4, 2010, Jacksonville, Florida USA

Experimental Investigation of Dynamic Load Estimation Using Small-Scale Testing

Kelly A. Salyards[1], Robert J. Firman III[2]

[1]Assistant Professor, Department of Civil and Environmental Engineering;
[2]Undergraduate Research Student, Department of Civil and Environmental Engineering
Bucknell University
Lewisburg, PA 17837

ABSTRACT

Vibration serviceability in large assembly-type structures such as stadiums and grandstands has become a design and management concern as crowds continue to impose significant dynamic loads on a structure through synchronized movement. However, there is little guidance on the appropriate dynamic design loading to be anticipated for the design of new structures or the assessment of existing structures. The guidance currently available for human-induced dynamic loading is limited to laboratory experiments and numerical models that fail to take into account the effects that many factors of the event atmosphere may have on the loading models. Because measurement of the forces exerted by a crowd under true event conditions is next to impossible, this project investigates the ability to estimate the dynamic loading with consideration of only the more reasonably obtainable acceleration response of the structure during the event. Experimental testing was performed on a simple floor structure subjected to dynamic forces generated by small groups to investigate the accuracy and sensitivity of the load estimation method for consideration when applying this method to large scale structures.

INTRODUCTION

The structural systems of stadiums and grandstands must be designed to resist loading conditions associated with wind, seismic, and live loads. Live loads typically result from human occupants acting either as a static or dynamic load on the structure. One of the most extreme loading conditions that a structure will experience due to live loads is the synchronized dynamic loading of a crowd. This type of loading has not been accurately specified in design standards and is not consistently established. However, it is critical to both strength and serviceability design. This study focuses on the dynamic load that a crowd of occupants is capable of producing and how to develop an estimate for this type of loading from actual events when it is impractical to measure the load itself.

Identification of the dynamic loading to be used in design is difficult as guidance is limited. Laboratory experiments have been performed to estimate the dynamic loading of small to medium sized groups [1]. However, the results of these experiments are limited in several respects. First, it is not practical or often feasible to conduct laboratory tests on a scale comparable to most events where a substantially large crowd may be present, such as in as in a stadium or concert venue. Also, laboratory tests are often unable to replicate some of the crowd variables present during an actual event, such as environmental factors including weather and crowd enthusiasm. For these reasons, it is important to determine a means of accurately estimating the dynamic load of a crowd in a more realistic environment so that the current limited recommendations for a rhythmic design loading can be confirmed or more specific recommendations developed.

BACKGROUND

Because it is next to impossible to measure the force exerted on a structure by a crowd, a method for estimating this load has been proposed [2]. This dynamic load estimation method (LEM) utilizes the more easily obtainable acceleration response of the structure during an event along with the dynamic properties of the structure including natural frequency, damping, and stiffness to estimate the dynamic crowd loading. The method assumes that the

T. Proulx (ed.), *Dynamics of Civil Structures, Volume 4*, Conference Proceedings of the Society for Experimental Mechanics Series 13,
DOI 10.1007/978-1-4419-9831-6_4, © The Society for Experimental Mechanics, Inc. 2011

loading can be reasonably represented by a periodic function. Figure 1 illustrates the general process used in this study, how it involves the load estimation method, and how verification of the results was approached.

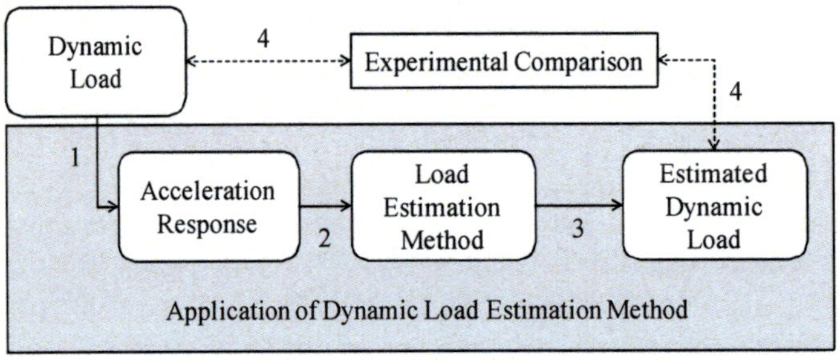

Figure 1: Overview of study, including load estimation method and verification of results

Dynamic Load Estimation Method Theory

With the assumption that the dynamic crowd loading can be represented by a periodic function, a Fourier series is often used to represent the coordinated, repetitive load. The appropriateness of modeling a dynamic force generated by an individual with a Fourier series has been well documented and the mathematical model is shown in Equation 1 [2].

$$F(t) = G_s\left(1 + \sum_{n=1}^{\infty} r_n \sin(2\pi nft + \phi_n)\right) \tag{1}$$

where $F(t)$ = time varying force
G_s = weight of the individual person
n = number of Fourier coefficient
r_n = n^{th} Fourier coefficient or dynamic load factor (DLF)
f = frequency of the applied rhythmic load
ϕ_n = n^{th} phase lag factor (related to the contact time between the occupant and the structure during impact)

It has been theorized that this model for an individual can be expanded for a crowd of people engaged in a coordinated, repetitive loading event as shown in Equation 2 [3].

$$F(x,y,t) = G(x,y)\left(1 + \sum_{n=1}^{\infty} r_{n,p} \sin(2\pi nft + \phi_n)\right) \tag{2}$$

where $F(x,y,t)$ = distributed force which varies with time
$G(x,y)$ = density and distribution of the human load in the crowd
$r_{n,p}$ = n^{th} Fourier coefficient or dynamic load factor (DLF) induced by *p* persons

The Fourier coefficients, r_n and $r_{n,p}$, can be derived from an examination of the acceleration response data measured during an actual event and an understanding of how the acceleration response is related to the dynamic load. The remainder of the variables involved in this derivation can be estimated from structural properties and observable loading conditions. The theory of the load estimation method is briefly discussed here but the derivation, in its entirety, can be found elsewhere [4].

The first step in utilizing the dynamic load estimation method is the determination of the dynamic properties of the unoccupied structure, which can be accomplished experimentally or through finite element modeling. A series of band-pass filters are then applied to the acceleration response data of the structure collected during an event. The band-pass filters are designed to isolate the portion of the acceleration response induced by each of the successive harmonic components of the forcing frequency, *f*. The first band-pass filter extracts the component of the response between 0.5*f* and 1.5*f*, the second between 1.5*f* and 2.5*f*, etc. For the purpose of this study, only three band-pass filters are utilized as it has been suggested that the first three harmonics are sufficient to estimate the quantities of interest: the force magnitude and frequency.

Once the original acceleration signal is filtered and the average peak acceleration, A_n, is identified for each harmonic component, Equation 3 can be used to estimate the Fourier coefficients of the load generating this acceleration response [3]. Equation 3 is derived from the peak of the theoretical acceleration response of a generalized single-degree-of-freedom system, which includes the dynamic magnification factor and the static deflection [4].

$$r_{n,p} = \frac{A_n\left(\sqrt{(1-n^2\beta^2)^2+(2\zeta n\beta)^2}\right)k}{G(2\pi nf)^2} \quad (3)$$

where A_n = average peak acceleration for the nth harmonic
β = ratio of applied rhythmic frequency to natural frequency of the structure
ζ = damping ratio of the structural system
k = static stiffness of the structure
G = generalized load of the crowd

An estimate of the dynamic force produced by the crowd can then be generated using the first three dynamic load factors estimated from Equation 3 and substituted into Equation 2.

Investigational Structure

The structure used to explore the LEM for real world structures is an interior bridge-type structure connecting two adjacent buildings and was selected because of its flexibility and the expected 'simple' vibrational modes. The structure itself consists of two wide-flange steel girders spanning thirty feet that are supported by a masonry wall at one end and frame into a steel girder of the adjacent building at the other end. Because the bridge is connected to the adjacent structure, this bay must be taken into consideration. Figure 2 shows images of the empty bridge structure.

Figure 2: (a) View of underside of the bridge (b) View from top of the bridge

The structural system includes various size wide-flange steel sections spanning between the two main girders to support metal deck and 5.25" concrete slab. In addition to the self-weight of the structural system, several other components add mass including hung ceilings, partitions, and office furnishings. The mass of partitions and furnishings is only considered for the bay adjacent to the bridge.

ESTIMATION OF DYNAMIC STRUCTURAL PROPERTIES

Finite Element Model

In attempt to understand the dynamic behavior of the test structure prior to experimental testing, a finite element model was created using SAP2000v12 [5]. The initial model consisted of only the bridge structure alone assuming simple support at both the masonry support and at the framed connection into the adjacent building structure. From this model, approximate nodal lines were determined in order to plan the placement of accelerometers during experimental testing. The results from the initial model indicated that the natural frequencies were greater than preliminary experimental tests would reveal. The model was subsequently modified to incorporate the adjacent bay. Figure 3 illustrates the first two modes of the structural system at 9.6 and 11.8 Hz. The model was also used to estimate the static stiffness of the bridge structure utilizing a unit load and corresponding deflection.

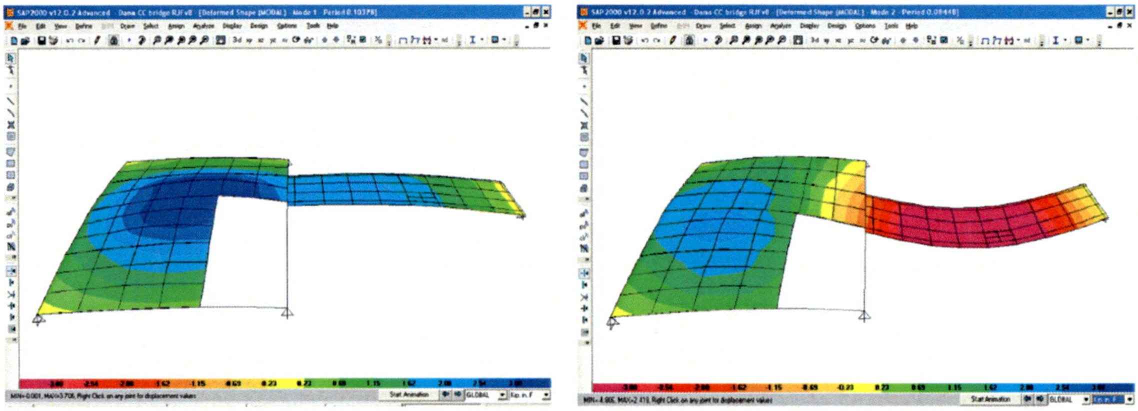

Figure 3: SAP2000 model of bridge (a) first mode, 9.6 Hz (b) second mode, 11.8 Hz

Experimental Procedures

The test structure described was used to explore the dynamic load estimation method at a small-scale with respect to large-scale applications involving significantly larger crowds and structures. Prior to the application of a rhythmic loading, the modal properties of the structure necessary for the load estimation method were experimentally determined. For all experimental testing, a WaveBook 516/E data acquisition system and WBK18 signal conditioner from IOtech were used. Preliminary tests to confirm equipment operation and quality of measurements were performed including checks of repeatability, reciprocity, and linearity [6].

Modal parameters were estimated using heel drop forces measured by force plates and acceleration responses of the structure measured using seismic accelerometers from PCB Piezotronics. From the force input measured by a force plate and the acceleration response measured by accelerometers, frequency response functions (FRFs) are generated within the EZ-analyst software by IOTech. Several experimental FRFs are shown in Figure 4 along with the coherence function for these measurements. From these FRFs, several definite peaks are apparent between 9-12 Hz but no other significant peaks are visible. As previously discussed, this preliminary measurement indicated the need for refinement of the initial finite element model which included only the bridge to the model shown in Figure 3 which includes the adjacent bay.

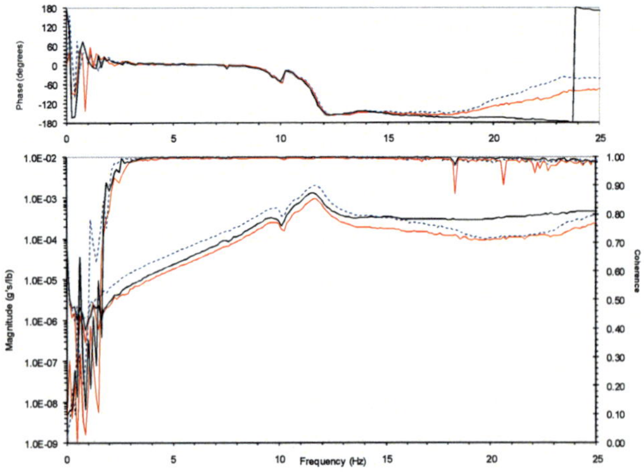

Figure 4: Frequency Response Functions indicating peaks in 9-12 Hz range

For the application of the load estimation method, only estimates of the structural natural frequency, damping, and stiffness are required. For this reason, basic manual modal analysis techniques, such as the quadrature response method applied to frequency response functions, log decrement applied to the time history, and half-power bandwidth method applied to frequency response functions, are utilized. These manual techniques yielded estimates of the natural frequencies and associated damping ratios. The stiffness established from the finite element model was determined to be acceptable after reasonable agreement of natural frequencies and mode shapes with the experimental results. The appropriateness of the finite element model was also evaluated by synthesizing FRFs from the finite element model and a reasonable comparison with the experimentally obtained FRFs was obtained. Structure properties, determined from these manual methods and the finite element model, used in the load estimation method, are summarized in Table 1.

Table 1: Dynamic properties of test structure estimated by experimental modal analysis techniques and finite element modeling

		Quadrature Response Analysis	Circle Fit	FE Model	Log Decrement	Half-Power Bandwidth
Mode 1	Frequency (Hz)	9.8	9.6	9.6		
	Damping				5.9%	3.3%
Mode 2	Frequency (Hz)	11.6	11.7	11.8		
	Damping					3.0%

RESPONSE DUE TO RHYTHMIC LOADING

Rhythmic loading tests were conducted on the test structure at several different jumping frequencies, with different types of audio signals, and with varying group sizes. Based on the finite element model, it was believed that the mode of the structural system to be excited by a rhythmic load on the test bridge would resemble the second mode or a combination of the first and second modes. For this reason, test subjects occupied the center of the bridge closest to the railing in order to maximize the structural response in this mode.

Summary of Rhythmic Loading Tests

During the rhythmic tests, each test subject jumped to a beat on an individual force plate and eight accelerometers were affixed to the structure in a grid pattern surrounding the test subjects. Between 9 and 12 channels of data were collected: 1 – 4 force plates and 8 accelerometers. The test subjects began jumping to the audio beat and once all subjects were reasonably synchronized (usually less than 10 seconds), the data acquisition was initiated and continued for 24 seconds. It was determined that 24 seconds is a realistic time frame to expect reasonably healthy test subjects to jump without becoming tired and is sufficiently long enough to allow post-processing filtering to yield a length of filtered acceleration data adequate for application in the load estimation method. The majority of rhythmic jump tests were performed twice and the data with the most consistent magnitude force was selected for analysis.

Table 2 displays the different combinations of test frequency and group size that were utilized. The frequencies of the songs were determined using beat analyzing software and were chosen to demonstrate the load estimation method across the frequency range typical for dynamic crowd loading [7]. The two songs at 2.25 and 2.37 Hz were selected because of their widespread and frequent use at sporting events. It was also established that the selected frequencies were not likely to result in a resonant condition with the structure. The only potential exception to this is with the 4th harmonic of Kernkraft 400 (2.37 Hz x 4 = 9.48 ~ 9.6 Hz). This potential will be discussed with regards to the results of the rhythmic loading tests.

Table 2: Summary of various rhythmic loading tests

		Group Size		
Frequency (Hz)	**Reference signal / Song (Artist)**	**1**	**2**	**4**
1.80	*Metronome*		X	X
2.00	*Metronome*		X	X
2.25	Sandstorm (Darude)	X	X	X
2.37	Kernkraft 400 (Zombie Nation)	X	X	X
2.80	Miracle (fast) (Cascada)	X	X	
2.90	Truly, Madly, Deeply (fast) (Cascada)	X	X	

The load estimation method previously outlined was then applied to the average of the eight acceleration signals and the estimated forcing function was compared to the measured forcing function, as illustrated in Step 4 in Figure 1. This comparison is presented here for validation of the method but would not be available for normal applications thus necessitating this estimation method.

Load Estimation Method Results

The dynamic load factors were first derived from the experimentally measured force data by filtering into harmonic components. The dynamic load factors were also estimated from the acceleration response of the structure utilizing the load estimation method applied in Matlab. Both sets of DLFs are summarized in Table 3 along with the recommended DLFs for design purposes specified in AISC's Design Guide 11 [8]. When the DLFs are utilized in conjunction with recommended phase lag factors to generate the forcing function, the results are graphically compared with the experimentally measured force as shown in Figure 5. The bumps in the estimated load at the expected time for the occupant to be in the air (zero expected dynamic force) are due in part to the limited number of harmonics, three, included in the estimated load.

Table 3: Summary of DLFs derived from LEM and experimentally measured force

Frequency (Hz)	Group Size	Experimental Force			Acceleration (LEM)		
		r_1	r_2	r_3	r_1	r_2	r_3
1.8	2	1.47	0.73	0.14	1.49	0.74	0.16
	4	1.43	0.64	0.10	1.51	0.68	0.13
2.0	2	1.52	0.82	0.21	1.48	0.78	0.21
	4	1.32	0.50	0.12	1.38	0.53	0.12
2.25	1	1.56	0.87	0.24	1.67	0.91	0.25
	2	1.41	0.48	0.09	1.37	0.47	0.08
	4	1.65	0.74	0.12	1.39	0.62	0.11
2.37	1	1.58	0.76	0.16	1.63	0.75	0.16
	2	1.58	0.76	0.19	1.64	0.75	0.17
	4	1.55	0.72	0.14	1.52	0.68	0.13
2.8	1	1.53	0.69	0.10	1.58	0.67	0.09
	2	1.49	0.62	0.08	1.60	0.61	0.07
2.9	1	1.47	0.70	0.11	1.47	0.65	0.08
	2	1.42	0.64	0.10	1.48	0.60	0.06
DG11 DLF Recommendations					1.50	0.60	0.10

DISCUSSION

The discussion of the results presented herein is limited by several factors associated with the test setup. These limitations must be taken into account when considering the application of this method to other structures and events. The rhythmic loading undertaken in this study is limited to groups of four or less people. Small groups of this size are not necessarily representative of larger groups and results cannot easily be extrapolated [7]. Despite this limitation, the application and acceptability of the results of this study do indicate that the method can produce meaningful and reasonably accurate results at a small-scale and gives reason to expand the preliminary testing to larger structures and crowds.

A second limitation of this study is the coordination and participation of the occupants performing the rhythmic activity. It has been shown that synchronization or coordination of a crowd is more difficult as the size of the crowd increases [7]. For small groups of four or less subjects, synchronization was achievable. The effects of lack of coordination of the crowd or activity that is out-of-phase with the primary rhythmic activity are not explored and may affect the accuracy of the load estimation method. In this study, all of the occupants of the floor actively participated in the rhythmic activity. This is certainly not the case with real-world crowd activity where certain members of the crowd are not participating in the rhythmic activity but rather acting as a passive occupant of the structure. It has been shown that passive occupants are more likely to maintain contact with the structure and this interaction, if present, has the potential to alter the properties of the dynamic system [1]. This

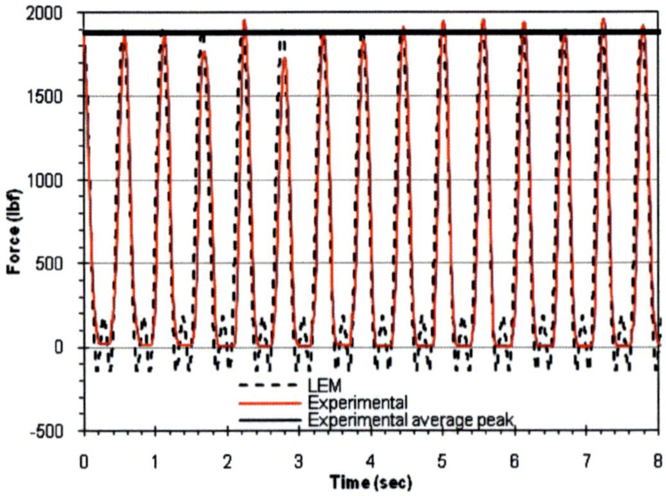

Figure 5: Typical comparison of estimated forcing function with experimentally measured force

phenomenon of human-structure interaction is not completely understood at this time and requires further research to evaluate the effects of the passive crowd.

In this study of a single structure, the natural frequency of the structure is a distinct data point which can affect the load estimation method. The natural frequency of the test structure is determined to be at 9.6 Hz, significantly higher than the possible forcing frequency of crowd activity. Because it is significantly higher, resonance with one of the first three harmonics of the forcing frequency is unlikely and the appropriateness of the method is confirmed. The application of this method to other structures must carefully consider the effect of the natural frequency and possible resonant conditions.

Despite these limitations, the results presented here are indicative of the applicability and beneficial nature of the load estimation in the determination of an appropriate design load for dynamic crowd excitation when considering both strength and serviceability. Expansion of the study with respect to these three areas of limitations is expected to reinforce these findings.

Dynamic Load Factor Estimation

Dynamic excitation due to human occupants is rarely concentrated at a single frequency but the energy associated with the dynamic force is distributed at the harmonics of the forcing frequency. This distribution of energy is represented by the dynamic load factors associated with each harmonic frequency. It is important that the distribution of the forcing energy, and therefore the dynamic load factors, be understood as they will affect the response of the structure. The DLFs estimated from the experimentally measured force, shown in Table 3, are compared to the recommendations for DLFs [8] to be used for serviceability design/assessment purposes. The average of the DLF for the first harmonic is exactly 1.50, equal to the recommended DLF. The average DLFs for the second and third harmonics are 0.69 and 0.13 respectively. These DLFs are slightly higher than the recommended DLFs of 0.60 and 0.10 indicating that the second and third harmonics of the forcing frequency may play a larger role in the excitation force than currently recommended. This could be particularly important when resonance with the second or third harmonic is possible. The energy of the crowd excitation at that particular frequency would be underestimated and therefore the structural response would also be underestimated.

The results in Table 3 indicate that the load estimation method produces reasonable estimates of the DLF for the first and second harmonics as determined from the actual measured dynamic force. The difference between the DLF estimated from the load estimation method was less than 8% for all rhythmic loading tests except one. For four people jumping at 2.25 Hz, the load estimation method underestimated the first DLF by more than 15%. Interestingly, the load estimation method also underestimated the second and third DLF significantly for this particular rhythmic test. The estimated DLFs for the third harmonic are much less accurate and it is interesting to note that the estimation of the third DLF seems to be more influenced by the excitation frequency as the accuracy decreases at the extremes of the forcing frequency range of this study: 1.8 or 2.8/2.9 Hz.

Force Magnitude Estimation

In addition to the relative distribution of the energy associated with the dynamic excitation, the magnitude of the force, albeit related to the distribution of the energy, is also important. The magnitude is also slightly influenced by the phase lag factors associated with each of the harmonics. The phase lags are related to the contact time between the occupant and the structure during impact. This experimentally determined contact time is in the range of the contact time recommended for high impact rhythmic activities such as jumping jacks [9]. Because exact calculation phase lag factors is not possible without measuring the applied force, the phase lags calculated for the recommended contact times are utilized in this study. The ratio of the contact time to the period of the forcing frequency is assumed to be $\alpha=1/2$ resulting in phase lag factors of $\phi_1 = 0$, $\phi_2 = -\pi/2$, and $\phi_3 = 0$.

The maximum force of the estimated load, derived with the estimated DLFs and the assumed phase lag factors, is compared to the average of the peak forces measured with each impact is shown in Figure 5 as the horizontal line at approximately 1900 lb for that particular test. The estimated load underestimates the measured load by approximately 5%. This underestimation can be attributed to any of a series of issues: the tolerances in the estimation of the DLFs, the differences between the actual and assumed phase lag factors, and the inclusion of only the first three harmonics of the forcing frequency. Despite the slight underestimation, the estimated dynamic load is still appropriate for serviceability design because the perception and related assessment thresholds are recommended levels or ranges, not limits, as perception and comfort levels differ from individual to individual.

With this in mind, the estimation of the magnitude of the dynamic load to within 5% of the maximum is deemed acceptable for serviceability design and assessment.

CONCLUSION AND RECOMMENDATIONS

The load estimation method, proposed by Ellis and Littler [2], is applied to a series of rhythmic loading scenarios on a small scale to confirm the method and identify any potential difficulties. The rhythmic loading tests were performed on an interior bridge-type structure with groups of 1 - 4 subjects jumping at a range of frequencies. To confirm the accuracy of the estimation procedures, the jumping loads were measured by individual force plates and compared with the estimated load. The comparison of the estimated and the measured rhythmic load exhibited nearly identical time histories and the peak estimated load was, on average, only 5% less than the measured load.

The distribution of the energy associated with the dynamic loading at each of the harmonics is critical to understanding a structure's response to the load as the response can be significantly impacted if resonance with one of the harmonics is possible. This distribution of energy is indicated by the dynamic load factors or Fourier coefficients for each harmonic. A comparison of the dynamic load factors from the measured dynamic load and the estimated load revealed differences of less than 8% for the first and second harmonic. Estimation of the dynamic load factor for the third harmonic is less accurate but also influences the dynamic load and structural response to a lesser degree.

The dynamic load factors determined from both the measured load and the estimated load are also compared with the recommended dynamic load factors for rhythmic excitation as recommended in the leading publication providing guidance for structural designers [8]. The recommended value for the first dynamic load factor is within the 95% confidence interval of this data, but the recommended values for the second and third dynamic load factors is slightly lower than the 95% confidence interval of this data. Because this test involves only small groups of 1-4 subjects and it has been shown that the dynamic load factors for larger groups are lower, the recommended values for the dynamic load factors are considered appropriate for design purposes. The load estimation method examined in this study provides a beneficial method of estimating the actual loading induced by a crowd and may offer insight into how other crowd variables such as coordination and human-structure interaction affect the loading or the response of the structure.

REFERENCES

1. Sachse, R., Pavic, A. and Reynolds, P. (2003), "Human-Structure Dynamic Interaction in Civil Engineering Dynamics : A Literature Review," *The Shock and Vibration Digest,* 35 January 2003, 3-18.
2. Ellis, B.R. and Littler, J.D. (2004), "Response of cantilever grandstands to crowd loads: Part 1: serviceability evaluation," *Proceedings of the Institution of Civil Engineers, Structures and Buildings,* v 157, n4, 235-241.
3. Ellis, B.R. and Littler, J.D. (2004), "Response of cantilever grandstands to crowd loads: Part 2: load estimation," *Proceedings of the Institution of Civil Engineers, Structures and Buildings,* v 157, n5, 297-307.
4. Ellis, B.R.; Ji, T. (2002); "Loads generated by jumping crowds: experimental assessment," BRE IP4/02, CRC Ltd., London, p 1-12.
5. Computers and Structures, Inc. (2005), *SAP2000 Manual: Linear and Nonlinear Static and Dynamic Analysis and Design of Three-dimensional Structures.*
6. Maia, N.M.M. and Silva, J.M.M. (ed.) (1997). Theoretical and Experimental Modal Analysis. Research Studies Press Ltd.
7. Ellis, B.R.; Ji, T.; Littler, J.D. (2000); "The Response of grandstands to dynamic crowd loads," *Proceedings of the Institution of Civil Engineers*, Structures and Buildings, v140, n4, Nov, 2000, 355-365.
8. Murray, T.M., Allen, D.E. and Ungar, E.E. (1997), "Floor Vibrations due to Human Activity," *Steel Design Guide Series, No. 11,* American Institute of Steel Construction (AISC).
9. Pernica, G. (1990), "Dynamic Load Factors for Pedestrian Movements and Rhythmic Exercises," *Canadian Acoustics,* 18(2), 3-18.

Proceedings of the IMAC-XXVIII
February 1–4, 2010, Jacksonville, Florida USA

PREDICTION OF PRESTRESS FORCE ON GROUTED TENDON BY EXPERIMENTAL MODAL ANALYSIS

Jung-Bum Jang[a], Hong-Pyo Lee[a], Kyeong-Min Hwang[a], Young-Chul Song[a]
[a] Nuclear Power Lab., KEPRI, 103-16 Munji-Dong, Yuseong-Gu, Daejeon, 305-380, Republic of Korea

ABSTRACT

The grouted tendon has been adopted to the containment building of some operating nuclear power plants in Korea and assessment of the prestress force on the grouted tendon is being issued as an important pending problem for the continuous operation beyond their design life. In order to assess the prestress force on the grouted tendon at present, indirect assessment techniques have being applied but they have limitations to accurately identify the real prestress force.

Therefore, the long-term research program was begun to assess the prestress force on the grouted tendon of the containment building using nondestructive techniques. As a first step of the research program, the experimental modal analysis was carried out using 6 post-tensioned concrete beams to obtain the variation of natural frequency according to the level of prestress force. The natural frequencies of post-tensioned concrete beams were calculated using numerical analysis with measured natural frequencies and the application possibility of natural frequency to predict the prestress forces of the post-tensioned concrete beams was confirmed.

1. EXPERIMENTAL MODAL ANALYSIS

1.1 Post-tensioned Concrete Beam

In order to investigate the variation of natural frequency according to the level of prestress force at the grouted tendon, post-tensioned concrete beams with the grouted tendon type were manufactured and they were representative of real wall and post-tensioning system of containment building at the nuclear power plant. The total number of post-tensioned concrete beams is 6 and their length is 8.0 m and area is 0.09(0.3×0.3 m) m^2. The 28-day compressive strength of concrete is 37.0 MPa, unit weight of concrete is 22.5 kN/m^3 and post-tensioning system is located at center of the post-tensioned concrete beam. Load cell was installed at the one end to measure accurately the prestress force which was applied to the post-tensioned concrete beam, as shown in Fig. 1. The anchorage type for prestressed strand is VSL Type P and the diameter of prestressed strand is 15.2 mm and 3 prestressed strands were used to apply the design prestress force. The final prestress forces which were applied to each post-tensioned concrete beam, are 0 kN, 146 kN, 264 kN, 356 kN, 465 kN, and 523 kN.

T. Proulx (ed.), *Dynamics of Civil Structures, Volume 4*, Conference Proceedings of the Society for Experimental Mechanics Series 13,
DOI 10.1007/978-1-4419-9831-6_5, © The Society for Experimental Mechanics, Inc. 2011

1.2 Experimental Modal Analysis

In order to get the natural frequency of post-tensioned concrete beams according to the level of prestress force, experimental modal tests were carried out. Among the experimental modal tests, impact hammer test and MIMO(Multi Input Multi Output) sweep test were adopted to get the natural frequency of post-tensioned concrete beams. The acceleration occurring from both tests, was measured by piezo-electrical sensors. In order to get the various mode shapes, 9 sensors were used as shown in Fig. 2. They were placed on top of the post-tensioned concrete beam and a sampling rate of 512 Hz was used.

Fig. 1. Post-tensioned concrete beam and load cell

Also, the loading point of an impact hammer and an exciter is P7 in Fig. 2. Impact hammer test was carried out 5 times to each post-tensioned concrete beam and the measured data were averaged. The frequency range of the impact hammer is from 0 Hz to 256 Hz. The loading magnitude of the exciter is 20 N and the frequency range of the exciter is from 2 Hz to 150 Hz.

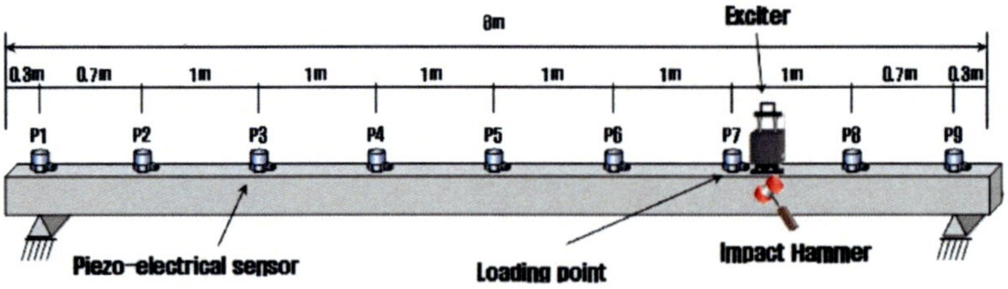

Fig. 2. Test setup

2. TEST RESULTS

2.1 Impact Hammer Test

Fig. 3 shows the impact signal of the impact hammer and Fig. 4 shows the Frequency Response Function(FRF) according to the level of prestress force. Table 1 indicates the natural frequencies corresponding to the each modal number and prestress force. The natural frequencies of post-tensioned concrete beams were generally proportional to the level of prestress force as shown in Table 1.

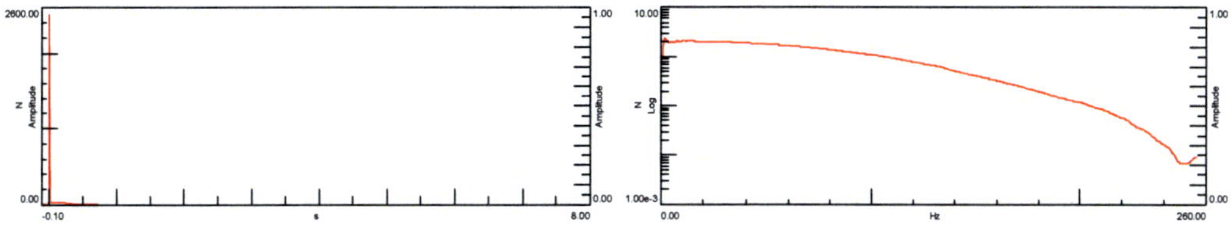

(a) Time domain (b) Frequency domain

Fig. 3. Impact signal

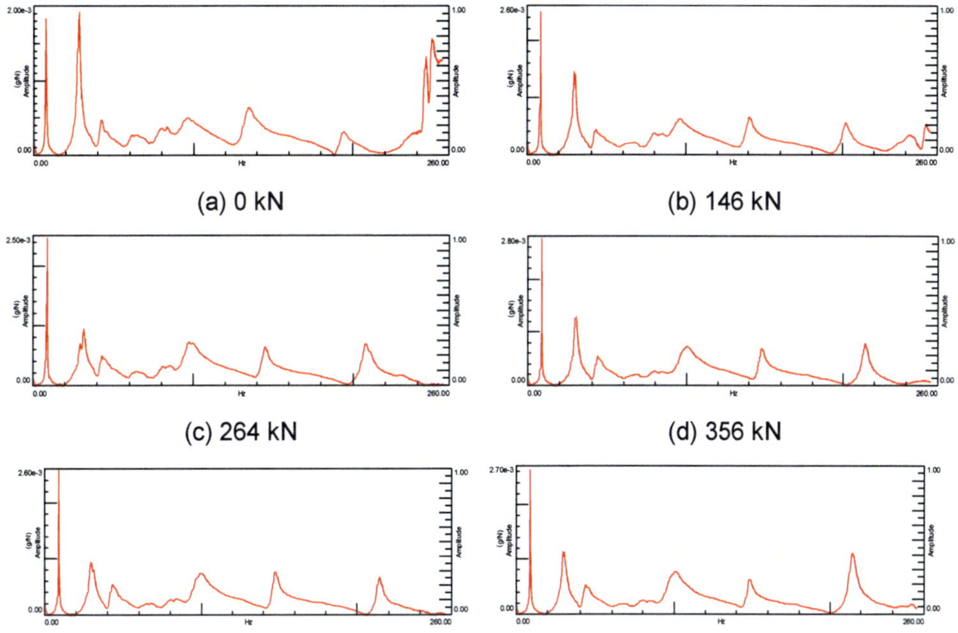

(a) 0 kN (b) 146 kN

(c) 264 kN (d) 356 kN

(e) 465 kN (f) 523 kN

Fig. 4. Frequency response functions by impact hammer test

Table 1. Natural frequencies by impact hammer test

Prestress force (kN)	1st mode (Hz)	2nd mode (Hz)	3rd mode (Hz)	4th mode (Hz)
0	7.513	28.183	94.876	132.999
146	8.216	29.399	95.052	140.023
264	8.539	30.976	97.334	145.087
356	8.724	29.997	99.418	147.913
465	8.697	30.126	98.812	147.480
523	8.794	29.850	99.466	148.334

2.2 MIMO Sweep Test

Fig. 5 shows the FRF according to the level of prestress force and Table 2 indicates the natural frequencies corresponding to the each modal number and prestress force. Also, the trend of natural frequencies to the level of prestress force by MIMO sweep test was almost same as that of natural frequencies by impact hammer test.

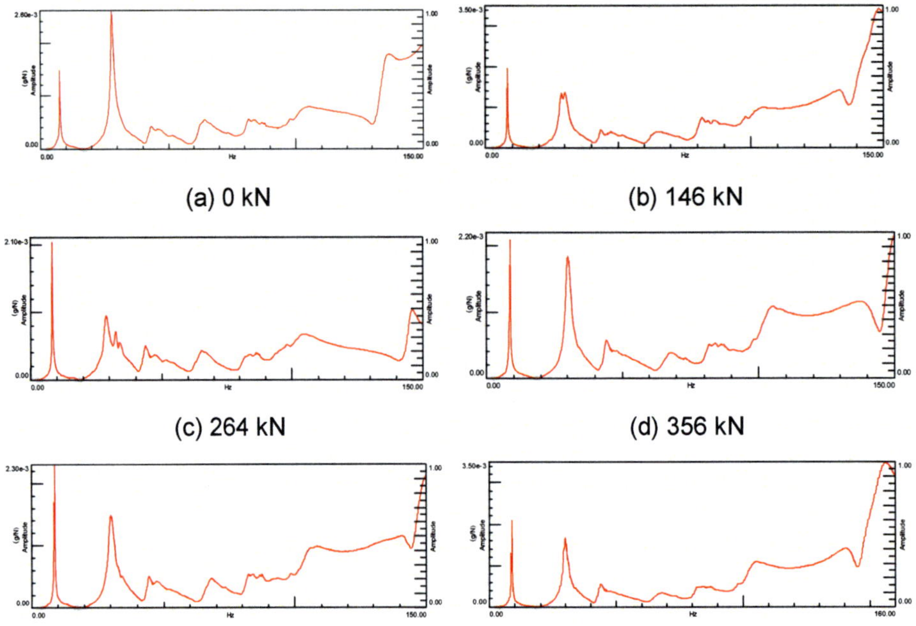

(a) 0 kN (b) 146 kN

(c) 264 kN (d) 356 kN

(e) 465 kN (f) 523 kN

Fig. 5. Frequency response functions by MIMO sweep test

Table 2. Natural frequencies by MIMO sweep test

Prestress force (kN)	1st mode (Hz)	2nd mode (Hz)	3rd mode (Hz)	4th mode (Hz)
0	7.567	28.223	101.095	133.812
146	8.190	26.609	100.727	135.912
264	8.498	29.583	102.160	145.208
356	8.672	30.269	102.814	144.053
465	8.690	30.331	104.474	144.280
523	8.757	30.241	103.231	144.522

In order to examine the similarity of natural frequencies got by both impact hammer test and MIMO sweep test, difference ratio of natural frequency classified by modal number according to the level of the prestress force, was compared. 1st modal natural frequency shows the highest similarity with minimum difference ratio of 0.71 % and the 2nd modal natural frequency shows the lowest similarity with maximum difference ratio of 10.5 %. Therefore, both tests showed the good accordance to the natural frequency according to the level of prestress force and both test data are available for predicting the prestress force of grouted tendon. As shown in Table 1 and 2, MIMO sweep test showed more consistent relationship between natural frequency and prestress force than impact hammer test and 1st modal natural frequency is the best data to predict the prestress force of grouted tendon among 4 modal natural frequencies.

3. EFFECTIVE FLEXURAL RIGIDITY

3.1 Effective Flexural Rigidity

It is very difficult to calculate the natural frequency of post-tensioned concrete beam with prestress force using the modal analysis by general structural analysis programs because the prestress force can't be inputted to the modal analysis of general structural analysis programs. Therefore, prestress force should be converted to the effective flexural rigidity for modal analysis to calculate the natural frequency of post-tensioned concrete beam.

In order to derive the effective flexural rigidity, the following equation for natural frequency of vibration was used for a simply supported prismatic beam.

$$\omega_n^2 = -\left(\frac{n\pi}{L}\right)^2 \frac{N}{m} + \left(\frac{n\pi}{L}\right)^4 \frac{EI}{m} \tag{1}$$

Where n = mode number, L = span length, N = axial compressive force, m = beam mass per unit length, E = modulus of elasticity, and I = moment of inertia for the beam section. The effective flexural rigidity EI can be expressed as a function of both natural frequency and prestress force from Eq. 1 and the resulting equation is as follows.

$$EI = \left(\frac{L}{n\pi}\right)^2 N + \left(\frac{L}{n\pi}\right)^4 m\omega_n^2 \tag{2}$$

3.2 Numerical Analysis Model

Numerical analyses were carried out to calculate the natural frequencies of 6 post-tensioned concrete beams using SAP 2000. Beam element was selected to model the post-tensioned concrete beam and the hinge supported condition was adopted as the boundary condition as shown in Fig. 6. Also, the prestress force was considered as the effective flexural rigidity using Eq. 2.

In order to verify the reliability of the numerical analysis model, modal analysis was performed to the case without prestress force. Table 3 indicates the numerical analysis results and analytical case 1 was in case that concrete elastic modulus of the post-tensioned concrete beam determined by concrete material test, was inputted to the modal analysis and analytical case 2 was in case that the effective flexural rigidity determined by Eq. 2, was inputted to the modal analysis and measured data are 1st modal natural frequency by MIMO sweep test.

Comparing natural frequencies by both analytical case 1 and 2 with measured natural frequency by MIMO sweep test, the numerical analysis model well estimated the natural frequency of post-tensioned concrete beams without prestress force within the difference ratio range from 0.2 % to 1.2 %. Therefore, the reliability of the numerical analysis model and the effective flexural rigidity were confirmed.

Fig. 6. Numerical analysis model

Table 3. Numerical analysis results without prestress force

	Analytical case 1	Analytical case 2	Measured data
Natural frequency (Hz)	7.477	7.552	7.567
Difference ratio (%)	-1.2	-0.2	

Using the numerical analysis model and the effective flexural rigidity, modal analyses were performed to calculate the natural frequencies of 5 post-tensioned concrete beams with prestress force. Table 4 indicates the numerical analysis results and difference ratios between numerical analysis and MIMO sweep test were very small. Also, difference ratio becomes larger in proportion to the level of prestress force but maximum difference ratio may be less than 5 % because the range of prestress force which has been applied to the real structure, is less than 465 kN. Therefore, natural frequency measured by experimental modal analysis is available for the prediction of prestress force on grouted tendon.

Table 4. Numerical analysis results with prestress force

Prestress force (kN)	Natural frequency (Hz)		Difference ratio (%)
	Numerical analysis	MIMO sweep test	
146	8.34	8.19	1.8
264	8.77	8.50	3.2
356	9.03	8.67	4.2
465	9.16	8.69	5.4
523	9.28	8.76	6.0

4. CONCLUSION

As a first step to develop the nondestructive techniques to identify the prestress force of the grouted tendon, the experimental modal analysis was carried out using 6 post-tensioned concrete beams with the grouted tendon type.

Also, in order to consider the prestress force in the modal analysis, the effective flexural rigidity was deduced from the equation for natural frequency of vibration and numerical analyses were performed to calculate the natural frequency of 6 post-tensioned concrete beams.

As a result of tests, both impact hammer test and MIMO sweep test showed the good accordance to the natural frequencies according to the level of prestress force and the natural frequencies of post-tensioned concrete beams by two tests were generally proportional to the level of prestress force. Also, it were confirmed that the reliability of both numerical analysis model and the effective flexural rigidity, and relatively good accordance between analytical and measured natural frequencies. Therefore, natural frequency measured by experimental modal analysis is available for the prediction of prestress force on grouted tendon.

ACKNOWLEDGEMENTS

This research is sponsored by the Korean Ministry of Knowledge Economy under long-term atomic energy R&D program.

REFERENCES

[1] Law, S.S., Lu, Z.R., “Time domain responses of a prestressed beam and prestress identification”, Journal of sound and vibration, 2005, 1011-1025.

[2] Hamed, E., Frostig, Y., “Natural frequencies of bonded and unbonded prestressed beams-prestress force effects”, Journal of sound and vibration, 2006, 28-39.

[3] Chen, H.L., Wissawapaisal, K., “Measurement of tensile forces in a seven-wire prestressing strand using stress waves”, Journal of engineering mechanics, 2001, 599-606.

[4] Saiidi, M., Douglas, B., Feng, S., “Prestress force effect on vibration frequency of concrete bridges”, Journal of structural engineering, 1994, 2233-2241

Proceedings of the IMAC-XXVIII
February 1–4, 2010, Jacksonville, Florida USA
©2010 Society for Experimental Mechanics Inc.

Finite element model selection using Particle Swarm Optimization

Linda Mthembu[1], Tshilidzi Marwala[2], Michael I. Friswell[3], Sondipon Adhikari[4]

[1]Visiting Researcher, Department of Electronic and Computer Engineering, Faculty of Engineering and Built Environment, University of Johannesburg, PO Box 17011, Doornfontein 2028, South Africa.

[2]Executive Dean Faculty of Engineering and Built Environment, University of Johannesburg, PO Box 17011, Doornfontein, 2028, South Africa, South Africa.

[3]Professor of Aerospace Structures, School of Engineering, Swansea University, Singleton Park, Swansea SA2 8PP, United Kingdom.

[4]Chair of Aerospace Engineering, School of Engineering, Swansea University, Singleton Park, Swansea SA2 8PP, United Kingdom

Abstract

This paper proposes the application of particle swarm optimization (PSO) to the problem of finite element model (FEM) *selection.* This problem arises when a choice of the best model for a system has to be made from set of competing models, each developed a priori from engineering judgment. PSO is a population-based stochastic search algorithm inspired by the behaviour of biological entities in nature when they are foraging for resources. Each potentially correct model is represented as a particle that exhibits both individualistic and group behaviour. Each particle moves within the model search space looking for the best solution by updating the parameters values that define it. The most important step in the particle swarm algorithm is the method of representing models which should take into account the number, location and variables of parameters to be updated. One example structural system is used to show the applicability of PSO in finding an optimal FEM. An optimal model is defined as the model that has the least number of updated parameters and has the smallest parameter variable variation from the mean material properties. Two different objective functions are used to compare performance of the PSO algorithm.

Nomenclature

FEM	Finite element model.
FEMU	Finite element model updating.
PSO	Particle Swarm Optimization.
d	Model Dimension.
m_{id}	i-th Finite element model position at parameter d.
v_{id}	i-th Finite element model velocity at parameter d.
p_{id}	Best position for the ith Finite element model at parameter d.
p_{gd}	Global best finite element model at parameter d.
w_k	The k-th dimension inertia weight.
M_{max}, M_{min}	Maximum and minimum model position respectively.
V_{max}, V_{min}	Maximum and minimum model velocity respectively.
E_i	Young's Modulus.
AIC	Akaike Information Criterion.
SSE	Sum of Squared Errors.
μ	Mean
$msrd_{data}$	Measured structural results.
$fem_{results}$	Finite element model results.

1. Introduction

The finite element model updating (FEMU) problem arises due to the mismatch between the initial finite element model results and the measure real system results [7, 9]. The modeller's problem is then to determine which aspects or feature of the initial model are uncertain or incorrectly modelled. This is effectively a system-identification problem [15]. In classic system-identification the real measured system/data is approximated by a set of mathematical equations, usually polynomial equations whose parameters or order is unknown and is to be identified. In this paper the system is a structural system which is described by a finite element model and what needs to be identified are the uncertain parameters of the initial FE model [8, 9].

In the literature there are two main directions to finite element model updating (FEMU); direct and indirect (iterative) methods [9]. In the direct model updating paradigm [4, 9] the measured modal data are directly equated to the model modal values thus freeing up the system matrices (the mass, damping and or stiffness matrix elements) for updating. This approach often results in unrealistic system matrix element magnitudes, for example, physically unrealisable mass elements. In the indirect or iterative model updating approach the updating problem is formulated as an optimization problem, often approached by the use of least squares, maximum likelihood, eigenvalue sensitivity, genetic algorithm optimization and Bayesian approaches [1, 2, 8, 9, 12, 13, 16, 17, 18, 20, 21 and 26]. This paper proposes an iterative scheme to model updating.

This paper continues addressing the FEMU problem with the approach proposed in [20]. This approach is based on directly answering three standard questions in FEM updating; (a) which aspects of the models do we need to update? (b) how are we to update these models? (c) is the updated model the best one? In [20, 21] a particular Bayesian approach to model updating was presented. The methods considered updating FE models using Bayesian stochastic techniques from which one can determine the best model in a given group. The approaches presented in [20, 21] and in this paper effectively propose performing finite element model updating within the model selection framework. This whole approach is based on the premise that different analysts would differ (for example the updating parameters of the GARTEUR structure in [14]) on which aspects of the initial finite element model are incorrectly modelled to answer question (a), and how to proceed with the model updating (when answering question (b)). This disagreement between researchers on the first two questions often makes it difficult to compare FEM updating results of the same structure let alone different updating techniques. To answer questions (b) and (c) we assume that a number of pre-existing potentially correct finite element models of a *particular* structure have been developed. These models could have been generated through engineering judgement or competing analysis by different people/techniques. Given the problem described in the previous paragraph, the problem is then to develop a method of both updating all the models and then selecting the best performing one. Ideally one would like an all-in-one procedure of both updating and selection. This would make it easier to compare the model updating results as they would have been updated 'similarly'. The particle swarm optimization framework proposed in this article allows this simultaneous updating of all competing models and the identification of the best model in the given group.

2. Particle Swarm optimization(PS0)

Particle swarm optimization was first developed by [10]. PSO is a population-based stochastic search algorithm inspired by the social-psychological behaviour of biological entities in nature when they are foraging for resources. The population/swarm of entities in nature could be that of birds, fish and or ants etc searching for food [5, 10, 11]. Each entity in the swarm is able to dynamically adapt individually and through group influence to the environment while in search of resources. The swarm adapts by stochastic moving towards previously good regions in the environment. This means the movement of the swarm in the search space has some random elements to it but this movement tends to converge to optimal points in the search space.

The swarm behaviour metaphor has been adopted by the evolutionary computation community [5, 11, 13] where the biological entities are called particles, the swarm is called a population, the environment is the solution space and the resource is the solution to the problem. One of the main differences between evolutionary and classic swarm based algorithms is the way the particles interact. In a typical evolutionary algorithm (e.g. Genetic algorithm [13]), particles combine and mutate within a population and over generations. In swarm based approaches for example, particles communicate instead of merging. There is no evidence of one method being superior to the other but consensus is that these methods are well suited to problems where the solution search space is too large to search exhaustively [6, 11, 24, 26].The practical interpretation of this analogy is that a number of particles stochastically 'move' through the problem search space searching for the minimum/maximum solution point. This means *each* considered particle can potentially find the optimal solution to the problem. In the FEMU context and approach proposed in this paper; each particle is a potentially correct model to the finite element modelling problem. The particle search space is defined by the number of updating parameters in the models; specifically the maximum number of updating parameters defines the complete search space. Obviously if the models do not have the same number of free parameters then some models will only search a subset of the full search space nonetheless they will be embedded in the full space (see section 4.1 for the chosen particle representation).

In the next section the mathematical operators of the PSO algorithm are presented.

2.1 PSO Operators

In the PSO algorithm each particle is described by its vector position in the search space as in equation 1 below;

$$m_i = \{m_{i1}, m_{i2}, m_{i3} \ldots m_{id}\} \quad (1)$$

where i is an arbitrary particle and d is the problem dimension defined by the maximum number of updating parameters for the models. The particle position features (m_{id}'s) are the potential solution variables. This means to evaluate each particle on the problem one substitutes the particle position to the model/function. The rate of the i-th particle's position change, the velocity, is represented by;

$$v_i = \{v_{i1}, v_{i2}, v_{i3} \ldots v_{id}\} \quad (2)$$

Each particle also stores its (local) best ever position as it searches the problem space. This is represented by;

$$p_i = \{p_{i1}, p_{i2}, p_{i3} \ldots p_{id}\} \quad (3)$$

The swarm also has a record of the best ever position by any particle, this is known as the global best solution (socially) which is represented by;

$$p_g = \{p_{g1}, p_{g2}, p_{g3} \ldots p_{gd}\} \quad (4)$$

On each iteration of the PSO algorithm the position and velocity of each particle are updated. The particle velocity is updated through the following equation;

$$v_{ik}(t) = w_k v_k(t-1) + c_1 r_1 (p_{ik} - m_{ik}) + c_2 r_2 (p_{gk} - m_{ik}) \quad (5)$$

and then the particle position is updated using equation 6;

$$m_{ik}(t) = m_{ik}(t-1) + v_{ik}(t) \quad (6)$$

where $\{i \in 1 \ldots m\}$ and $\{k \in 1 \ldots d\}$ which means each particle's position and velocity parameters /dimensions are updated on each iteration of the algorithm. See section 4 for the PSO algorithm pseudo code. In equation 6, c_1 and $c_2 \in \mathbb{R}$ are constants weighting that normally vary between 2 and 4 [10]. The random constants $r_1, r_2 \sim U[0,1]$ introduce randomness to the search process. In order to prevent the tendency of the particle position and velocity to explode in magnitude, M_{max}, M_{min}, V_{max} and V_{min} are defined for each particle dimension. Thus if

$m_{ik} > M_{max}$ then $m_{ik} = M_{max}$
$m_{ik} < M_{min}$ then $m_{ik} = M_{min}$

Similarly if

$v_{ik} > V_{max}$ then $v_{ik} = V_{max}$
$v_{ik} < V_{min}$ then $v_{ik} = V_{min}$;

The setting of these limits on each dimension of the problem would depend on the analyst's understanding of the problem. The types of constraints are also very much dependent on the problem, for example some particle dimensions might be known to be constrained to be positive values, and in that case the absolute $|M_{max}|$ or $|V_{max}|$ might be applicable [10,11, 25].

A modification to the original PSO algorithm by [24] introduced w_k, the inertia weight variable. It controls the influence of the previous velocity on the current velocity value. An adaptive inertia weight is often used to improve the algorithm's search from an initially explorative (global) search to a more local search as this variable decreases. This also has a tendency to improve the algorithm's convergence rate [11, 24]. This variable is specified by the starting weight (w_{start}), w_f is the fraction of iterations over which the inertia weight is decreased and w_{end} the final inertia weight value The initial w_k in equation 5 is w_{start} and it is decrease by $w_k = w_k - w_{dec}$ where

$$w_{dec} = \frac{w_{start} - w_{end}}{N - w_f} \tag{7}$$

from the first iteration up to iteration N x w_f, thereafter w_k is w_{end}. The (p_{ik}- m_{ik}) term in equation 5 measures how far each particle is currently from its personal best position (local) and (p_{gk}- m_{ik}) measures how far each particle is from the global (social) best particle in the swarm. This means the middle term in equation 6 tends to control the particle's velocity based on the particles own best position while the last term allows the particle to be influenced by the best performing particle in the swarm.

In the next section we present the finite element models and propose a particular representation of these in the particle swarm context.

3. Finite Element Models

The finite element models updated in this paper were all developed from the unsymmetrical h-beam structure In [20] and previously used in [19]. The details of the modelled beam are described in the next section.

3.1 Unsymmetrical H-Beam

A simple unsymmetrical h-beam shown in figure 1 is modelled. This unsymmetrical h-beam is suspended on rubber bands (see [19] for more details on the structure and experimental set-up). The measured natural frequencies of interest of this structure occur at; 53.9Hz, 117.3Hz, 208.4Hz, 254Hz and 445Hz which correspond to modes 7, 8, 10, 11 and 13 respectively. The aluminium beam material has a Young's Modulus of $7.2\text{x}10^{10}$ Pa, the beam length is 600mm with a width of 32.2mm and a section thickness of 9.8 mm. The left edge has a length of 400mm, the right edge length 200mm and a density of 2793 kg/m^3.

Figure 1 shows that the beam is divided into elements numbered from one to twelve. Each cross-sectional area is 9.8mm by 32.2 mm. Each finite element model used standard isotropic material properties and Euler Bernoulli beam elements to approximate the beam sections of the structure. The beam is free to move in all six degrees.

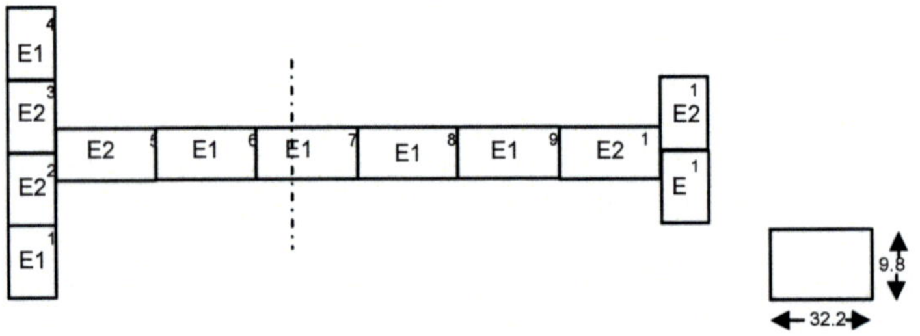

Figure 1: Model M_2 of a 12 Element Unsymmetrical H beam.

3.2 The mathematical models of the beam

All models in this example assume the only uncertain beam property is its Young's Modulus (E) value. To design different models of the beam, beam elements are grouped differently. The beam is modelled by eight competing models, m_i, $i = 1...8$. Model m_1 assumes the whole beam's Young's modulus is the updating parameter to be updated from the average given material value. Model m_2 has two parameter, E_1 and E_2; the elements numbered 1,4, 6,7,8,9 (all forming parameter E_1) are to be varied equally while elements 2, 3, 5, 10, 11, 12 (all E_2) are to be varied equally (see figure 1 for the element numberings).

Model m_2 models the elements connected near the structural joints as one parameter and those away from the joints as another. Model m_8 assumes the left edge together with the first horizontal element, the horizontal section and right edge together with the last horizontal element are best updated differently, thus the three parameter arrangement. Table 3 lists the rest of the models and their parameterizations.

Model Identity	Number of model parameters	Parameter Labels	Element grouping
m_1	1	E_1	{1-12}
m_2	2	E_1 & E_2	{1,4,6-9} & { 2,3,5,10-12}
m_3	3	E_1 E_2 E_3	{1,4,6-9}, {2,3,11,12} & {5,10}
m_4	4	E_1 E_2 E_3 E_4	{1,4,6-9}, {2,3} {11,12} & {5,10}
m_5	5	E_1 E_2 E_3 E_4 E_5	{1,4,6-9}, {2,3} {11,12},{5} & {10}
m_6	2	E_1 E_2	{1,2,3,4} & {5-12}
m_7	2	E_1 E_2	{1- 6} & { 7-12}
m_8	3	E_1 E_2 E_3	{1-5}, {6-9} & (10,11,12}

Table 1. Model Parameterization

Perhaps the most important step in the implementation of the PSO algorithm is the choice of particle representation. This fundamentally dictates the problem search space and the ease of algorithm implementation. The next section presents the model representation adopted in the current finite element model updating procedure.

4. PSO Algorithm

4.1 Particle Representation

Each particle or finite element model ($m_{i...m}$) is described by the following E_i vector arrangement:

$m_1 = [E_1,0,0,0,0]$; $m_2 = [E_1,E_2,0,0,0]$; $m_3 = [E_1,E_2,E_3,0,0]$; $m_4 = [E_1,E_2,E_3,E_4,0]$;

$m_5 = [E_1,E_2,E_3,E_4,E_5]$; $m_6 = [E_1,E_2,0,0,0]$; $m_7 = [E_1,E_2,0,0,0]$; $m_8 = [E_1,E_2,E_3,0,0]$;

where in each case

$$E_{i...5} = \mu + q\sqrt{0.5e20}\text{N.m}^{-2}. \tag{8}$$

and the mean

$$\mu = 7.2e10\text{N.m}^{-2}.$$

The q variable samples random numbers from a normal distribution between [-∞, ∞].The parameter *location* of the models as described in Table 1 nullifies the concern that models m_2, m_6 and m_7 seem to be described by the same parameter vector.

This choice of model representation sets the problem search space to five dimensions. Even though all the models search the five dimensional space, each is actually constrained to only searching a particular manifold of the space. This contextually means each subspace is assumed to have an optimum somewhere which the particle is suppose to find guided by individualistic and social performance. This means if model m_5 finds the best solution within the group (thus it is p_g) at some coordinate/parameter values, all the other particles will adapt towards model m_5's parameter values. In this particle representation, all the particle parameters will incrementally change towards model m_5's values, even the zeros in the particle vector description. But since, for example, model m_2, m_6 and m_7 are only dependent on the first and second updating values, it does not matter what happens to the zero features! Each model will somewhat also resist moving towards model m_5's coordinates by also incrementally moving towards their own previous best positions.

4.2 PSO Pseudo-Code

In this section the FEM-PSO pseudo-code algorithm together with the parameter settings used in the simulations is presented.

FEM-PSO Algorithm Pseudo Code

```
%Set constants
N = 500;        The max number of iterations
c1 = c2 = 2;    Individual and Group Influence
m = 8;          Number of potentially correct models
wstart = 1.2;   Initial inertia weight
wf= 0.5;        Inertia decrement factor
wend = 0.4;     Final inertia weight
%Initialise
wdec   - Using equation 7.
mi..m  - Randomly initialise the models
pi...m - Randomly initialise particle best solution (In this paper initial pi=mi)
% Compute
F(mi..m) - Calc model Fitness using the objective function in section 4.4
Identify Pg
Iteration =1
Repeat
  While iteration< N do
        for all mi..m do
          for each dimension d
               Calculate particle velocity(vi) using equation 5
               Update particle position (mi) using equation 6
          end for
          Compute F(mi)
          Update pi  if mi(t) > pi(t-1)
        end for
        Update pg if any pi > pg(t-1)
        If iteration < |N wf|
           wk = wk-wdec
         end if
     iteration = iteration +1;
  end while
return pg
```

Table 2. The FEM-PSO algorithm pseudo code

Obviously if the original PSO algorithm is implemented i.e. the inertia variable is eliminated, the last If statement condition is not executed and the velocity equation does not have w_k on the first term. The next section describes the particle fitness functions used in all the simulations.

4.3 Model parameter constraints

In our FEMU problem the constraints placed on the particle velocity and position in the algorithm were as follows; the maximum parameter magnitude M_{max} for each dimension was set at 7.5e10 N.m^{-2} and the minimum, M_{min}, was set at 5.5e10. The maximum velocity magnitude was set to the difference between M_{max} and M_{min}. The minimum velocity was set to 1e9 N.m^{-2}. This means the velocity in this algorithm was tracing a factor of the standard deviation of the parameter values (i.e. the second term in equation 8) and the particle position was determining the mean parameter value (E_i) in equation 8.

4.4 Objective Functions

A number of fitness or objective functions are available in the scientific literature. In the FEMU problem Occam's razor is very much applicable. This means one seeks a model with the fewest updating parameters that will produce FE model results closest to measured lab results. In this paper we compare two objective/fitness functions; the Akaike Information Criterion (AIC) [25] and the Squared Sum of Errors (SSE). The AIC function is given by the following equation:

$$AIC = n\log(\sigma^2) + 2d \tag{9}$$

where,

$$\sigma^2 = \frac{\sum_{i=1}^{n}(msrd_{data} - fem_{results})^2}{n}, \tag{10}$$

and d is the number of model parameters, n is the number of measured modes, $msrd_{data}$ is the measured data and $fem_{results}$ are the finite element model results. The squared sum of errors is given by

$$SSE = \frac{\sum_{i=1}^{n}(msrd_{data} - fem_{results})^2}{2} \tag{11}$$

As it can be seen from equations 9 to 11, the first term of equation 9 is effectively the SSE and is commonly referred to as the data-fit term [3, 20, 21, 23] and the second term is known as the model complexity penalty term. This assumes a models' complexity is determined by the number of free variables. This is not always the correct way to define model complexity as argued in [22, 23]. The minimization/maximization of the squared sum of error objective function does not account for model complexity but only measures how well a given model fits the data. We would expect the implementation of the AIC function as the objective function in the PSO algorithm to be biased to models with fewer parameters.

5. Simulation Results

The simulations presented in this section are all modelled using Version 6.0 of the Structural Dynamics Toolbox (SDT®) for Matlab. A number of simulations were run using different setting of the PSO algorithm parameters. Initially the number of iterations in all settings was set to *N*=1000 but it was found that the algorithm consistently converged before N= 500 iterations. In each of the experiments the convergence figures will only focus on the main convergence part of the graph, where necessary the figure will be expanded to show 500 iterations.

5.1 Simulation settings

PSO Parameter	Simulation No.1	Simulation No. 2	Simulation No. 3	Simulation No. 4
C_1 Local influence	2	2	2	2
C_2 Global Influence	2	2	2	2
w	0	0	Adaptive	Adaptive
Objective function	AIC	SSE	AIC	SSE

Table 3. PSO Simulation parameter settings

5.1.1 Simulation number 1

In this simulation the original PSO algorithm is implemented on the FEMU problem, this means there is no inertia variable as shown in table 3. Figure 2 shows the convergence plot of the AIC objective function over the 200 iterations of the algorithm. Figure 3 shows the variation of the best particle (p_g) in the swarm over 10 iterations.

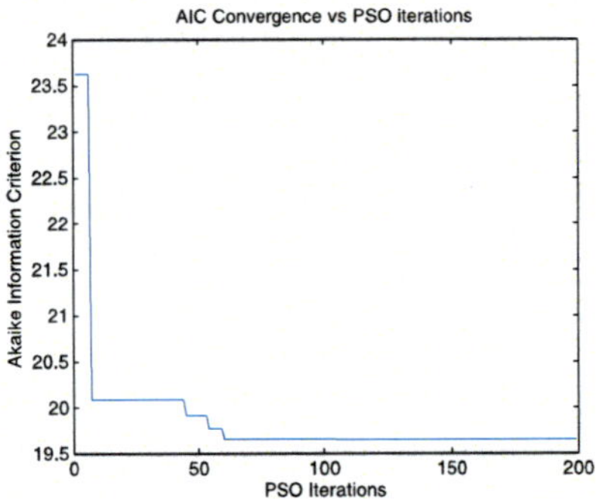

Figure 2. Akaike Information Criterion (AIC) convergence vs. PSO iterations.

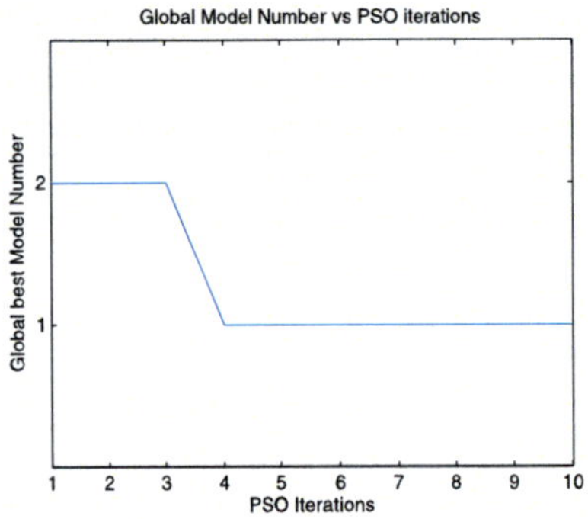

Figure 3. Global Model Number vs. PSO iterations

Figure 2 illustrates that the PSO algorithm rapidly converged close to the ultimate minimum error within the first 80 algorithm iterations. The global best model in this simulation started off as model m_2 but after 3 iterations it changed to model m_1 and remains unchanged for the rest of the simulation. The model order, based on the minimum of the objective function for these PSO settings was m_1, m_6, m_2, m_7, m_8, m_3, m_4 and then m_5.

5.1.2 Simulation number 2

This simulation is the same as simulation 1 except the objective function has been changed to SSE. Figure 4 shows the convergence plot of the SSE objective function over the first 100 algorithm iterations. Figure 5 illustrate the convergence behaviour of global best model over 100 iterations.

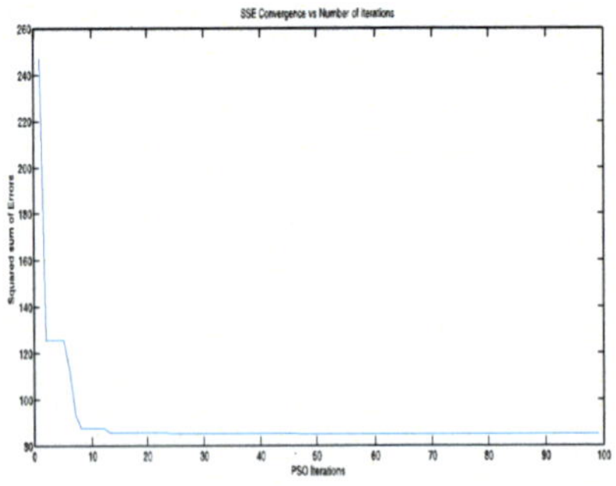

Figure 4. Squared sum of Errors (SSE) vs. PSO iterations

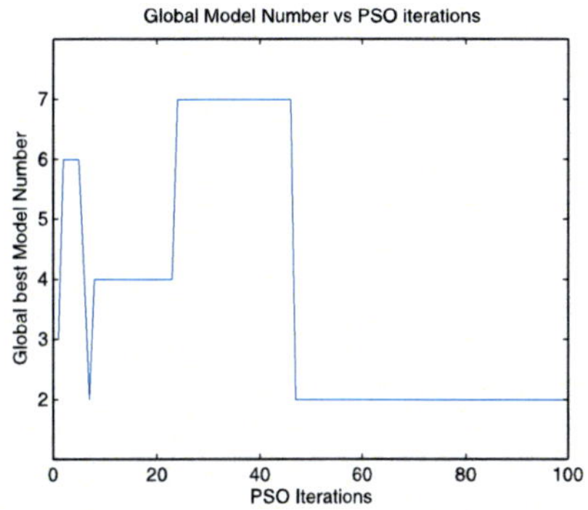

Figure 5. SSE Global best model for the first 100 PSO iterations.

The objective function (in Figure 4) did not improve much after 100 iterations even thou the global model changed. A relatively small improvement occurred (not shown) due to model m_6 becoming the global best model after 300 iterations. It is clear from figure 5 that the objective function had a significant role in the updating of the model parameters. Initially model m_3 was the p_g then the global best changed to being m_6, m_2, m_4, m_7, m_2 then finally model m_6 was again the p_g. The final objective function based model order in this simulation was m_6, m_1, m_2, m_5, m_4, m_8, m_3 and then m_7. This is different to the results in simulation 1 where the less complex models attained lowest errors.

5.1.3 Simulation number 3

In simulation number 3 the PSO algorithm has been changed by introducing the adaptive inertia weight. As mentioned in section 2.1 the inertia parameter allows the algorithm to initially explore a wider search area and near the end to exploit the local search space.

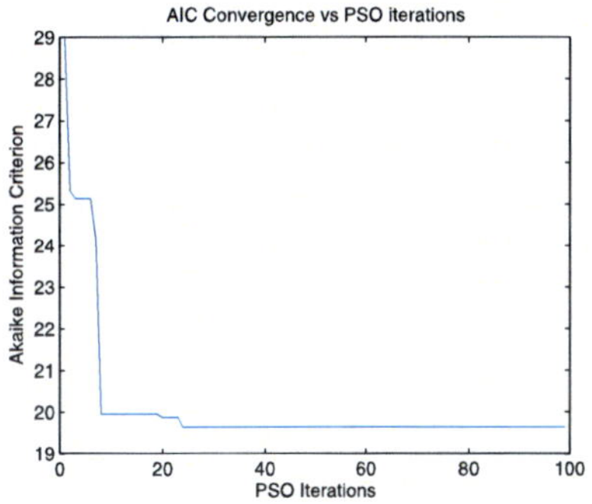

Figure 6. Akaike Information Criterion (AIC) convergence vs. PSO iterations

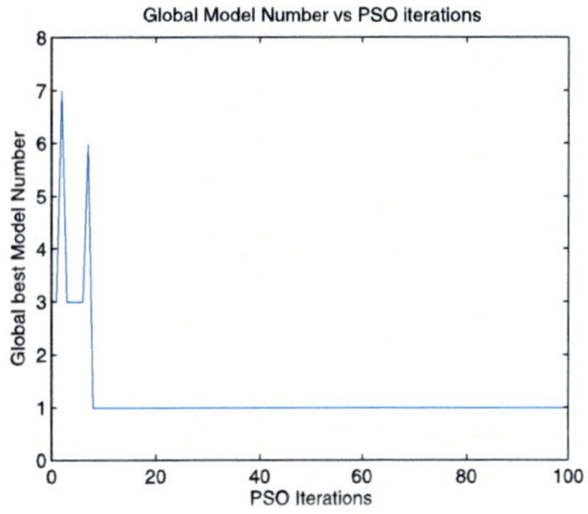

Figure 7.AIC Global model convergence vs. PSO iterations

This is evident from figure 6, where the search converged much quicker than in figure 2 whilst using the same objective function. Figure 7 also supports the initial explore to local exploitation concept because a number of models were initially p_g as opposed to figure 3 but towards the end a firm favourite was converged on. The final model order in this simulation was m_1, m_6, m_7, m_2, m_3, m_8, m_4 and then m_5. Perhaps the AIC objective function is too critical of the model complexity as it seems to always select models according to it.

5.1.3 Simulation number 4

In simulation number 4 the PSO algorithm has the adaptive inertia weight but uses the SSE as the objective function.

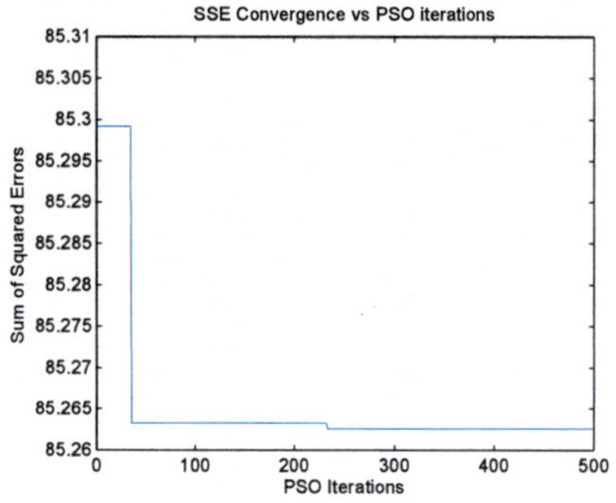

Figure 8. Squared sum of Errors (SSE) vs. PSO iterations

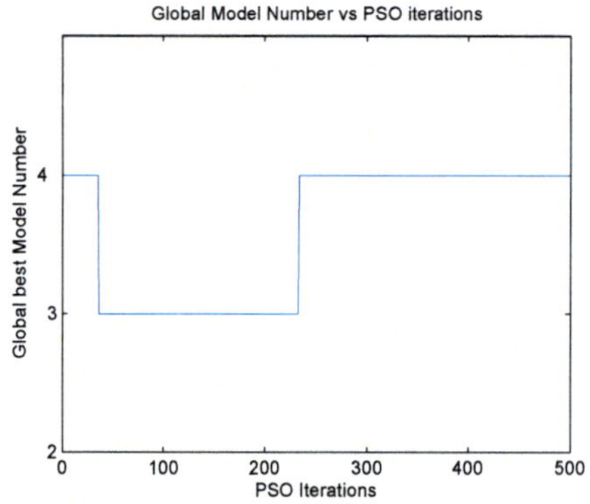

Figure 9.SSE Global model convergence vs. PSO iterations

The final model order in this simulation was m_4, m_1, m_3, m_6, m_2, m_7, m_5 and then m_8. Clearly the SSE objective function is not concerned with the complexity of the finite element model. The best model in this case is one of the most complex. It is not easy to directly compare the objective function magnitudes the algorithms converge to. This would have allowed for better analysis of why different models behave so differently under different objective functions. More conclusive decisions on the choice of objective function in this type of updating methodology can only be made with further experiments on different types of objective functions.

6. Conclusions

A particle swarm based method of finite element model updating and selection is presented. The method updates finite element model parameters using a stochastic-population based optimization procedure. Each potentially correct model of a structure is treated as an adaptive particle in the finite element model updating problem space. This space is defined by the number of potentially updatable model parameters. A number of simulations, using different objective function, are performed on eight competing models of a particular structure. The particle swarm based optimization approach to finite element model updating offers the researcher an ability to simultaneously update and select the best model in a given group. This is desirable in the cases where multiple competing model updating models of one structure exist. This paper has also highlighted that the choice of objective function is crucial in obtaining a reasonable best model.

7. Acknowledgements

We would like to acknowledge the support of National Research Foundation of the Republic of South Africa.

7. References

[1] J.L. Beck, K-V. Yuen. **Model selection using Response Measurements: Bayesian Probabilistic Approach.** Journal of Engineering Mechanics, Vol. 130, pages 192-203, 2004.

[2] J.L. Beck, L.S. Katafygiostis. **Updating Models and their uncertainties. I:Bayesian Statistical framework.** Journal of Engineering Mechanics, Vol. 124, No4, pages 455-461,1998.

[3] C. Bishop. **Pattern Recognition and Machine Learning**. Springer, 2006.

[4] N. Datta. **Finite element Model Updating, Eigenstructure Assignment and Eigenvalue Embedding Techniques for vibrating Systems**. Mechanical Systems and Signal Processing, Vol. 16, pages 83-96, 2002.

[5] M. Dorigo, C. Bhlum. **Ant colony optimization theory: A survey**. Theoretical Computer Science, Vol. 344, pages 243-278, 2005.

[6] H-J Escalante,M. Montes, L-E,Sucar. **Particle Swarm Model Selection**. Journal of Machine Learning Research, Vol. 10 (Feb), pages 405-440, 2009.

[7] D.J. Ewins. **Modal Testing: Theory and Practice**. Publisher Letchworth, Research Studies Press, 1984.

[8] J.R. Fonseca, M.I.Friswell, J.E. Mottershead, A.W. Lees. **Uncertainty identification by the maximum likelihood method**. Journal of Sound and Vibration, Vol. 288, pages 587-599, 2005.

[9] M.I. Friswell, J.E. Mottershead. **Finite element model updating in structural dynamics**. Kluwer Academic Publishers: Boston, 1995.

[10] J. Kennedy, R. Eberhart. **Particle Swarm Optimization**. In Proceedings of the International Conference on Neural Networks, Vol. IV, pages 1942-1948, 1995.IEEE.

[11] J.Kennedy,R. Eberhart. **Swarm Intelligence**. Morgan Kaufmann Publishers, Inc., San Francisco,CA,20001.

[12] Kim G-H, Park Y-S. **An automated parameter selection procedure for finite element model updating and its applications**. Journal of Sound and Vibration. Vol. 309, No. 3-5, pages 778-793, 2008.

[13] M.I. Levin, N.A.J. Lieven. **Dynamic finite element model updating using simulated annealing and genetic algorithms.** Mechanical Systems and Signal Processing, Vol. 12, No. 1, pages 91–120, 1998.

[14] M.Link, M,Friswell. **Working Group 1: Generation of validated structural dynamic models-Results of a benchmark study utilising the GARTEUR SM-AG19 testbed.** Mechanical Systems and Signal Processing, 2003, Vol.17, No. 1, pages 9-20.

[15] Ljung L. **System identification: Theory for the user**. Englewood Cliffs, NJ: Prentice hall, 1987.

[16] C.Mares, B.Dratz, J.E. Mottershead, M.I. Friswell. **Model Updating using Bayesian Estimation**. ISAM 2006, Leuven, Belgium, pages 2607-2616, 18-20 September 2006.

[17] C.Mares, J.E. Mottershead, M.I. Friswell. Selection **and Updating of parameters for the GARTEUR SM-AG19 testbed.** 3rd International Conference on Identification in Engineering Systems, Swansea, pages 130-141, April 2002.

[18] T. Marwala. **Finite element model updating using particle swarm optimization**. International Journal of Engineering Simulation, Vol. 6(2), pages 25-30, 2005.

[19] T.Marwala, S.Sibisi. **Finite Element Model Updating using Bayesian Framework and Modal Properties**. Journal of Aircraft, Vol. 42, No. 1, pages 275-278. January-February 2005.

[20] L.Mthembu, T.Marwala, M.I. Friswell, S.Adhikari. **Finite element model updating using a Bayesian Approach**. Submitted Mechanical Systems and Signal Processing, August 2009.

[21] M. Muto, J.L Beck. B**ayesian Updating and Model Class Selection for Hysteretic Structural Models Using Stochastic Simulation**. Journal of Vibration and Control, Vol. 14, No. 1-2, pp 7-34, 2008.

[22] I. Murray , Z. Ghahramani . **A note on the evidence and Bayesian Occam's razor**. Technical Report Gatsby Computational Neuroscience Unit GCNU-TR 2005-003. August 2005

[23] I.J. Myung. The importance of complexity in model selection. Journal of Mathematical Psychology, Vol. 44, pages 190-204, 2000.

[24] Y. Shi,R.C. Eberhart. **Empirical Study of particle swarm optimization**. In Proceedings of the Congress on Evolutionary Computation, pages 1945-1949, Piscataway, NJ, USA, 1999. IEEE.

[25] M.S Voss, X. Feng. Arma **model selection using particle swarm optimization and AIC criterion**. In Proceedings of the 15th IFAC Triennial World Congress on Automation Control, Barcelona, Spain, 2002.

[26] Titurus B and Friswell MI. **Regularization in model updating**. International Journal for numerical methods in engineering, Vol. 75, pages 440-478, 2008.

Proceedings of the IMAC-XXVIII
February 1–4, 2010, Jacksonville, Florida USA

A Force Identification Approach for Multiple-Degree-of-Freedom Systems

Lage, Y. E., Maia, N. M. M., Neves, M. M., Ribeiro, A. M. R.

Department of Mechanical Engineering, Instituto Superior Técnico,
Av. Rovisco Pais, 1049-001 Lisboa, Portugal.

ABSTRACT

This paper presents an inverse procedure for the determination of forces applied to a structure from the measurement of its dynamic response. The authors propose an approach to the inverse problem in two main points: identification of the force position in the structure and its quantification. Once the numerical model is established, the measurement of the dynamic responses at a chosen set of co-ordinates along the structure should allow for the estimation of the applied loads. For this study, the behavior of a beam is simulated numerically and validated later on experimentally in laboratory, for a given type of load. The proposed algorithm combines the finite element model using the Bernoulli-Euler beam theory with the dynamic responses of the structure measured in laboratory. Several methodologies are studied and discussed, varying the position of the sensors, in order to verify the ability of the method to identify the force location and magnitude.

Keywords: Force identification, Frequency Response Function, inverse problem.

1 Introduction

Force identification is an important engineering issue in vibro-acoustics, especially when operational forces cannot be measured directly as it happens at inaccessible locations in installed machinery [1, 2]. Moreover, transducers cannot usually be introduced in the structure in order to allow experimental determination of external loading and only a limited number of sensors and positions are available.

For these reasons, detection of unknown forces from vibrations measured at accessible locations is of interest. The calculation of these forces is usually reduced to an inverse problem, also named as inverse force determination.

The theory of inverse methods has been actively developed in many research areas presenting in common with other inverse problems the effects of matrix ill-conditioning due the ill-posedness of the inverse problem. These can be overcome by methods as pseudo-inversion for over-determined systems, singular value rejection, and Tikhonov regularization [3-5].

Various research works in this area can be found in the literature as, for example, those related to the identification of impact forces, implementation of prediction models based on reflected waves or simply from the dynamic response [6 – 13]; prediction of forces in plates for systems with time dependent properties [6] and identification of harmonic forces [8].

Wang and Lu [14] present a study on the problem of force identification with the help of Eddy current sensors (ECS) where the localization and the amplitude of the forces were achieved using an optimization method relating the predicted and measured responses of the sensors. The numerical and experimental results are given for a cantilevered beam.

DOI 10.1007/978-1-4419-9831-6_7, © The Society for Experimental Mechanics, Inc. 2011

Michaels and Pao [6] presented a method for the inverse problem to identify oblique concentrated forces applied on a surface; which includes the identification of the localization, orientation, amplitude and time dependence. The identification of impact forces through the analysis of wave propagation along the structure was studied by several authors and can be found in, e.g., Martin and Doyle [7]. They mainly study the problem of reflections and their role in the response and reveal the difficulties in the evaluation of the force in the frequency domain due to those reflections, which tend to cause lost of information in the response, especially close to the joints. As mentioned by those authors, combining analytical and experimental model helps to minimize the problems.

Wang and Chiu [13] studied the identification of impact forces. The prediction model here assumes that the acceleration response and the modal properties of the structure (i.e., natural frequencies, damping and mode shapes) are known. They developed two methods, one in the time domain and the other in the frequency domain, trying to solve the problem using an error function between the estimated and the measured responses.

A practical case of force identification is presented by Otsuka and Okada [10], related to forces in components of a ship engine. As in many other real problems it may be not possible to measure the dynamic response of the structure in certain points that are inaccessible; in such a situation it may be necessary to estimate those responses. The solution of this problem is proposed by Kammer [11], where the prediction is based on the measurements undertaken in other locations and then related to the desired ones through a transformation matrix that uses Markov parameters; these parameters represent the response to a unitary impulsive force and contain the dynamic properties of the system.

Other studies involve sensitivity analyses with respect to the force and response locations, which can provide valuable information for the force identification, e.g., O'Callahan and Piergentili [12]. They assume the complete knowledge of the model of the structure and study specifically the influence of the sensitivity of the prediction on the force location.

This paper presents a study on the ability of an inverse procedure for the identification of applied forces (position and magnitude are unknown) using an approach based on the combination of a finite element model with the dynamic responses measured in the laboratory. The test specimen is a free-free beam. Three possible inverse procedures are presented and discussed. Based on the three procedures, a brief comparison on the efficacy to locate the unknown force for an application case is presented and discussed. Finally, as this is an ongoing study, directions for future work are mentioned.

2 Theorical considerations

The finite element method is used to build the model of our structure, which is a Bernoulli-Euler free-free beam. The dynamic stiffness matrix $[Z(\omega)]$ is built and will be used for the force identification. One shall suppose that the model of the beam is completely known (and validated through some kind of model updating) and that only part of the measured responses is available; this constitutes the basis for the estimation of the localization and intensity of the applied forces. The computer program is developed in MatLab, building the dynamic stiffness matrix by discretizing the beam into N elements (*NN=N+1* nodes) in free-free conditions. Each node has three degrees of freedom (*x, y, θ*) and so the matrix dimensions will be *(3xNN)x(3xNN)* ($[Z]\in\Re^{3NNx3NN}$). At this stage one shall neglect the damping. With this simple methodology, the dynamic system is completely defined. For the proposed problem, one shall work with the inverse of the dynamic stiffness $[Z]$, the receptance matrix $[H]$.
One shall also assume that the only measured displacements are the vertical ones (along *y*), meaning that from a receptance matrix $[H]^{3NNx3NN}$ one only has a sub-matrix, $[H]^{NNxNN}$. The model will be like it follows:

$$\{y\}=[H]\{f\}\Leftrightarrow\begin{Bmatrix} y_1 \\ y_2 \\ y_3 \\ \vdots \\ y_{NN} \end{Bmatrix}=\begin{bmatrix} H_{1,1} & H_{1,2} & H_{1,3} & \cdots & H_{1,NN} \\ H_{2,1} & H_{2,2} & H_{2,3} & \cdots & H_{2,NN} \\ H_{3,1} & H_{3,2} & H_{3,3} & \cdots & H_{3,NN} \\ \vdots & \vdots & \vdots & \ddots & \vdots \\ H_{NN,1} & H_{NN,2} & H_{NN,3} & \cdots & H_{NN,NN} \end{bmatrix}\begin{Bmatrix} F_1 \\ F_2 \\ F_3 \\ \vdots \\ F_{NN} \end{Bmatrix} \tag{1}$$

Therefore, the vector $\{y\}$ represents the measured responses, $[H]$ represents the finite element response model (from MatLab) from which two degrees of freedom are eliminated (*x* and *θ*) and the vector $\{F\}$ contains the unknown imposed loading that one needs to estimate. It is easy to understand that if all the responses are known, corresponding to the finite element discretization, the solution will be simply given by:

$$\{F\}=[H]^{-1}\{y\} \tag{2}$$

It happens, however, that in the majority of the cases the number of known responses is insufficient to complete the vector $\{y\}$, due to the very large number of coordinates, to the complexity of the structure or even to the lack of access to some locations in the structure. Being impossible the determination of the forces in such circumstances, it is proposed here to undertake the calculations in two steps:

1. Localization of the forces
2. Force reconstruction

First of all, one shall identify the possible locations for the applied forces and after that, the reconstruction of the forces will be undertaken. The localization of the forces is realized through the maximum occurrence counting, i.e., by counting of the number of times that their maxima are achieved along the frequency range. Supposing that the response vector $\{y\}$ is incomplete, say, with only 2 accessible co-ordinates for measurement, one has, for example, the following situation:

$$\begin{Bmatrix} y_3 \\ y_9 \end{Bmatrix}=\begin{bmatrix} H_{3,1} & H_{3,2} & H_{3,3} & \cdots & H_{3,8} & H_{3,9} \\ H_{9,1} & H_{9,2} & H_{9,3} & \cdots & H_{9,8} & H_{9,9} \end{bmatrix}\begin{Bmatrix} F_1 \\ F_2 \\ F_3 \\ \vdots \\ F_9 \end{Bmatrix} \tag{3}$$

As $[H]$ is not invertible, one proposes three possibilities to obtain the force. The first possibility consists in completing equation (3) using several frequencies, so that $[H]$ becomes at least a square matrix. This implies, of course, that one assumes that the amplitudes of the force vector remains constant along those chosen frequencies. If one wishes to contemplate the general possibility of localizing a force vector whose amplitudes vary along the frequency, this approximation can only include a very narrow frequency band for each inversion. One shall have to solve a system of equations for each frequency band. Let us suppose, from eq. (3) that one considers five consecutive frequencies from the frequency spectrum, $\{\omega_1\ \omega_2\ \omega_3\ \omega_4\ \omega_5\}$. The problem to solve is:

$$\begin{Bmatrix} y_3^{\omega 1} \\ y_9^{\omega 1} \\ y_3^{\omega 2} \\ y_9^{\omega 2} \\ \vdots \\ y_3^{\omega 5} \\ y_9^{\omega 5} \end{Bmatrix}=\begin{bmatrix} H_{3,1}^{\omega 1} & H_{3,2}^{\omega 1} & H_{3,3}^{\omega 1} & \cdots & H_{3,8}^{\omega 1} & H_{3,9}^{\omega 1} \\ H_{9,1}^{\omega 1} & H_{9,2}^{\omega 1} & H_{9,3}^{\omega 1} & \cdots & H_{9,8}^{\omega 1} & H_{9,9}^{\omega 1} \\ H_{3,1}^{\omega 2} & H_{3,2}^{\omega 2} & H_{3,3}^{\omega 2} & \cdots & H_{3,8}^{\omega 2} & H_{3,9}^{\omega 2} \\ H_{9,1}^{\omega 2} & H_{9,2}^{\omega 2} & H_{9,3}^{\omega 2} & \cdots & H_{9,8}^{\omega 2} & H_{9,9}^{\omega 2} \\ \vdots & \vdots & \vdots & \ddots & \vdots & \vdots \\ H_{3,1}^{\omega 5} & H_{3,2}^{\omega 5} & H_{3,3}^{\omega 5} & \cdots & H_{3,8}^{\omega 5} & H_{3,9}^{\omega 5} \\ H_{9,1}^{\omega 5} & H_{9,2}^{\omega 5} & H_{9,3}^{\omega 5} & \cdots & H_{9,8}^{\omega 5} & H_{9,9}^{\omega 5} \end{bmatrix}\begin{Bmatrix} F_1 \\ F_2 \\ F_3 \\ \vdots \\ F_9 \end{Bmatrix} \Rightarrow \{F\}=[H]^{+}\{y\} \tag{4}$$

where "+" stands for pseudo-inverse. If *n* is the number of available measurements, *p* the number of frequencies and *q* the number of possible force locations, one has:

$$\{y\}=[H]\{F\} \quad where \quad y\in\Re^{(pxn)x1}, H\in\Re^{(pxn)xq}\ ,\ F\in\Re^{qx1}\ \text{and}\ pxn\geq q \tag{5}$$

The second and third possibilities consist of inverting $[H]$ in the first place and only after that eliminate the matrix columns corresponding to the unavailable information. For instance, if y_3 and y_9 were the only measured responses, the equation would be:

$$\begin{Bmatrix} F_1 \\ F_2 \\ F_3 \\ \vdots \\ F_9 \end{Bmatrix} = \begin{bmatrix} Z_{1,3} & Z_{1,9} \\ Z_{2,3} & Z_{2,9} \\ Z_{3,3} & Z_{3,9} \\ \vdots & \vdots \\ Z_{8,3} & Z_{8,9} \\ Z_{9,3} & Z_{9,9} \end{bmatrix} \begin{Bmatrix} y_3 \\ y_9 \end{Bmatrix} \tag{6}$$

The difference between the second and third possibilities is in the way the inverse is calculated. In the second hypothesis it is the inverse of the complete matrix, including the three degrees of freedom per node, while in the third one the matrix to invert only has the *y* degrees of freedom. In both these options and contrarily to what happens in the first case (eq. 4), here the forces are calculated at each frequency.

3 Numerical and experimental applications

A free-free steel beam subjected to an applied force at node 7 will be tested, as schematically shown in figure 1. The frequency range is 0-800 Hz and the responses in the y direction are partial- or totally known, depending on the test case. The beam properties are given in table 1 and the amplitude of the applied force (resulting from a multi-sine signal) is shown if figure 2.

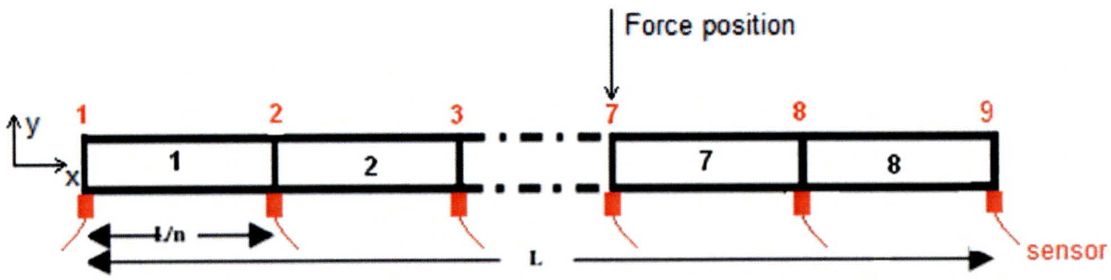

Figure 1 Free-free steel beam showing the placement of the sensors and force.

Table 1 Beam properties

Density –ρ	7860 Kg/m^3
Young's modulus – E	210 GPa
Length – L	0.8 m
Section area	$8.\times10^{-5}\,m^2$
Second area moment	$1.6667\times10^{-10}\,m^4$

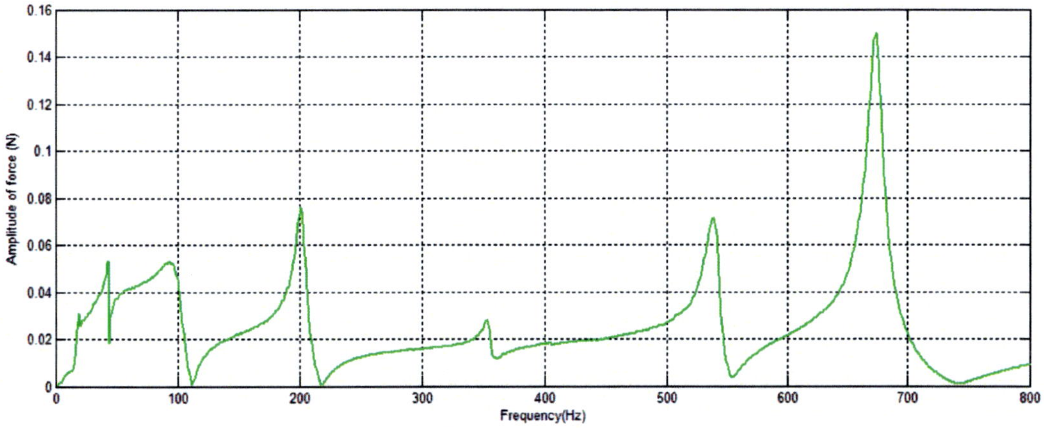

Figure 2 – Measured applied force at node 7.

One intends to reproduce the graph of figure 2 from the knowledge of the finite element model and of the measured displacements *y*.

4 Results and discussion

4.1 Theoretical model *versus* experimental model

The numerical model is undamped and therefore it is expected that the peaks of the frequency response functions do not totally agree in magnitude. However, it is important that they agree as much as possible in frequency; for that purpose, some extra concentrated masses were added, in order to model the effect of the accelerometers. In figure 3 the numerical and experimental amplitudes of the FRFs corresponding to the response at node 8 for the force at node 7 are superimposed, showing a considerably good match.

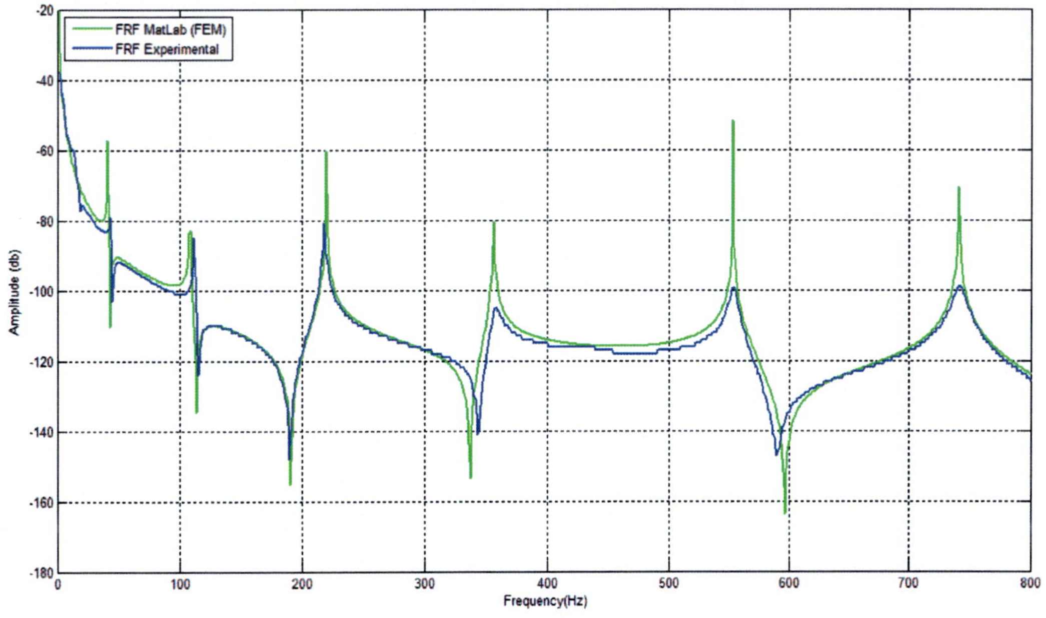

Figure 3 – Numerical and experimental frequency response functions H_{87}

4.2 Force localization – theoretical analysis

For the force localization one shall use the methods presented before, in the frequency range 0-800 Hz. Using the force spectrum of figure 2, the response is generated for all nodes and frequencies. The force localization will be taken as the location where the loading vector shows more often its maximum value, along the given frequency range.

In order to evaluate the performance of each method for the force localization, an algorithm has been developed, capable of analizing the results from different combinations of the nine possible accelerometer positions, with a minimum of 3 accelerometers, i.e., one has tested all the position combinations for 3 sensors, for 4, etc, up to 9 (total of 465 possible cases). This shall allow for assessing the robustness of each algorithm when there is no knowledge about the best location for the sensors or the locations are so restricted in terms of availability that there is no real choice; the following analysis assumes that any set of positions for the sensors is possible and there is no possibility to know, *a priori*, which set would be the best.

The results for the force localization are given in table 2, corresponding to the 9 possible locations, and for the three methods. Each entry of table 2 corresponds to the total number of maximum values found at that force location and the respective method, for all combinations of sensors and for the entire frequency range.

Table 2 Results for the force localization, for each method

	Force location								
	Nodes (force in node 7)								
Method	1	2	3	4	5	6	7	8	9
1	4	92	43	36	14	43	133	99	1
2	0	161	13	11	4	5	147	123	1
3	1	28	23	16	12	19	338	28	0

From table 2 one can observe that the highest result is given by method 3, localizing correctly the force at node 7 in the large majority of the cases. One has taken all the combinations, from 3 to 9 sensors. Some positions of the sensors are not, obviously, the most appropriate. To understand better this point, one presents in table 3 the results obtained by method 3, for various positions in groups of 4 sensors (there are a total of 126 possible combinations).

Table 3 - Force localization with method 3

Counting without considering the node where the force is applied (70 possible cases)										Counting considering the node where the force is applied (126 possible cases)									
Nodes	1	2	3	4	5	6	7	8	9	Nodes	1	2	3	4	5	6	7	8	9
1459	17	58	191	23	75	114	7	170	144	1467	0	33	41	86	38	183	389	29	0
1468	13	6	30	142	55	51	300	200	2	1468	13	6	30	142	55	51	300	200	2
1469	0	25	18	157	97	76	141	181	104	1469	0	25	18	157	97	76	141	181	104
1489	23	87	46	237	32	0	268	29	77	1478	19	25	11	122	35	96	397	82	12
1568	87	0	0	96	52	45	296	222	1	1479	4	60	8	106	33	104	387	97	0
1569	110	0	5	111	56	45	227	160	85	1489	23	87	46	237	32	0	268	29	77
1589	76	46	0	20	265	56	270	2	64	1567	43	33	12	131	88	74	399	19	0
1689	96	43	0	0	28	183	384	14	51	1568	87	0	0	96	52	45	296	222	1
2345	203	58	0	2	342	194	0	0	0	1569	110	0	5	111	56	45	227	160	85
2346	33	22	0	141	177	323	103	0	0	1578	0	26	0	65	103	53	434	118	0
2348	73	31	12	179	67	0	129	308	0	1579	9	63	0	60	86	65	411	102	3
2349	139	31	11	225	71	0	0	158	164	1589	76	46	0	20	265	56	270	2	64

Table 3 - Force localization with method 3 (Continued)

Counting without considering the node where the force is applied (70 possible cases)										Counting considering the node where the force is applied (126 possible cases)									
Nodes	1	2	3	4	5	6	7	8	9	Nodes	1	2	3	4	5	6	7	8	9
2356	0	78	25	218	27	249	202	0	0	1678	0	1	0	0	96	136	490	65	11
2358	0	5	61	141	28	126	132	306	0	1679	0	20	0	0	74	148	398	159	0
2359	0	4	65	175	129	143	0	145	138	1689	96	43	0	0	28	183	384	14	51
2368	13	16	116	32	48	106	285	181	2	1789	70	103	0	0	0	195	431	0	0
2369	27	49	128	30	53	126	121	169	96	2367	2	20	117	33	38	141	398	50	0
2389	0	47	284	99	0	0	277	16	76	2368	13	16	116	32	48	106	285	181	2
2456	0	223	104	73	23	257	119	0	0	3456	0	106	146	0	15	319	213	0	0
2458	0	223	0	42	69	36	131	298	0	3457	0	28	104	0	32	191	374	70	0
2459	0	300	0	34	101	59	13	166	126	3478	0	1	52	127	58	50	420	85	6
2468	0	198	1	68	31	53	297	143	8	3479	0	0	70	115	46	74	384	110	0
2589	0	273	22	0	220	22	182	22	58	3589	0	21	237	28	115	72	262	0	64
2689	0	219	3	0	29	141	328	22	57	3678	0	0	154	71	14	63	450	47	0
3456	0	106	146	0	15	319	213	0	0	3679	0	0	99	66	18	85	383	148	0
3458	0	8	134	0	61	132	108	356	0	3689	0	0	248	35	33	63	364	4	52
3459	0	3	242	0	49	175	14	145	171	3789	0	0	252	36	0	113	398	0	0
3468	0	5	90	34	123	63	248	236	0	4567	0	0	76	126	0	68	432	97	0
3469	0	0	172	36	164	69	86	164	108	4568	0	0	46	102	0	26	382	243	0
3489	0	23	212	203	47	0	237	4	73	4569	0	0	52	169	0	48	310	146	74
3568	0	24	127	113	19	41	300	172	3	4578	0	0	38	52	68	85	430	126	0
3569	0	27	147	142	16	44	201	136	86	4579	0	0	23	70	59	104	432	105	6
3589	0	21	237	28	115	72	262	0	64	4678	0	0	32	122	40	108	439	49	9
3689	0	0	248	35	33	63	364	4	52	4679	0	0	54	108	7	122	376	132	0
4568	0	0	46	102	0	26	382	243	0	4789	0	0	1	126	44	173	442	13	0
4569	0	0	52	169	0	48	310	146	74	5678	0	0	0	105	73	0	488	96	37
4589	0	0	53	183	49	98	345	16	55	5679	0	0	0	109	103	14	399	174	0
4689	0	0	18	211	67	34	409	15	45	5789	0	0	0	80	117	130	454	18	0
5689	0	0	0	69	117	23	560	0	30	6789	0	0	0	0	146	158	495	0	0
No. of correct cases	**0**	**10**	**10**	**5**	**2**	**7**	**25**	**11**	**0**	**No. of correct cases**	**0**	**10**	**10**	**5**	**2**	**7**	**81**	**11**	**0**

As it can be observed from table 3, not including the node where the force is located (number 7), the total number of correct cases is 25 out of 70; when the sensors placement include that node, the number of correct cases increases to 81 out of 126. In both cases the right force location has been clearly identified.

4.3 Force localization – experimental tests

Using the third method (the one that gave the best results) for the experimental identification, the force is almost always correctly localized. The experimental identification uses the displacement measurements at the various nodes, together with the numerical finite element model. The results are presented in table 4, showing various combinations of sensors and the counting of maxima, frequency by frequency (from 0 to 800 Hz).

Table 4 – Counting results (method 3)

Sensor positions	Counting results from 0-800Hz								
2,5,8	0	124	40	52	194	52	215	122	0
3,6,8	0	0	126	58	71	32	463	47	2
4,5,8	0	0	24	118	140	94	208	215	0
6,8,9	0	0	0	0	36	99	428	224	12
1,4,6,8	0	14	36	64	90	34	473	87	1
2,4,6,8	0	141	0	37	50	37	487	40	7

4.4 Force reconstruction

Once the location of the force is made, its reconstruction is a simple calculation, using the measured displacement at one of the nodes and the corresponding FRF relating that node to the force location. As in the given example the force is located at node 7, taking the measurement at node 8, for instance, it follows that:

$$y_8 = H_{8,7}\, F_7 \quad \Leftrightarrow \quad F_7 = \frac{y_8}{H_{8,7}} \tag{7}$$

Figure 4 shows the measured and reconstructed forces; the result is quite good, except for the high peaks, although these have to do with the antiresonances of H_{87} and are therefore expected.

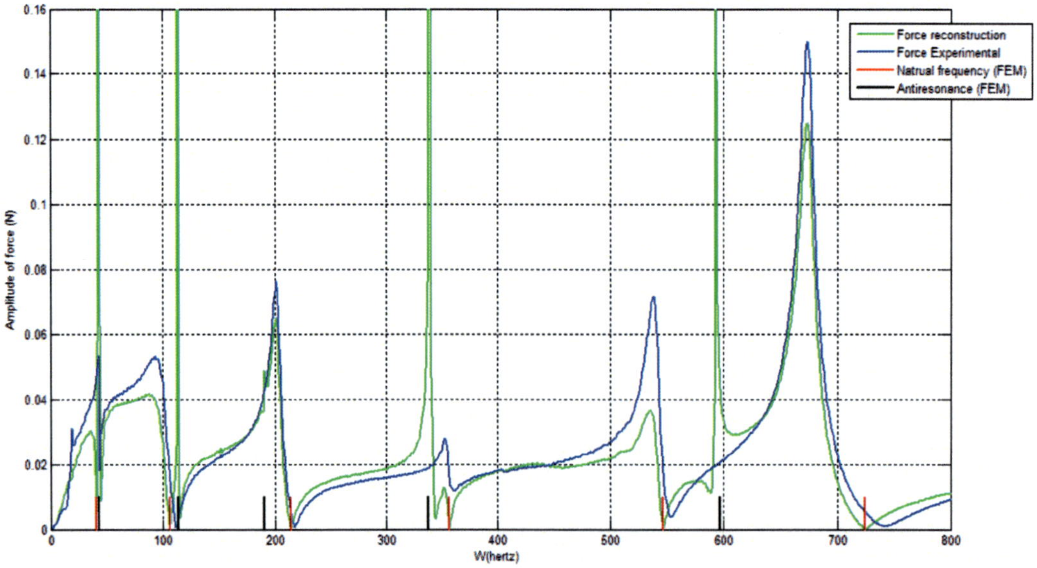

Figure 4 – Comparison between measured and reconstructed force at node 7. The high peaks correspond to the anti-resonances of FRF $H_{8,7}$ (Figure 3)

5 Conclusions

From this study one can draw the following conclusions:

- It seems possible to determine the localization of forces acting on a structure, based on an updated numerical model and on measured displacements along the structure and over a given frequency range. Nevertheless, the possibility of obtaining false positives due to the positioning of the sensors is far from negligible.
- The localization step can be undertaken, a priori, using different approaches. From those, one has selected the most performant, for the specimen under study.
- After the localization stage, one can estimate the magnitude of the forces in a simple way.
- In the future the subject of force identification will be pursued and further refinements will be introduced to improve the accuracy of the results.

Acknowlegements

The current investigation had the support of FCT, under the project PTDC/EME-PME/71488/2006, and IDMEC/IST.

References

[1] Mas P, Sas P, Wyckaert K., Indirect force identification based on impedance matrix inversion: a study on statistical and deterministic accuracy. In: Proceedings of 19th international seminar on modal analysis, Leuven, Belgium, 1994, pp. 1049–65.

[2] Dobson BJ, Rider E., A review of the indirect calculation of excitation forces from measured structural response data, Proc Inst Mech Eng C, 1990, 204, pp. 69–75.

[3] Thite AN, Thompson DJ., The quantification of structure-borne transmission paths by inverse methods. Part 1: improved singular value rejection methods, J Sound Vibr 2003, 264, pp. 411–31.

[4] Thite AN, Thompson DJ., The quantification of structure-borne transmission paths by inverse methods. Part 2: use of regularization methods, J Sound Vibr, 2003, 264, pp. 433–51.

[5] H.G. Choi, A.N. Thite, D.J. Thompson, A threshold for the use of Tikhonov regularization in inverse force determination, Applied Acoustics 67, 2006, pp. 700–719.

[6] Jennifer E. Michaels and Yih-Hsing Pao, The inverse source problem for an oblique force on an elastic plate, J. Acoust. Soc. Am 77 (6), June 1985, pp. 2005-2011.

[7] M. T. MARTIN and J. F. DOYLE, Impact force identification from wave propagation responses, International Journal of Impact Engineering Volume 18, Issue 1, January 1996, pp. 65-77.

[8] C.-H. Huang, An inverse non-linear force vibration problem of estimating the external forces in a damped system with time-dependent system parameters, Journal of Sound and Vibration 2001, 242(5), pp. 749-765.

[9] C.D. Johnson, Identification of unknown, time-varying forces/moments in dynamics and vibration problems using a new approach to deconvolution, Shock and Vibration 5, 1998, pp. 181-197.

[10] Tetsuya Otsuka, Takeshi Okada, Tetsuro Ikeno, Kazuyuki Shiomi, Masaaki Okuma, Force identification of an outboard engine by experimental means of linear structural modeling and equivalent force transformation, Journal of Sound and Vibration 3008, 2007, pp. 541-547.

[11] Kammer, D. C., Estimation of Structural Response using Remote Sensor Locations, Proceedings of SPIE, the International Society for Optical Engineering 1997, vol. 3089 (2), pp. 1379-1385.

[12] John O'Callahan and Fabio Piergentili, Force estimation using operational data, Proc. SPIE Vol. 2768: Proceedings of the 14th International Modal Analysis Conference.1996, pp. 1586-1592.

[13] Bor-Tsuen Wang and Chun-Hsien Chiu, Determination of unknown impact force acting on arbitrary structures, International modal analysis conference N°17, vol. 3727 (2), pp. 1653-1659.

[14] Wang, B. T., and Lu, S. H., The Use of Eddy Current Sensor for Harmonic Force Prediction, Proceedings of the 23rd International Modal Analysis Conference, 2005, Paper No. 68, Session 39.

Proceedings of the IMAC-XXVIII
February 1–4, 2010, Jacksonville, Florida USA

Model Updating by Combining Substructure Energy Functions and Subset Selection

Haishan WU
Tongji University.
State Key Laboratory for Disaster Reduction in Civil Engineering.
Siping Rd. 1239. Shanghai.China
East China Jiaotong University
Department of Civil Engineering
Shuanggang Rd. Nanchang. China

Limin SUN
Tongji University.
State Key Laboratory for Disaster Reduction in Civil Engineering.
Siping Rd. 1239. Shanghai.China

ABSTRACT

Since a reliable finite element model has broad applications in structural response prediction, control, health monitoring and condition assessment, model updating techniques are often applied to improve the performance of an analytical model for its intended use. The sensitivity-based iterative model updating approach becomes preferable because of its physically meaningful results, however, the selection of updating parameters remains a difficult problem to handle. In this paper, a synthesized approach, which combined substructure energy functions and subset selection, was proposed recognizing that the former can be used to locate the area containing dominant modeling errors and the latter to ensure the least number of effective parameters be chosen to reduce the modal residue adequately. This synthesized method can potentially reduce the burden of calculation but have more credible results. Its effectiveness was demonstrated systematically by numerical simulation with encouraging results.

NOMENCLATURE

K, K_j	stiffness matrix, substructure stiffness matrix;
M, M_j	mass matrix, substructure mass matrix;
$\omega, \omega_A, \omega_E$	circular frequency, analytical circular frequency, experimental circular frequency;
ψ, ψ_A, ψ_E	mode shape vector, analytical mode shape vector, experimental mode shape vector;
N	number of measured mode shape;

DOI 10.1007/978-1-4419-9831-6_8,

$\Delta\prod_j^s, \Delta\prod_j^k$	strain energy, kinetic energy functions of the J th substructure;
S	sensitivity matrix;
b	modal parameter residue between measured and analytical results;
θ	physical parameters vector for updating;
Φ, Φ_E	mode shape matrix, experimental mode shape matrix;
Φ_{UA}	analytical mode shape components corresponding to unmeasured degrees of freedom;
Φ_{MA}, Φ_{MA}^+	analytical mode shape components corresponding to measured degrees of freedom and its pseudoinverse matrix;
$\lambda, \lambda_{i,j}$	eigenvalue, the derivative of the i th eigenvalue with respect to the j th physical parameter;
$\psi, \psi_{i,j}$	eigenvector, the derivative of the i th eigenvector with respect to the j th physical parameter;
$K_{,j}$	the derivative of stiffness matrix with respect to j th physical parameter;
$M_{,j}$	the derivative of mass matrix with respect to j th physical parameter;
W	weight matrix;

1. INTRODUCTION

Since a reliable finite element model has broad applications in structural response prediction, control, health monitoring and condition assessment, model updating techniques are often applied to improve the performance of an analytical model for its intended use. Model updating is a sequential process which including several steps, and the selection of parameters for updating is probably the most important task [1]. The parameters should be chosen which associated with features of the model which are in doubt for physical reasons, thus these recognized uncertainties can be corrected. So parameter selection means modeling error localization. However, this is a nontrivial task. It needs structural analysts' application of considerable physical insights. Several parameter selecting methods have been proposed during the past decades, and they have relation to parameterization method used in model updating.

In direct model updating, the potential updating parameters are individual terms of stiffness and mass matrixes. Lallement and Piranda proposed a method of balancing the eigenvalue equation $[K-\omega^2 M]\psi=0$, in which the analytical eigensolution $(\omega^2, \psi)=(\omega^2{}_A, \psi_A)$ is replaced by eigenvalues and properly expanded eigenvectors from experiments (ω_E^2, ψ_E), to give a localization matrix $[K-\omega^2{}_E M]\psi_E=L$ [2]. The dominant modeling errors are reflected by those degrees of freedom associated with large values of $q_i=\sum_{h=1}^{n} p_h L_{ih}^2, i=1,\cdots,n$, where P_h is weighting scalars.

In iterative model updating, one parameterization method is substructure parameter, which is associated with individual finite element or groups of finite elements, the other parameterization method adopts physical parameters, such as Young's modulus, mass density and so on. With regard to the former case, Link and Santiago [3] proposed the use of energy functions based upon the substructures to locate regions which possibly have modeling error. In the latter case, Lallement and Piranda proposed a method named best subspace to choose those updating parameters which can reduce the discrepancy more effectively between analytical and experimental results [2]. Selecting updating parameters from all possible ones is in fact a subset selection, and the theory of which has been well established in the statistical literature [4]. And the best subspace method adopts a forward subset selection approach. Fritzen studied QR orthogonalisation strategy in subset selection, and compared it with Efroymson's criterion method, saying that both performed successful [5] [6]. Golub and Van loan described SVD-based subset selection procedure [7]. Friswell extended the subset selection method to circumstance of multiple testing data [8]. In subset selection, all kinds of methods try to select a subset of parameters which are least in correlation thus reduce the residue most effectively.

Updating parameters should be selected with the aim of correcting modeling errors. Subset selection method which is not based upon physical insights choose the parameters which most effectively improve the correlation of analytical and experimental results, However, this is not a sufficient condition to assure that these parameters must have dominant errors. Substructure energy functions is a method based upon physical insights with the potential to find modeling error regions, but not all parameters in error regions are suited to be updated due to the property of sensitivity-based model updating approach. These two methods can complement each other to obtain potentially the parameters selected not only based upon physical insights but with high effectiveness in updating process.

In this paper, a synthesized approach based on the discussion above was proposed which begins with the use of substructure energy function aiming at obtain the likely regions with modeling errors, then only the physical property parameters in these regions are filtered by subset selection, the best subspace method is used in subset selection for example, with parameters obtained as final updating parameters. Section 2 outlines some important theory used in this synthesized approach. A simulated cantilever beam example is used to demonstrate the properties of the method in section 3, coming to conclusions in section 4.

2. THEORETICAL BACKGROUND

In this synthesized approach, the measured mode shapes are firstly expanded by the system equivalent reduction expansion process (section 2.3), and the likely regions with modeling error are obtained according to the theory of substructure energy functions (section 2.1), then physical parameters in these likely error regions are filtered by the subset selection theory (section 2.2), finally, those parameters chosen are used to be updated by a sensitivity-based model updating approach (section 2.5).

2.1 Substructure Energy Functions

Link and Santiago proposed the use of energy functions based upon the substructure stiffness and mass matrixes [3]. The substructure strain energy function can be expressed as,

$$\Delta \Pi_j^s = \sum_{h=1}^{N} (\psi_{ah} - \psi_{Eh})^T K_j (\psi_{ah} - \psi_{Eh}) \tag{1}$$

And substructure kinetic energy function as,

$$\Delta \Pi_j^K = \sum_{h=1}^{N} (\psi_{ah} - \psi_{Eh})^T M_j (\psi_{ah} - \psi_{Eh}) \omega_{Eh}^2 \tag{2}$$

Poorly modeled regions are believed to be detected by large residual substructure strain energy function or kinetic energy function values. Since in both residual energy functions, full experimental eigenvectors are obtained from a few component values of mode shape corresponding to places with sensors by modal expansion process in which the analytical eigenvectors with errors are used in most practices, the detected regions with modeling error can just be likely.

2.2 Subset selection

In sensitivity-based model updating, the following equation is obtained,

$$S \cdot \theta = b \tag{3}$$

where S is the sensitivity matrix and contains the first order derivatives of modal residual between experimental and analytical results with respect to the candidate updating parameters. It is assumed that only a subset of these parameters are non-zero. Lallement and Piranda proposed a subset selection procedure, commonly known as forward subset selection, to produce a sub-optimal solution [2].

Firstly, a single column is sought which can best represent the residual vector b. If the columns of S are given by S_j, namely $S = [S_1\, S_2\ \ \dots\ \ S_p]$, then the parameter which minimizes the residual

$$J = \left\| b - S_j\, \tilde{\theta}_j \right\|_2^2 \tag{4}$$

is selected, where $\tilde{\theta}_j$ is the least squares estimate of the j th parameter, with the expression of

$$\tilde{\theta}_j = S_j^T b \,/\, S_j^T S_j \tag{5}$$

Substituting equation (5) into (4), The result is obtained that the j th parameter is selected that maximizes

$$(S_j^T b)^2 \,/ (S_j^T S_j) \tag{6}$$

Since b is a constant vector,

$$\cos^2 \phi_j = (S_j^T b)^2 \,/ (S_j^T S_j)(b^T b) \tag{7}$$

is maximized equivalently when the j th parameter is selected, where ϕ_j is the angle between the vectors b and S_j. In other words, the j th parameter is selected that minimizes the angle between the vector b and S_j.

Then, the forward subset selection method starts to select the second parameter, together with the already selected one, to make a combination of two columns of S, which constitutes the best sub-basis for the representation of b. Let j_1 represents the first parameter selected and the corresponding column of S be S_{j1}. To select the subsequent parameter following j_1, the vector b and columns of S must be transformed to be orthogonal to vector S_{j1}. Thus the columns of S and vector b are replaced with

$$S_j \rightarrow S_j - S_{j1}(S_{j1}^T S_j \,/\, S_{j1}^T S_{j1}) \tag{8}$$

$$b \rightarrow b - S_{j1}(S_{j1}^T b \,/\, S_{j1}^T S_{j1}) \tag{9}$$

In this transformed problem, the second parameter is similarly obtained $j_2 (\neq j_1)$ as the first one , which minimizes the angle between the vector b and S_j. So far, an iterative process is produced to find a subset collection from all candidate parameters, provided a stop rule is specified.

2.3 Modal Expansion Process

In model updating, problems arise from comparing measured experimental modal data with numerical analytical modal data because of incompleteness. There are fewer freedoms of response measured by sensors in experiments than those calculated in analysis. Aiming to make this comparison done, one way is to reduce the number of degrees-of-freedom (DOFs) of analytical model, the other is to expand the DOFs of measured modal vectors. In this paper, the updating parameters are defined in the analytical model, which is to be updated, thus, the measured incomplete mode shapes vectors are expanded to the full set of DOFs of the analytical model.

Several modal expansion techniques are available [1], and in terms of preservation of eigeninformation, the system equivalent reduction expansion process (SEREP) is adopted here considering its improved accuracy resulting from preserving the dynamics before and after the process in the least-squares sense [9]. The expanded mode shape matrix is given by

$$\tilde{\Phi} = T \cdot \Phi_E \tag{10}$$

where the transformation matrix T is given by

$$T = \left[\frac{\Phi_{MA}}{\Phi_{UA}} \right] \cdot \Phi_{MA}^{+} \tag{11}$$

2.4 Eigenpair Sensitivities

To obtain sensitivities of eignvalues and eigenvectors with respect to physical parameters, the method proposed by Fox and Kapoor's [10] is used in this paper. The expression of sensitivity of eigenvalue is

$$\lambda_{i,j} = \psi_i^T [K_{,j} - \lambda_i M_{,j}] \psi_i \tag{12}$$

And the sensitivity of eigenvector is

$$\psi_{i,j} = \sum_{k=1}^{N} a_{ijk} \psi_k \tag{13}$$

where coefficients a_{ijk} is determined by

$$a_{ijk} = -\psi_l^T [K_{,j} - \lambda_i M_{,j}] \psi_i \,/ (\lambda_i - \lambda_l) \qquad l \neq i \tag{14}$$

$$a_{iji} = -(\psi_i^T M_{,j} \psi_i \,/ 2) \tag{15}$$

2.5 Sensitivity-based model updating methodology

In sensitivity-based finite element model updating using measured modal parameters, the identification of structural parameters is formulated in an optimization problem where the structural parameters are sought so that the updated finite element model can reproduce as closely as possible the measured modal parameters [11]. The objective function measuring the residual of modal data (natural frequencies (NF) and mode shapes) between experiment and numerical analysis is expressed by

$$MinJ_X = \varepsilon^T W \varepsilon = \left\| W^{1/2} (\tilde{z} - z(\theta) \right\|_2^2 \tag{16}$$

where ε is the output error of the modal parameters, $\tilde{z}$ and $z(\theta) \in R^n$ are the experimental and analytical modal vectors with $n = n_f \cdot \ (n_m + 1)$, n_f and n_m are the numbers of measured modal frequencies and measured coordinates for each mode respectively.

The solution to equation (16) is obtained from a sensitivity-based iterative optimization approach. Beginning with an initial estimate of updating parameters, the solution for updating parameters at the Kth iteration step is obtained by the expressions of

$$\theta^{K+1} = \theta^K + \Delta\theta^K \tag{17}$$

$$\Delta\theta^K = [S^T S]^{-1} S^T (\tilde{z} - z(\theta^K)) \tag{18}$$

where S is sensitivity matrix obtained by the theory described in section 2.4.

3. NUMERICAL EXAMPLE

In this illustrative numerical example, a refined FE reference model is built to represent the true behavior of the experimental structure (see Figure 1), and mode shape data with partial DOFs is obtained by a limited number of sensor measurements from this reference model. Accordingly, a FE updating model with a relatively coarse FE mesh is built which has modeling errors to be updated (see Figure 1). The structural type used is a 1D cantilever beam with the following properties: density 2,500 Kg/m^3 and elastic modulus 3.0e+10N/㎡; length 6.4 m, cross section area $b \cdot h = 0.2 \cdot 0.3m^2$. Matlab is used for building the finite element model and performing the whole numerical calculation process.

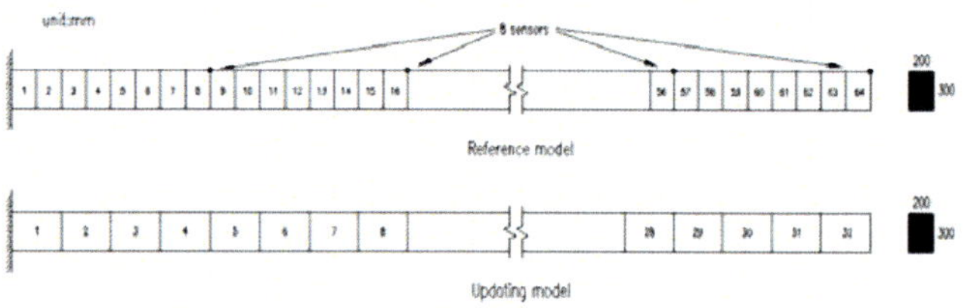

Figure 1: The reference model and updating model

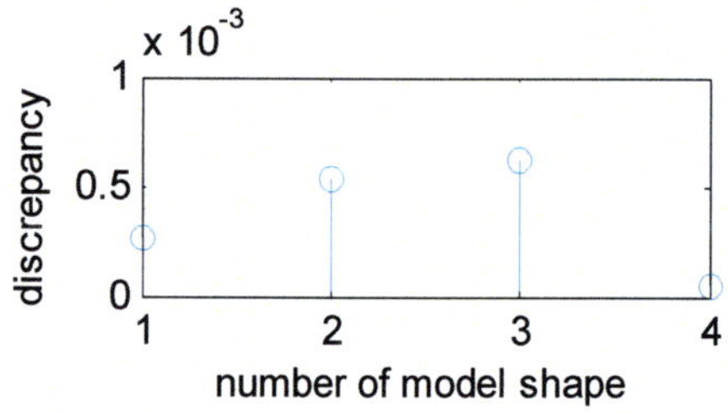

Figure 2: The first four NF discrepancy between reference model and theoretical value

The reference model(RM) has 64 Euler-Bernouli beam elements, and the updating model(UM) 32 elements. Modeling errors are simulated with reduction of elastic moduli E_i of 40% and 50% of the 8th and 18th elements respectively in updating model. Eight equally spaced responses (representing sensor measurements) are obtained from the reference model, and only first five modes with 9 DOFs (eight from sensors and one from fixed end of cantilever) are used in the model updating process to represent realistic situation.

The reference model is firstly inspected whether it can represent the true physical structure. The first four natural frequencies are compared to their theoretical values [12], with the result of the largest discrepancy below 0.1 percent (see Figure 2). Thus the reference model can be assumed to represent the experimental data accurately, and the discretization error ignored. Before updating, the discrepancy between the reference model and updating model are reflected by natural frequencies discrepancy percent J and modal assurance criteria(MAC) values (see Table 1), where

$$J = \left| f_{ref} - f_{upd} \right| / f_{ref} \cdot 100\% \tag{19}$$

$$MAC = \left| \Phi_E^T \Phi_A \right|^2 / (\left| \Phi_E^T \Phi_E \right| \cdot \left| \Phi_A^T \Phi_A \right|) \tag{20}$$

Table 1. The First Five Natural Frequencies (NF) and MAC Comparisons before and after Updating.

Modes	NF-RM (Hz)	NF-UM (Hz)(before)	NF-UM (Hz)(after)	J(%) (before)	J(%) (after)	MAC (before)	MAC (after)
1	4.0947	3.9947	4.0875	2.44	0.18	1.0000	0.9989
2	25.6650	24.8689	25.6395	3.10	0.10	0.9998	1.0000
3	71.8840	70.7533	71.8963	1.57	0.02	0.9987	1.0000
4	140.9275	136.0643	140.9898	3.45	0.04	0.9988	1.0000
5	233.1025	229.9906	233.9547	1.34	0.36	0.9985	1.0000

In order to use the substructure energy function to localize the regions with modeling errors, the updating model is organized into 16 substructures with every substructure composed of two contiguous FE elements, and the first five mode shapes are expanded using SEREP method. In this example, structural mass is assumed unchanged, and substructure strain energies of all 16 substructures are calculated and normalized to the greatest one of them (see Figure 3).

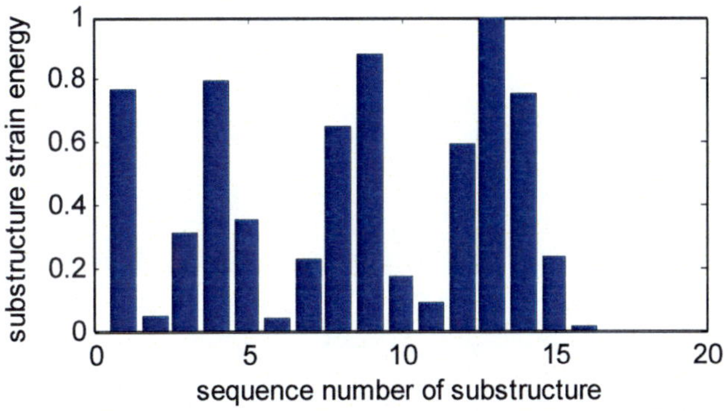

Figure 3: Normalized substructure strain energy

Four substructures with largest strain energy are taken to be likely regions with modeling errors, which are sequentially the 13th, 9th, 4th and 1st substructure. Thus the moduli of 1, 2, 7, 8, 17, 18, 25, 26 elements of updating model will be provided as candidate parameters to be filtered by subset selection. After implementing a best subspace subset selection, the first four elements chosen are 2, 17, 18, 8 elements. Based on the results of these two methods, the moduli of 17,18,8 elements are finally chosen to be updated by a sensitivity-based approach with good results shown in the Table 1 and Table 2.

Table 2. The Values of updating parameters before and after Updating. Unit: N/m²

Element	Before updating	After updating	True values	discrepancy
8	1.80e+10	2.97e+10	3.00e+10	-1.00%
17	3.00e+10	3.07e+10	3.00e+10	+2.33%
18	1.50e+10	2.95e+10	3.00e+10	-1.67%

If the updating parameters are chosen by subset selection alone as usually, all of moduli of 30 elements in updating model will be candidate parameters, leading to that not only greater quantity of calculation, especially in step-wise regression subset selection [8], but also relatively large number of elements will be finally used in order to include the two true error parameters. From this example, the fact that it is very hard to accurately choose the updating parameters is also discovered. In order to obtain good updating result, one needs make trade-offs between using more updating parameters hoping to contain all parameters with errors and the negative influence by those parameters which should not be included indeed in practice.

4. SUMMARY

The selection of updating parameters is a very important question in model updating, but it seems that there is no an exact method to solve it. It is recommended to use several different methods and only to pick out those parameters which have been selected by more than one technique, and the approach proposed here is a try. From the results of this synthesized approach composed of two complementing methods, one can potentially be more sure of the feasibility of the rightness in choosing updating parameters and reduce the burden of calculation though it is still very difficult to obtain the exact parameters with modeling error.

5. ACKNOWLEDGMENTS

The writers are grateful to the National Science Foundation for financing this research work through Grant No. 50538020, Science and Technology Commission of Shanghai Municipality through Grant No. 062112007 and State Key Laboratory for Disaster Reduction in Civil Engineering through Grant No.SLDRCE-08-A-05.

6. REFERENCES

[1]. Friswell, M.I., Mottershead, J.E., Finite Element Model Updating in Structural Dynamics, Kluwer Academic Publisher, Dordrecht, 1995.

[2]. Lallement, G., Piranda, J.,. "Localization Methods for Parameter Updating of Finite Element Models in Elastodynamics", Proc. 8th Int. Modal Analysis Conf., Orlando, FL. pp. 579-585, 1990.

[3]. Link, M., Santiago, O.F., "Updating and Localizing Structural Errors Based on Minimization of Equation Errors", Proc. Int. Conf.: "Spacecraft Structures and Mechanical Testing", ESA/ESTEC, Noordwijk, Holland,1991.

[4]. Miller, A., Subset Selection in Regression, Second Edition, Chapman & Hall/CRC, 2002.

[5]. Fritzen, C.P., Jennewein, D., "Damage Detection Based on Model Updating Methods", Mechanical System and Signal Processing, 12(1), pp. 163-186, 1998.

[6]. Fritzen, C.P., Bohle, K., "Parameter Selection Strategies in Model-Based Damage Detection", Proceedings of the International Workshop on Structural Health Monitoring, Stanford University, pp.901-911, 2000.

[7]. Golub, G.H., Van Loan, C.F., Matrix Computations, Third Edition, John Hopkins Univ. Press, Baltimore, Maryland, 1996.

[8]. Friswell, M.I., Penny, J.E.T., Garvey, S.D., "Parameter Subset Selection in Damage Location", Inverse Problems in Engineering, 5(3), pp. 189-225, 1997.

[9]. Callahan, O., Avitable, P., and Riemer, R, "System Equivalent Reduction Expansion Process (SEREP)", Pro. 7th Int. Modal Analysis Conf., pp. 29-37, 1989.

[10]. Fox., R.L., Kapoor, M.P., "Rates of Changes of Eigenvalues and Eigenvectors", AIAA Journal, 6(12), pp. 2426-2429, 1968.

[11]. Hua, X.G., Ni, Y.Q., Chen, Z.Q., and Ko, J.M., "An Improved Perturbation Method for Stochastic Finite Element Model Updating", Int. J. Numer. Meth. Engng, 73, pp. 1845-1864. 2008.

[12]. Clough, R.W., Penzien, J., " Dynamics of Structures", Third Edition, Computers & Structures Inc, Berkeley, CA, 2003.

Proceedings of the IMAC-XXVIII
February 1–4, 2010, Jacksonville, Florida USA

Dynamic Identification and FE Updating of S. Torcato Church, Portugal

Luís F. Ramos, PhD in Civil Engineering, Assistant Professor, Department of Civil Engineering – ISISE, University of Minho, Campus de Azurém 4800 - 058 Guimarães, Portugal

Murat Alaboz, MSc in Civil Engineering, Department of Civil Engineering – ISISE, University of Minho, Campus de Azurém 4800 - 058 Guimarães, Portugal

Rafael Aguilar, Civil Engineering PhD student, Department of Civil Engineering – ISISE, University of Minho, Campus de Azurém 4800 - 058 Guimarães, Portugal

Paulo B. Lourenço, PhD in Civil Engineering, Full Professor, Department of Civil Engineering – ISISE, University of Minho, Campus de Azurém 4800 - 058 Guimarães, Portugal

NOMENCLATURE

ω Natural frequency value
φ Mode shape vector

ABSTRACT

San Torcato Church is located near the city of Guimarães, Portugal. The construction of the church started in 1871 and completed in recent years. The church combines several architectonic styles, like Classic, Gothic, Renaissance and Romantic. This "hybrid" style is also called in Portugal as "Neo-Manuelino". At the moment, the church has significant structural problems due to soil settlements of the main façade. Cracks can be observed in the main and lateral facades. The cracks are visible from the outside and inside of the temple. The bell-towers are leaning, and the arches in the main nave present vertical deformations. Non-stabilized phenomena are present in the structure. To stabilize the damage, a structural intervention is planned to occur soon and the church is already monitored to assist the intervention. The paper clearly presents the problem with emphasis to the dynamic analysis carried out before the structural strengthening, namely: the experimental tests with output-only techniques for frequencies, damping and mode shapes estimation, and FE model updating analysis.

1. INTRODUCTION

Before any intervention to a structure, identification of all structural properties and damage sources has a significant importance. Although numerical model analyses methods allow one to simulate various cases, estimation or assumptions of the structural properties are not easy. To alter these difficulties many non-destructive methods have being developed but the information of those techniques rather local or insufficient.

Experimental dynamic identification techniques, which are generally based on the acceleration records of a structure, allow us to obtain natural frequencies, mode shapes and damping coefficients. These data represent the dynamic response of a structure, as a result of its physical properties that are unknown or difficult to obtain.

T. Proulx (ed.), *Dynamics of Civil Structures, Volume 4*, Conference Proceedings of the Society for Experimental Mechanics Series 13,
DOI 10.1007/978-1-4419-9831-6_9,

By this way, the real response of the structure under specified or unknown excitations can be used to calibrate Finite Element (FE) models.

In this context, modal identification tests were carried out on San Torcato Church to tune a FE model used for safety analysis. The work includes the experimental campaign for estimation the modal parameters (natural frequencies, mode shapes and damping coefficients) and procedures to tune accurately the FE model.

2. SAN TORCATO CHURCH

San Torcato Church is located in the village of St. Torcato, 7 km North from the city of Guimarães, Portugal. Construction of the church started in 1871 and completed in recent years. In Figure 1 early stages of the construction can be seen. The church combines several architectonic styles, like Classic, Gothic, Renaissance and Romantic. This "hybrid" style is also called in Portugal as "Neo-Manuelino".

(a) (b)

Figure 1 - Early stages of the construction (San Torcato Museum): (a) mail façade without towers, and (b) the first tower construction.

Architectural plan shape of San Torcato Church exhibits a classical cross basilica scheme (Figure 2) with the main nave covered with a barrel vault, transept that is crowned with a dome in the middle and the apse part. Besides the main façade, two towers are located. Between towers and transepts, one story with low height was added to the church from both sides of the nave. Accesses to the towers are provided from inside those buildings.

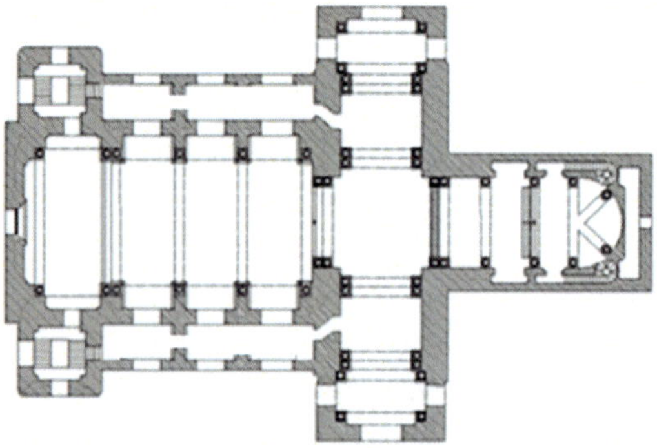

Figure 2 - Plan of San Torcato Church.

The nave of the church has 25 m length and 15 m width. The walls have 1.30 m of thickness. The nave is covered with a barrel vault which is supported by stone arches. The vault stands on stone columns that are supported with buttresses adjacent to body walls of the nave. At the transition with transept, main dome is carried by a masonry double arch.

The bell towers are located beside the main nave, adjacent to the facade. They have rectangular plan shape with 7.50 m in length and 6.20 m width, measured from outer surface. Load bearing walls of the towers are 1.45 m in thickness and have 32.70 m high until the belfry level. Belfry parts of the towers have two arched openings in all directions and have 8 m of elevation. The towers are covered with conical roofs.

Due to the construction long duration, structural techniques and materials vary within whole structure. The main façade, towers, lateral façades beside the towers and main nave are prior in construction phase and are made of granite stone with dry joints. The apse is combined with reinforced concrete walls after the springing level of the vault.

The significant structural problem of the structure is the crack pattern on the main facade due foundation settlements. The cracks than can be seen on the main façade, starting from the mid arch at the main entrance and goes through the rose window and reaches to the left and right corners of tympanum (Figure 3). The continuity of cracks inside the church and tilting of the towers indicates a settlement due to high stress level of towers and soft filling layer of soil.

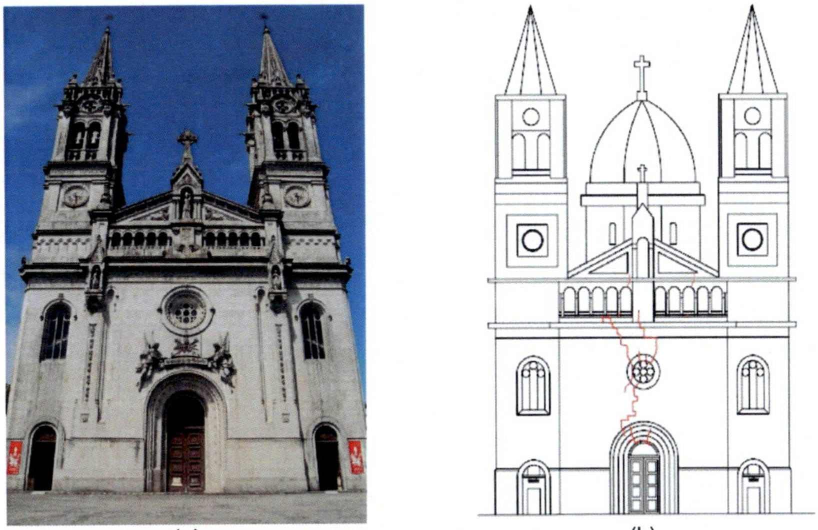

(a) (b)

Figure 3 - Facade view of San Torcato Church and the damage pattern: (a) general view; and (b) damage map.

3. DYNAMIC IDENTIFICATION

3.1 Test Planning

In experimental modal analysis of San Torcato Church, 10 uniaxial PCB 393B12 model piezoelectric accelerometers with a bandwidth ranging from 0.15 to 1000 Hz (5%), a dynamic range of ±0.5 g, sensitivity of 10 V/g, 8 µg of resolution and 210 gr of weight were used. For data acquiring, a 16 channel Digital to Analogue Converter (ADC) was used (Figure 4).

(a) (b)

Figure 4 - Conventional equipments used for dynamic identification: (a) accelerometers; and (b) and data acquisition system.

For the expected frequency range, preliminary test results [1] were taken into account. The first four natural frequencies were ranging from 2 to 4 Hz. To select the measuring, mode shapes of the FE model previously studied for safety analysis was used.

Records were taken in 35 points within 9 test setups. Schematic layout of sensors is given in Figure 5. Reference accelerometers were decided to be placed at the top of the towers (two accelerometers in each tower in two perpendicular directions) because of their high amplitude and modal contribution in each mode. Towers and main facade had to be measured accurately as they have serious damages. Thus, the amount of measurement points was chosen dense in this area. The nave of the structure was decided to be measured in 13 points in the vault; at the supports and at the top. Transept was decided to be measured at the corners. In the apse part, two points were selected at the corners.

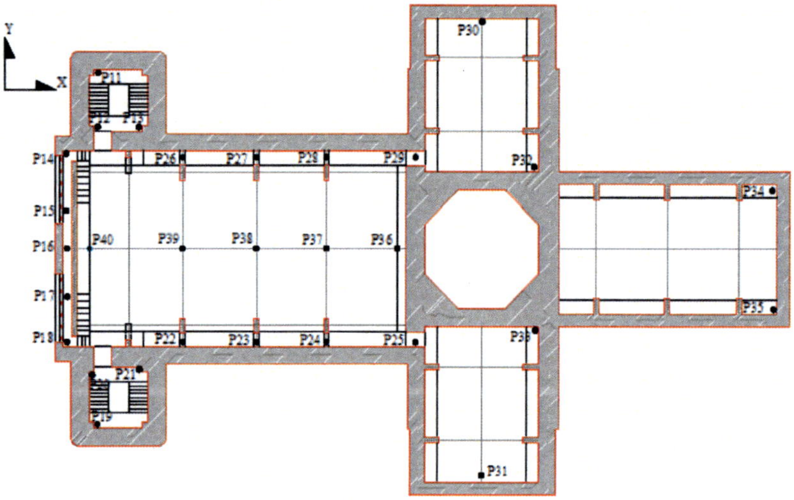

Figure 5 - Plan scheme of measured DOFs.

Accelerations were recorded with 200 Hz sampling rate for 10 minutes. Ambient vibration was used as excitation. The wind and traffic excitation was enough to obtain, at least, the first four modes for all structure.

3.2 Dynamic Results

By using Frequency Domain Decomposition (FDD) [2] technique each mode is estimated as a decomposition of the system's response spectral densities into several single-degrees-of-freedom (SDOF) systems. For frequency domain analysis, the frequency range was between 0 and 2 Hz, with 1024 points window length, and with 66.7% overlap.

Frequency response graphs are very useful to interpret the data roughly. When the graphs are examined, in each setup, between 2 to 3 Hz clear peaks are identified (Figure 6a). In different setups some other peaks are visible in the range of 4 to 5 Hz. However, to be able to identify global frequencies, all the setups were processed together.

To compare the results with other frequency domain techniques, the Enhanced FDD [3] and Curve-Fit FDD [4] were used. Also the Stochastic Sub-Space Identification (SSI) method [5], based on the solution of state space matrices, was used to check the results. For the SSI analysis, SSI-PC (Stochastic Subspace Identification-principal component) method was used (Figure 6b). In Table 1 the frequency estimation of natural frequencies within the chosen methods is given.

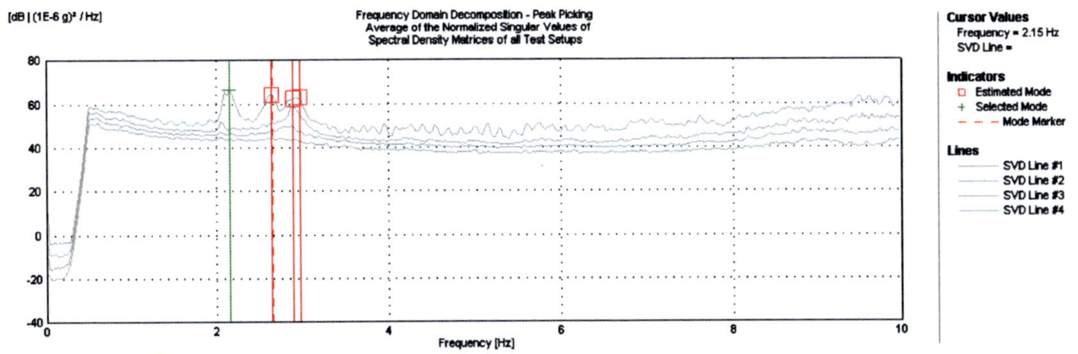

(a)

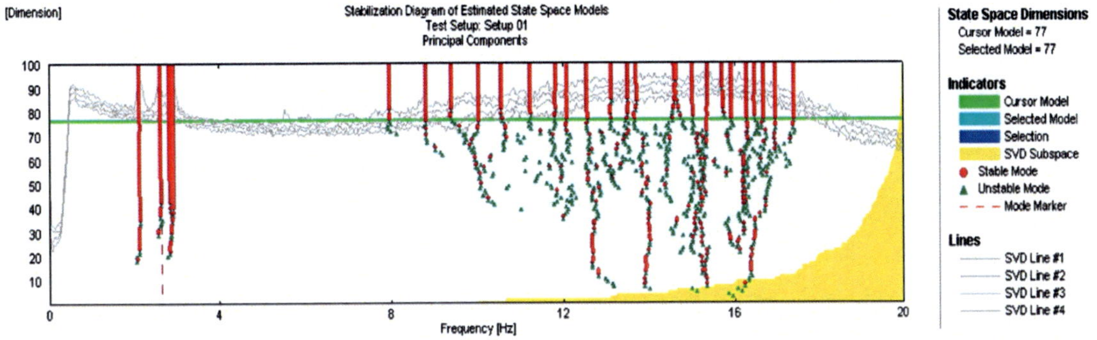

(b)

Figure 6 – Estimation methos: (a) FDD method, and (b) SSI method.

Table 1: Comparison of frequency domain methods with SSI method

	FDD		EFDD		CFDD		SSI
	Frequency [Hz]	Error [%]	Frequency [Hz]	Error [%]	Frequency [Hz]	Error [%]	Frequency [Hz]
Mode 1	2.15	0.4	2.14	0.1	2.13	0.3	2.14
Mode 2	2.64	0.4	2.62	0.1	2.62	0.1	2.63
Mode 3	2.89	1.3	2.89	1.1	2.86	0.3	2.85
Mode 4	2.97	1.4	2.94	0.3	2.93	0.1	2.93

The estimated frequencies with all different methods gave very close results. Also the results of the previous dynamic analysis [1] which were carried out in towers and on the balcony in separated setups exhibit similar results, with the first four frequencies equal to 2.13, 2.60, 2.81, and 2.92 Hz for the North tower, and 2.13, 2.62, 2.85, and 2.90 Hz for the South tower.

Figure 7 presents the estimated mode shapes. In the first two modes it is observed translational modal displacements in X and Y directions. Third and forth mode exhibits torsion modes. Towers move diagonally in opposite directions while the nave is bending.

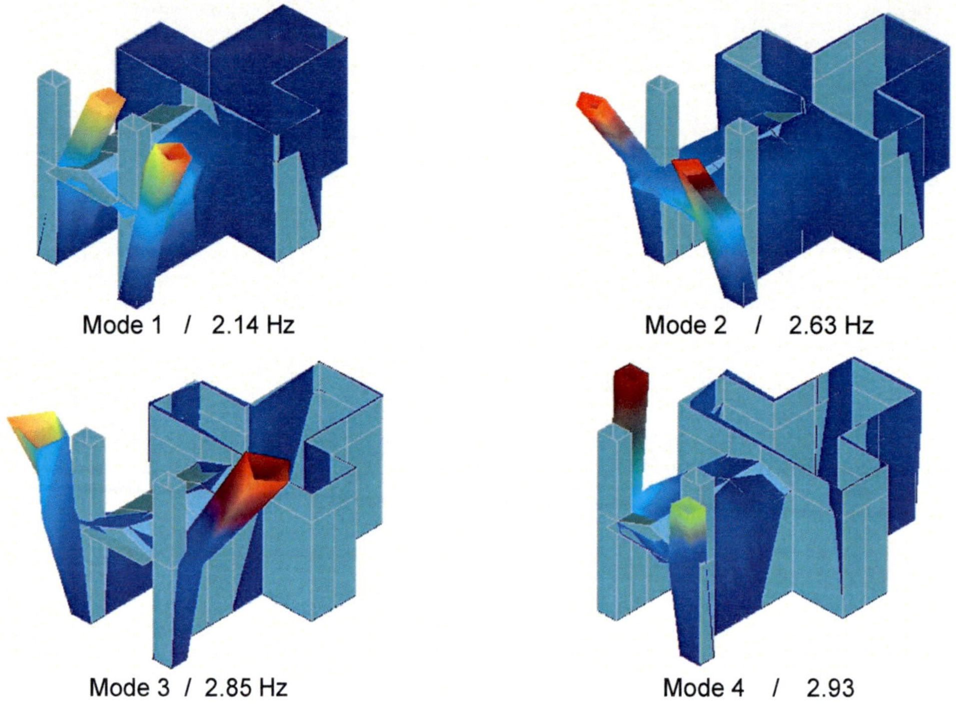

Figure 7 – Mode shapes of the structure

4. MODAL UPDATING

4.1 FE Model

The numerical model of San Torcato church was built for the static non-linear analysis. The model was built in iDiana Release 9.3 Software [6]. Only the nave, façade and towers were modeled.

The model was built with 20 nodes quadratic solid elements CHX60 and with 15 nodes quadratic wedge elements CTP45. The architectural details were neglected and only structural solid parts were modeled. Totally, the model has 3044 solid elements with 3153 DOF.

For soil structure interaction, 16 nodes quadrilateral interface elements CQ48 were used. Soil properties defined elastic and masonry elements were considered homogeneous in all parts of the structure (Figure 8).

Figure 8 – 3D visualization of numerical model

The following hypothesis and assumptions were used in the model:

- Soil-structure interaction is defined with interface elements which allow definition of horizontal and vertical stiffness properties of soil. For the numerical assumption of the properties, a previous soil investigation results were used [1];
- Weight of the masonry assumed as 25 kN/m^3;
- The Poisson's ratio of the masonry defined as 0.2;
- Homogeneous masonry material was used in all parts of the structure;
- Due to incomplete model, the missing part of the structure was simulated by introducing interface elements at the intersection with transepts, to simulate the stiffness of the missing parts.

4.2 Model Updating

The finite element model updating procedure was carried out with four updating parameters with the initial values given in the Table 2. The Young's modulus of the masonry E_m, the normal stiffness of the first soil-structure region E_s, the normal and shear stiffness properties of interface elements at the missing part, E_{tn} and E_{ts}, respectively.

Table 2: Initial values for the updating parameters

E_m	15	GPa
E_s	3900	kPa
E_{tn}	Rigid	-
E_{ts}	Rigid	-

For the comparison between experimental and numerical data frequencies and modes shapes were used. The modes were compared by the Modal Assurance Criterion (MAC) [7]. The MAC is the most commonly used method to investigate the consistency of two different models. It depends on the correlation of modal vectors and varies from zero to one, depending on the consistence of the models. A MAC value equal to one indicates 100% match of both mode shape vectors. MAC is defined by:

$$MAC_{e,n} = \frac{\left| \sum_{i=1}^{n} \varphi_i^e \varphi_i^n \right|^2}{\sum_{i=1}^{n} \left(\varphi_i^e\right)^2 \sum_{i=1}^{n} \left(\varphi_i^n\right)^2} \qquad (2)$$

where φ^e and φ^n are the mode shape vectors, experimental and numerical, respectively.

The structural updating parameters and their effect on the updating results were first studied manually by comparing the frequencies and MAC values. The contribution of each parameter was studied independent from the other variables. To the best combination was chosen after several trials. The results of the starting point for the robust updating analysis are presented in Figure 9, in terms of mode shape comparison, according to the different parts of the structures, and the MAC and frequency comparison plot. The numerical model was too stiff according to the experimental natural frequencies.

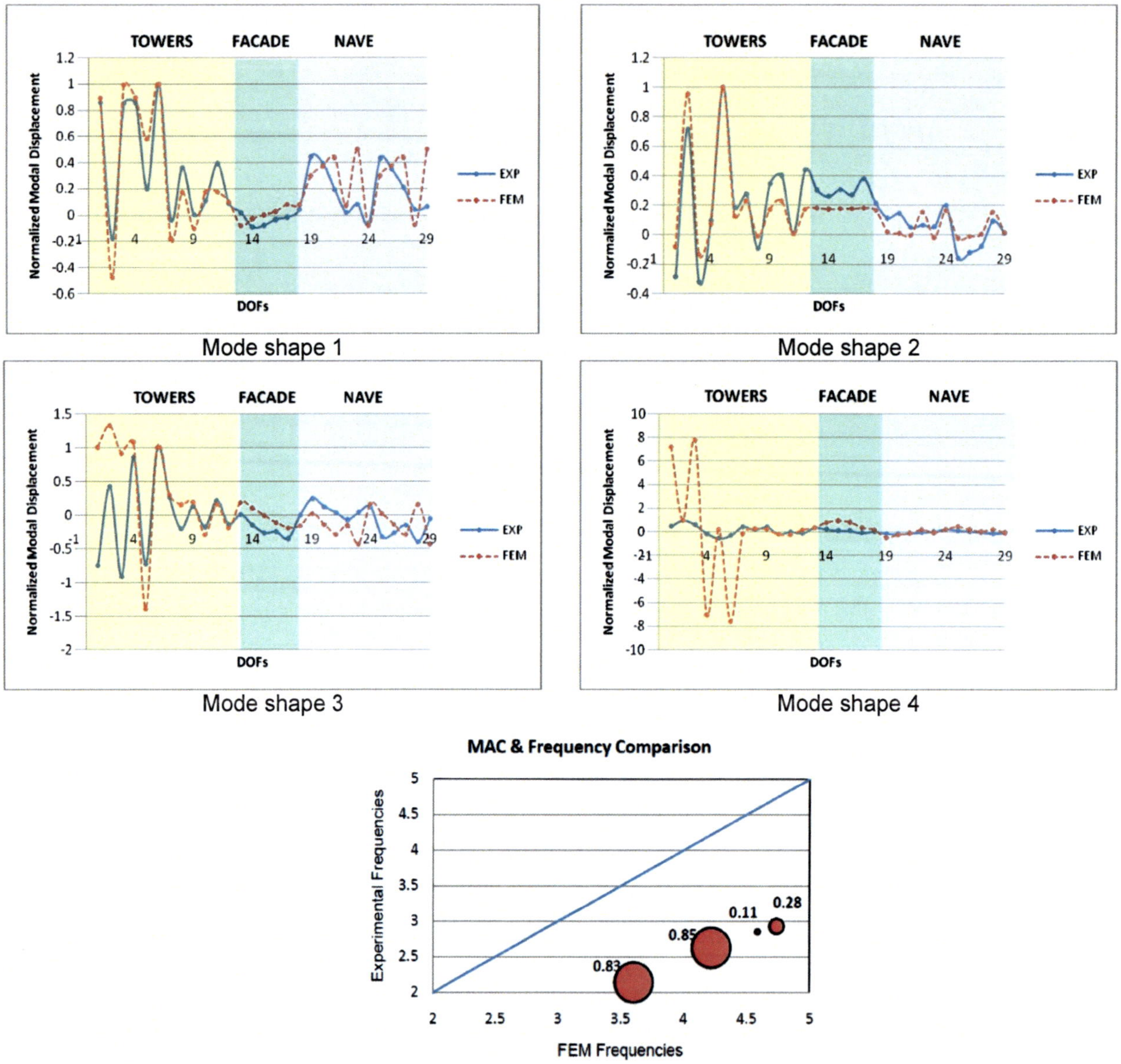

Figure 9 - Comparison between experimental and numerical results before robust updating analysis

For the robust updating analysis, the Douglas-Reid method was used [8]. Lower and upper boundaries for the updating variables were defined according to engineering judgment. The updating results are presented in Table 3. The frequency errors are lower than 3% and the MAC values higher than 77%. The updated values for the structural parameters are consistent from the engineering point of view.

Table 3: Results of the robust updating analysis

Name	Optimum	Modes	ω_{exp} [Hz]	ω_{FE} [Hz]	ω Error [%]	MAC
E_m	5.642 GPa	1st (Transversal Y)	2.14	2.14	0.09	0.92
E_s	0.629 GPa	2nd (Longitudinal X)	2.63	2.55	2.86	0.86
E_{tn}	0.046 GPa	3rd (Torsional)	2.85	2.93	2.70	0.83
E_{ts}	21.591 GPa	4th (Torsional)	2.93	2.94	0.41	0.77

The mode shapes of the updated model exhibit high consistence with experimental mode shapes (Figure 10). In the first and second mode shapes (translational modes), no significant difference is observed between the modes. The third mode shape of the numerical model present similar behavior with experimental model, but when the north tower is examined, it can be seen that the torsion effect is higher than the numerical model. The forth mode shape also exhibit similar behavior, but the reason for having a low MAC value can be related with amplitudes of DOF's and the damage pattern which was not introduced in the analysis.

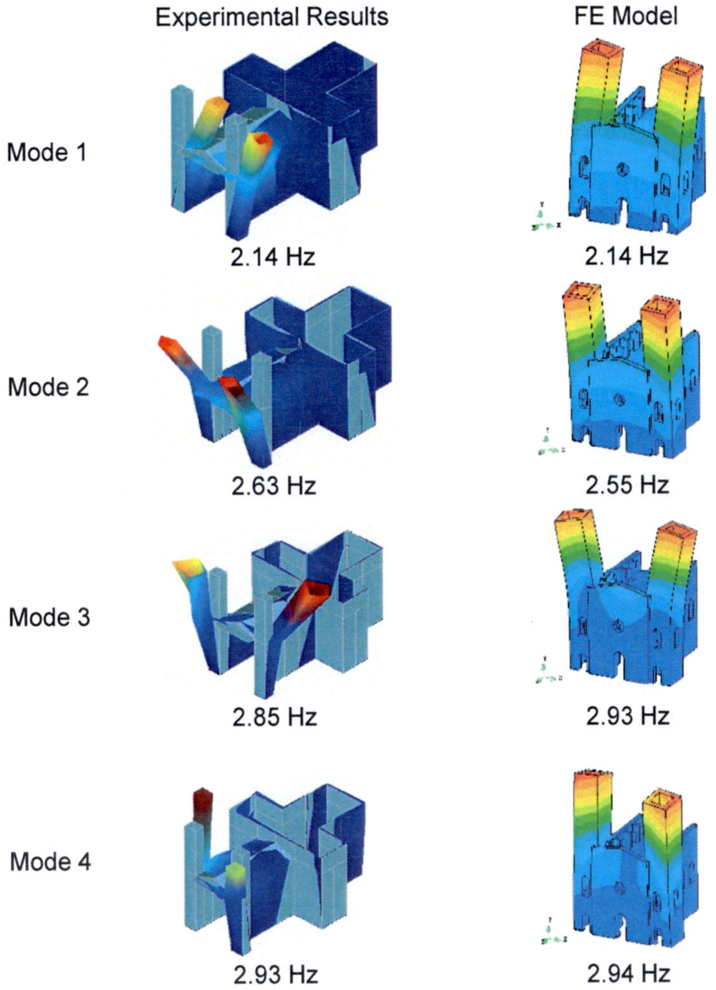

Figure 10 – Mode shape comparison

Figure 11a presents the MAC and frequency comparison plot, where now it can be seen that the frequencies are almost in the 45° line. Figure 11b shows the evolution of the results during the manual tuning and robust updating method. The effect of all the modifications is plotted by means of frequency ratio and MAC values. The final point presents a high contribution of the modifications relative to the starting point. However, the evaluation of results and still being far from the target point, which is desired to be one for MAC values and one for frequency. This indicates that the model needs of other modifications to achieve better results. Considering the crack's widths observed on the main façade, the simulation of the cracks by means of interface elements are advised for the next step.

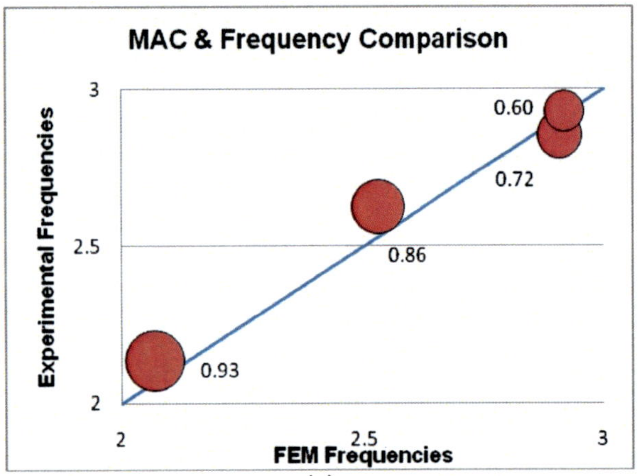

(a)

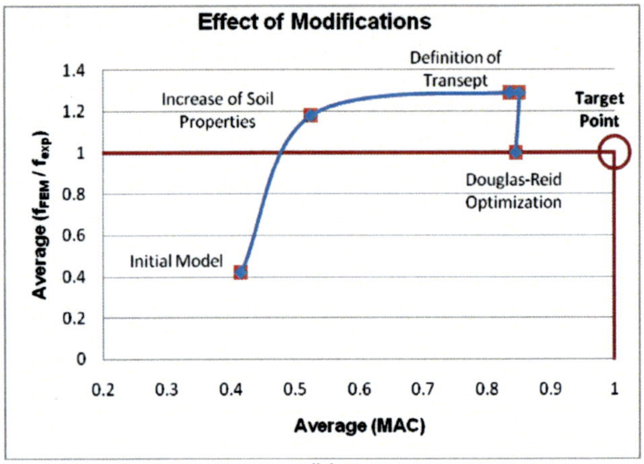

(b)

Figure 11 – Final updating results: (a) MAC and frequency comparison plot; and (b) evolution of the updating analysis.

5. CONCLUSIONS

The paper presents the modal identification analysis of San Torcato Church, together with one updating analysis of a FE model.

The data collected in the field was checked and processed with different methods. Due to the low level of excitation, only the first four global modes of the structure were estimated. The obtained results are high consistency with preliminary dynamic identification results, in terms of frequency and mode shapes.

A FE model, which was previously built for safety analysis, was updated by using manual and robust optimization procedures. Modifications were carried out on homogeneous elastic properties of masonry, elastic soil properties and stiffness parameters of the interface elements which were added to the model in order to simulate the missing part of the transept and apse. High consistence of the numerical model was obtained with manual modifications and non-linear algorithms, but further analysis considering the damage (cracks) should be carried out to achieve better results.

6. REFERENCES

[1] Ramos L.F., Aguilar R. (2007). Dynamic Identification of St. Torcato's Church: Preliminary Tests. Guimarãess, Portugal, University of Minho, Technical Report, 30p.

[2] Brincker R., Zhang L, Andersen P. (2000). Output-only Modal Analysis by Frequency Domain Decomposition. Proceedings of ISMA25 Volume 2

[3] Rodrigues J., Brincker. R., Andersen P. (2001). Improvement of Frequency Domain Output-Only Modal Identification from the Application of the Random Decrement Technique.

[4] Jacobsen, N., J., Andersen P., Brincker R. (2008). Applications of Frequency Domain Curve-fitting in the EFDD Technique

[5] Peeters B., De Roeck G. (2001). Stochastic System Identification for Operational Modal Analysis:A Review. Journal of Dynamic Systems, Measurement, and Control , 659-667.

[6] TNO (2008), "DIsplacement method ANAlyser", version 9.3, Cd-Rom, Netherland.

[7] Allemang J. R. (2003). The Modal Assurance Criterion – Twenty Years of Use and Abuse. Sound and Vibration , 14-21.

[8] Douglas-Reid, B.M., W.H. (1982). Dynamic tests and system identification of bridges, Journal Struct. Div., ASCE,108, 2295-2312

Proceedings of the IMAC-XXVIII
February 1–4, 2010, Jacksonville, Florida USA
©2010 Society for Experimental Mechanics Inc.

Modal Properties of a High Rise Building Under Construction

Tomás R. Nuñez[1], Rubén L. Boroschek[2]
[1]Msc. Earthquake Engineering, University of Chile, Chile.
tnunez@ing.uchile.cl
[2]Associate Professor, Dept. of Civil Engineering University of Chile, Chile.
rborosch@ing.uchile.cl
Blanco Encalada 2002, Santiago. Chile.

1 Nomenclature

$\{x_{k+1}\}$: State vector on $k+1$.

$\{x_k\}$: State vector on k.

$[A]$: Discrete System Matrix.

$[C]$: Output Matrix.

$\{w_k\}$: Process noise vector.

$\{v_k\}$: Measurement noise vector.

i : *i-th* mode.

λ_i : Eigen value from continuous system matrix.

μ_i : Eigen value from discrete system matrix.

Δt : Sample time.

ω_i : Angular frequency.

ξ_i : Damping ratio.

$[\phi]$: Modal shape.

$[\Psi]$: Eigen vector from discrete system matrix.

2 Abstract

This article presents the results of the continuous vibration monitoring and modal parameter identification of a 55 story shear-wall concrete building under construction. The building has a near elliptical floor shape with 61x34 meters maximum plan dimensions and 196 meters in height. Metal hysteretic energy dissipating devices have been incorporated into the building to reduce the vibration due to earthquake excitations. The continuous monitoring network is composed by 3 accelerometers on the 20th story and had the objective of identifying the variations of modal parameters induced by the construction process. Ambient and earthquake records have been obtained during monitoring period. The dynamic characteristics of the building were identified using parametric and non-parametric techniques such as the Spectrogram and the Stochastic Subspace Identification (SSI) methods. The procedures obtain the variations of modal parameters with a resolution of ten minutes. The recorded motions were used to validate 15 different Finite Element Models (FEM), created according to the construction progress of the building. The mathematical model frequencies were correlated with the results of the identification procedures; maximum differences between 2% and 14% were obtained. The difficulties of monitoring vibration and modal identification during construction process are indicated and causes of main differences between design and real structure are commented.

T. Proulx (ed.), *Dynamics of Civil Structures, Volume 4*, Conference Proceedings of the Society for Experimental Mechanics Series 13,
DOI 10.1007/978-1-4419-9831-6_10, © The Society for Experimental Mechanics, Inc. 2011

3 Introduction

Many studies to determine modal properties have been performed over the past decades, with most of applications focus on structural health monitoring [5, 7, 9], calibrated mathematical models [3, 16], diagnosis [1,2, 12], damage detection [11, 13], variations on dynamic properties [4, 10], among others. The objective of the continuous identification during the construction process was to track part of the dynamic properties variations, to detect anomalies and to compare the identified values with different Finite Adaptative Element Models (FAEM).

Two types of monitoring procedures were implemented: a) along the building height, when the construction process was located on floors 11, 17, 22, 28 and 55 and b) continuously under-construction for a period of 5 months. This article presents the results of the continuous monitoring only, [10].

The Titanium La Portada building, developed by ASL Sencorp and Structural Engineering from Alfonso Larrain Vial. ALV & Associated, is the tallest building in Chile, and it is currently in its latest stage of construction. The building has 55 stories, 7 underground levels and 126,000 m2. Structurally, the building consists of a reinforced concrete walls core surrounded by frames and columns located in the perimeter, as shown in Figure 1.

The construction process consists in the concreting of the core and perimeter columns and beams, after which the prefabricated slabs are supported by this system. In order to control earthquake vibration several steel dampers are included along the height of the structure. During the study, the dampers were lock so they don't participate in the energy dissipation of the system.

Figure 1. General views.

4 Modal Parameters from ambient vibration.

Continuous ambient vibration records were obtained during a 5 month period from the three uniaxial accelerometers located in the 20th floor, as shown in Figure 2. Due to this instrument arrangement only frequency and damping, and partial mode shapes could be obtained. Channels were sampled at 100 [Hz]. Data was recorded continuously, and modal properties were obtained every 10 minutes.

The permanent variation of the dynamic properties due to constructive process was monitored and compared at different stages of construction using finite element adaptive models, according to the progress of the building construction shown in Figure 3.

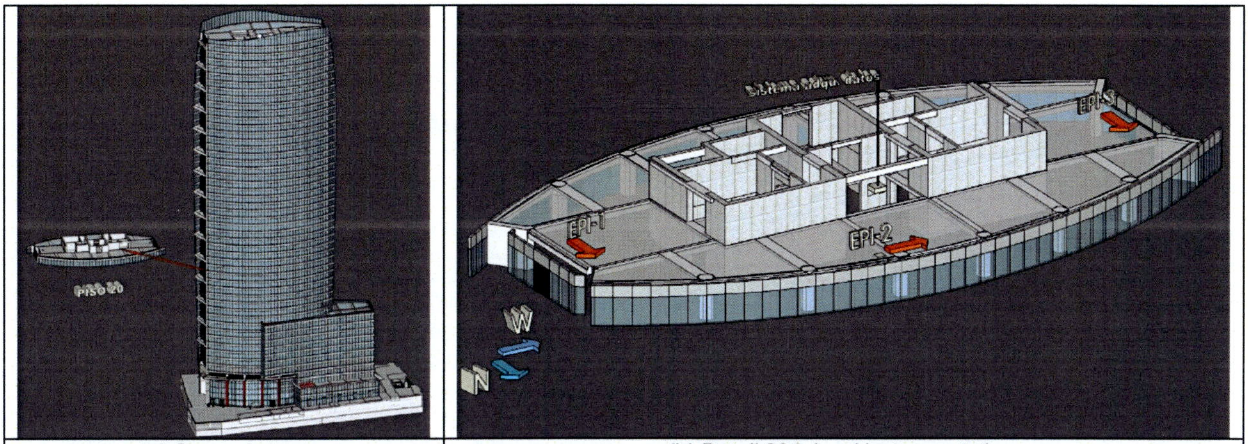

(a) General view. | (b) Detail 20th level instrumentation.

Figure 2. Accelerometer location during construction.

(08-03-06) | (08-05-12) | (08-06-05) | (08-07-10) | (08-08-09) | (08-09-01) | (08-10-15) | (08-11-10) | (08-12-01) | (08-12-15) | (08-12-30)

Figure 3. Construction sequence. Date (yy-mm-dd).

5 System Identification Technique

The Stochastic Subspace Identification Technique developed by Van Overschee and De Moor [14, 15] is used to estimate modal frequency, damping and shapes. This technique uses the stochastic state space model, described by equation (1.1) to identified modal parameters from output only response signals.

$$\begin{aligned} \{x_{k+1}\} &= [A]\cdot\{x_k\}+\{w_k\} \\ \{y_k\} &= [C]\cdot\{x_k\}+\{v_k\} \end{aligned} \tag{1.1}$$

Eq. (1.1) constitutes the basis for time-domain modal identification through ambient vibration measurements. There are several techniques and algorithms to obtain modal parameters from stochastic subspace model [14], equation (1.2). The mathematical background for many of such methods is similar, differing only in implementation aspects. The algorithms identify the state–space matrices ($[A],[C]$) based on the measurements by using robust numerical techniques, such as *QR* factorization, singular value decomposition (SVD), and least squares [15].

Once the mathematical description of the structure is found, modal parameters such as frequency, ω_i, damping ratio, ξ_i, and operational mode shapes, $[\phi]$ are determined as:

$$\lambda_i = \frac{\ln(\mu_i)}{\Delta t} \qquad \omega_i = |\lambda_i| = \sqrt{\lambda_i \cdot \lambda_i^*} \qquad \xi_i = \frac{real(\lambda_i)}{|\lambda_i|} \qquad [\phi]=[C]\cdot[\Psi] \tag{1.2}$$

To validate results the Short Time Fourier Transform, Spectrogram, was also used, [6].

6 Results of continuous monitoring

Several difficulties are present when ambient vibration monitoring is done during construction. They are related with high transient vibration due to impact, continuous variations of mass and stiffness, operation of small and large machinery like cranes and others. All this activities affects the system identification processes producing non stationary signals. Figure 4 shows vibrations during concrete pumping and Figure 5 a comparison between workday and holyday.

One important observation from time series is the difference between acceleration amplitude at workdays compared to holyday. While the workdays the acceleration amplitude was in the order of 10^{-3}[g] the holydays showed two order less, 10^{-5}[g], Figure 5.

This construction process did not affect the identification of predominant frequencies. Nevertheless, they did affect considerably the damping values, producing a larger scatter of values for each consecutive 10 minute window. For better estimates an average of six consecutive values was used.

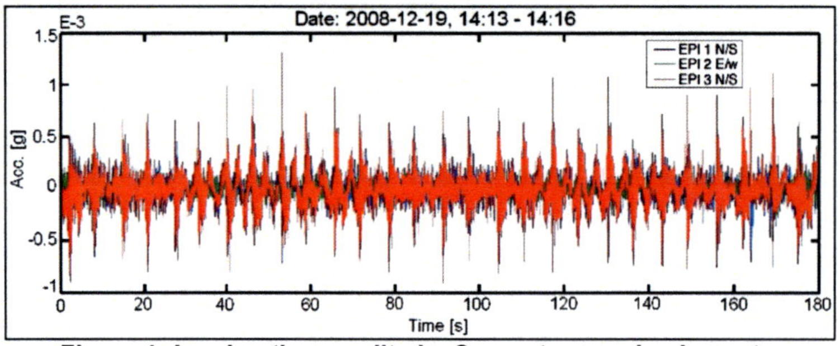

Figure 4. Acceleration amplitude. Concrete pumping impacts.

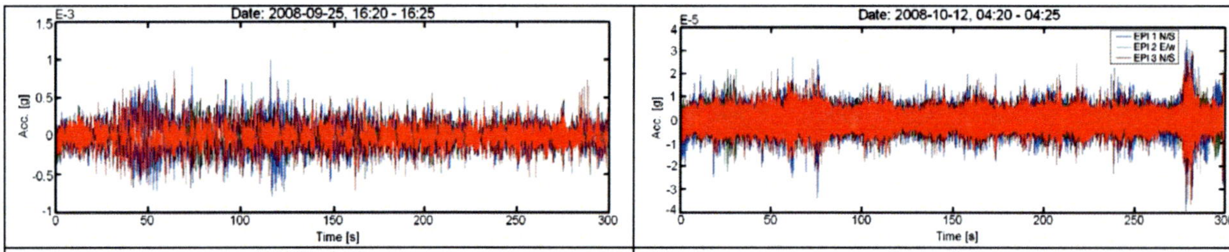

(a) Work day	(b) Holiday

Figure 5. Acceleration amplitude comparison between (a) holyday and (b) work day. The difference is up to one order magnitude.

From continuous monitoring, 13 natural frequencies and their associated damping ratios were estimated. The last 4 modes were identified during the last months of monitoring.

6.1 Frequency

Figure 6 present the variation of frequency for all identified modes during the five month continuous monitoring. The results show a continuous reduction in frequency and a daily variation, this last effect is the one that produces the apparent thick line in the Figure 6. The variations in frequency are mainly caused by two characteristic and repetitive patterns: 1) construction progress (mass and stiffness) and 2) daily ambient and usage variations.

To separate both tendencies a power spectral density analysis was performed to identify weekly and daily frequency variations. The results obtained from the power spectral peaks show predominant variations of 23.9 hours and 7.4 days in average. A band-pass filter was applied around these patterns, separating the observe frequency variations. The result of the filtering for the 7.4 day pattern is shown in Figure 7 (a) and the nearly 24 hour pattern in Figure 7 (b).

The most important frequency variation is associated to the addition of mass after construction of a new story. According to the construction program, the average global progress of the structure corresponds to 3.75 stories per month. These nearly weakly patterns are the consecuence of the concreting of the wall-core and perimeter frame, and the second due to the slabs placement. This last procedure provides greater amount of mass and therefore the changes in frequency are more drastic.

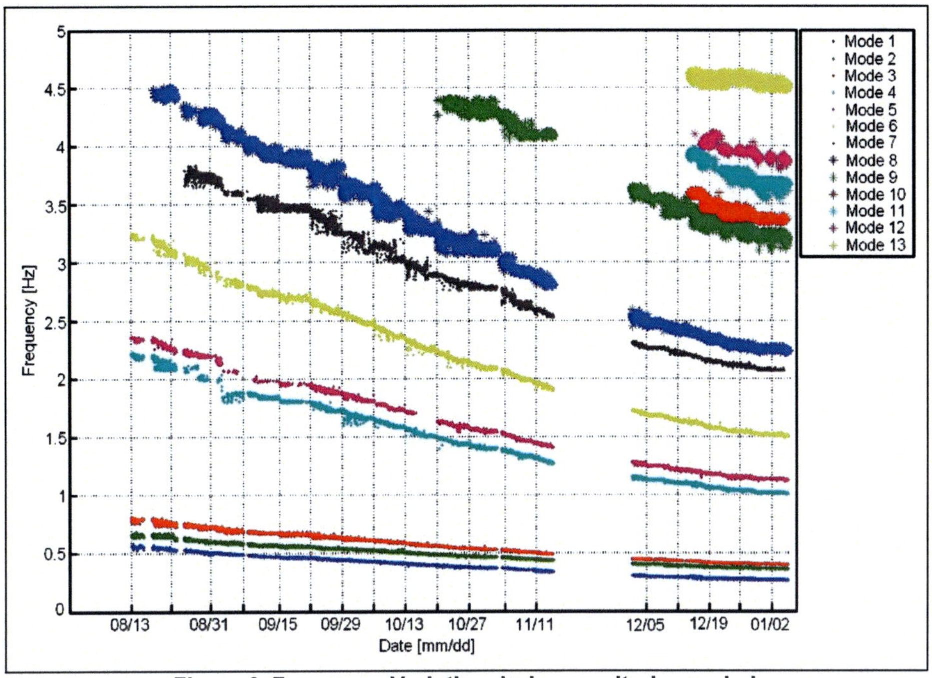

Figure 6. Frequency Variation during monitoring period.

The weekly pattern generated variations of frequency in a range between 1.5% and 5.5% respect to the frequency at the beginning of each week, Figure 7 (a). This behavior was observed for the first eight building vibration modes.

Their maximum values due to daily frequency variations (deviation from daily mean value) are in the order of 1.6%, Figure 7 (b). These variations were directly related to environmental day-night cycles and the presence of workers, machinery and materials in the building.

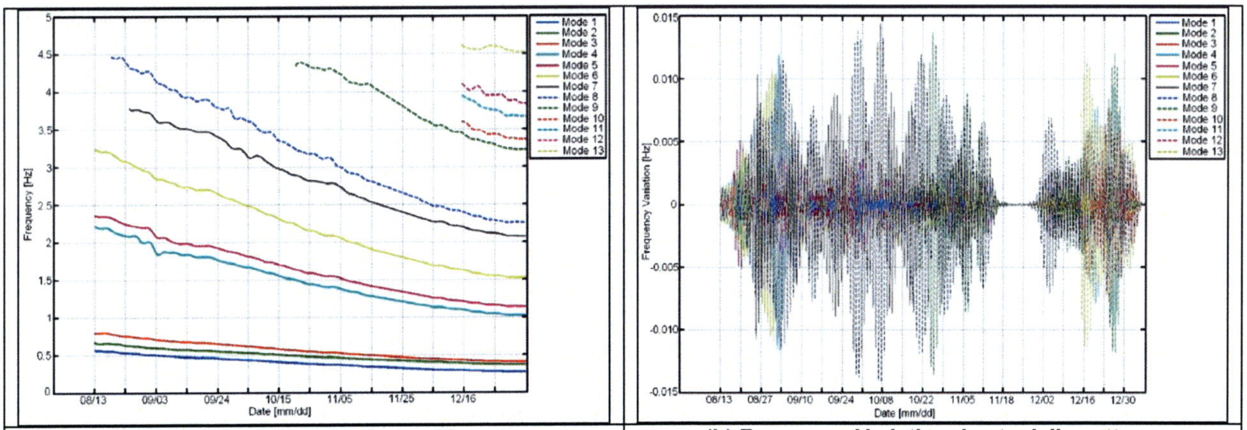

(a) Frequency Variation due to weekly pattern. **(b) Frequency Variation due to daily pattern.**

Figure 7. Filtered Frequency variation.

A similar pattern is possible to appreciate using the spectrogram method, Figure 8. An important result is associated to observe the diary cycles, producing high energy amplitude during working period (day) and low energy at night or weekends (vertical lines on spectrogram, Figure 8).

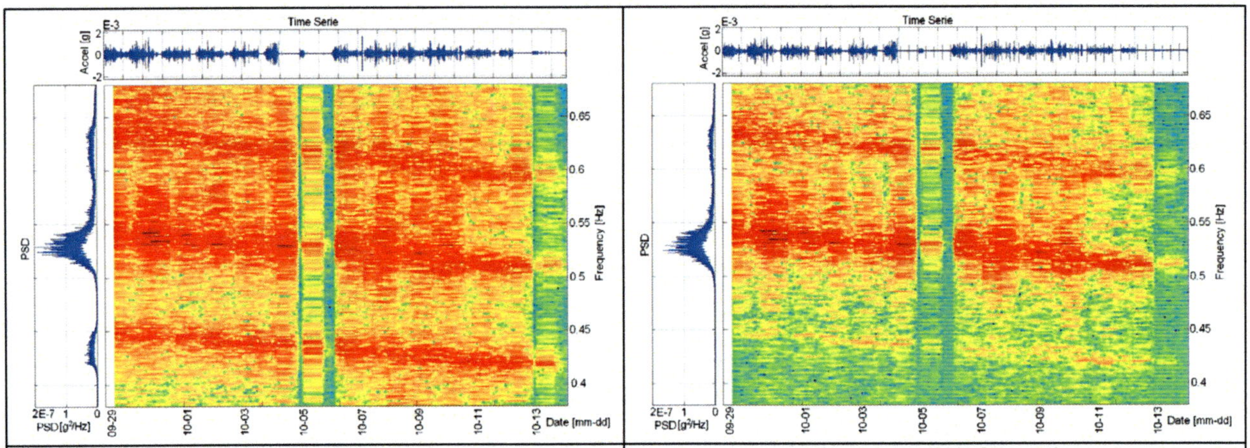

(a) Channel 1. N/S Orientation	(a) Channel 2. E/W Orientation.

Figure 8. Spectrogram.

6.2 Damping

Damping ratio estimations show high scatter between consecutive 10 minute windows. Average values were obtained for six consecutive non overlapping windows. Result of the average, smooth, data are presented in Figure 9 and Figure 10 for the first four modes. The damping values showed a increasing magnitude with increasing structural height for the first predominant frequency and a decreasing magnitude with increasing structural height for the higher modes, nevertheless daily variation are large overall variation during the five month period is small.

Damping ratio is one of the most difficult to determine, since it was observed that is highly dependent on the amplitude of the excitation. For that reason, the filtering procedure applied to characterize the frequency variation was performed to damping ratio, identifying a damping variation pattern spanning around 18.4 days, period required to build about 3 stories, Figure 11.

From the analysis of the records it was established that the damping ratio, under ambient vibration and construction process, varies in a range between 0.5% and 1.5% [10], Figure 11.

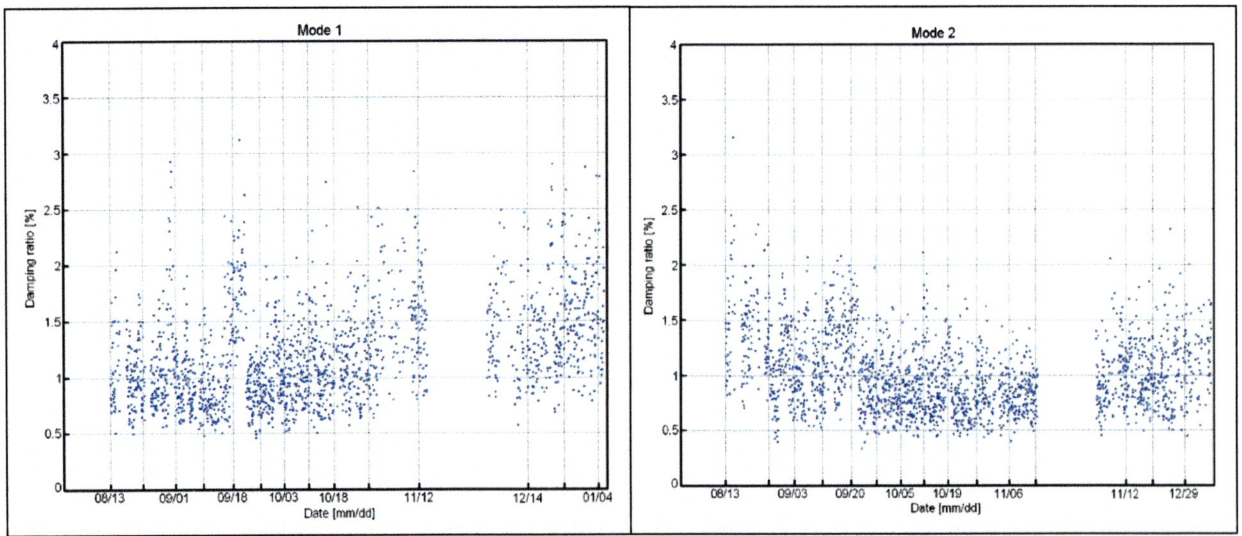

Figure 9. Damping Ratio variation during monitoring period. Modes 1 and 2. Average values obtained for six consecutive non overlapping windows.

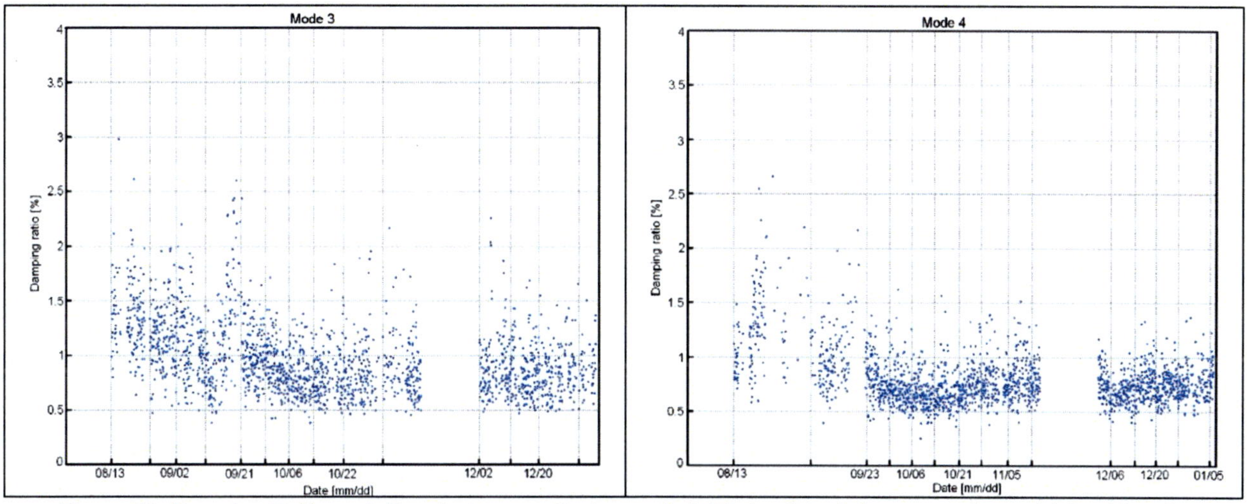

Figure 10. Damping Ratio variation during monitoring period. Modes 3 to 4. Average values obtained for six consecutive non overlapping windows.

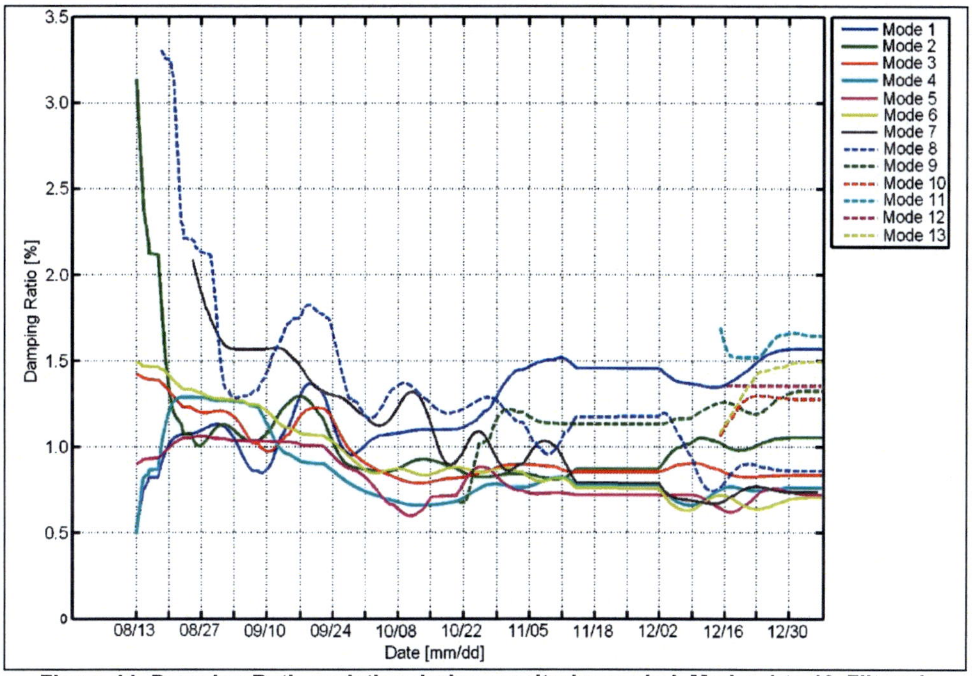

Figure 11. Damping Ratio variation during monitoring period. Modes 1 to 13. Filtered.

7 Mathematical Structural Models

The results of the structural monitoring were compared using several finite elements models created for different construction stages, according to the information provided by the contractor. The structure was modeled considering fixed base, with two types of Young's module: 331.248 [kgf/cm2] and 409.765 [kgf/cm2]. As mass source, the model considers only dead load, because permanent loads of endings and live loads were not present during the monitoring period. Gross section was considered for all elements and beam did not considered the contribution of the slab.

Maximum differences between 2% and 14% on the modal frequency are found when experimental and finite element models results are compared. Figure 12 shows correlation between real and predicted frequencies for the first and second mode. The differences between the dynamic properties of the real structure and the finite element adaptive model has not been studied in detail yet, but could initially be associated with the restraint present for underground levels and additional construction masses not considered in the analysis model (forklifts, cranes, etc.). Further studies are under way.

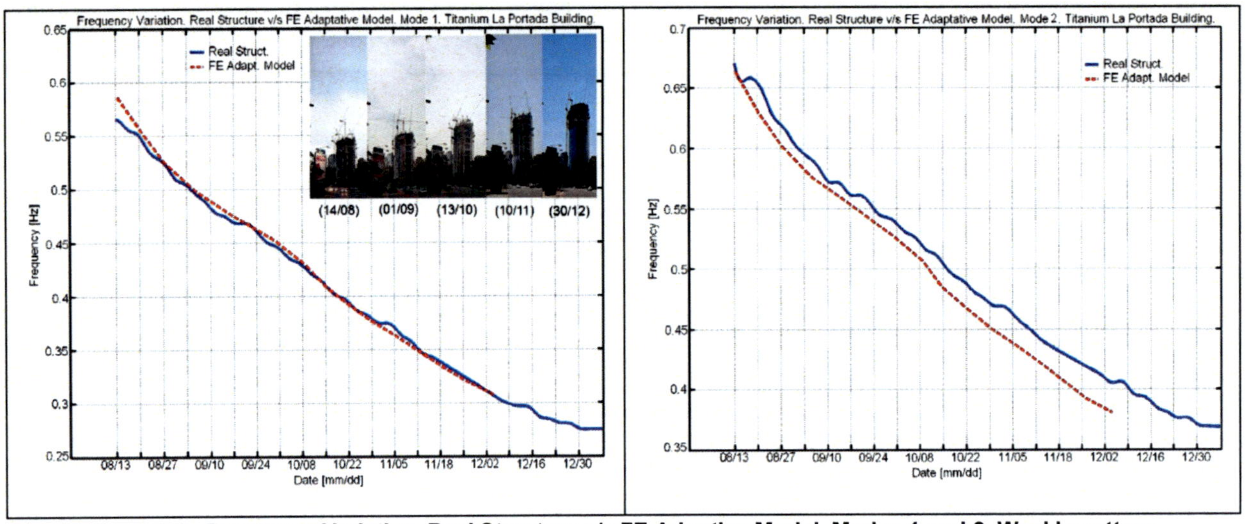

Figure 12. Frequency Variation. Real Structure v/s FE Adaptive Model. Modes 1 and 2. Weekly pattern.

8 Earthquake Records

During the five month continue vibration monitoring 11 low level earthquakes were recorded, Figure 13.

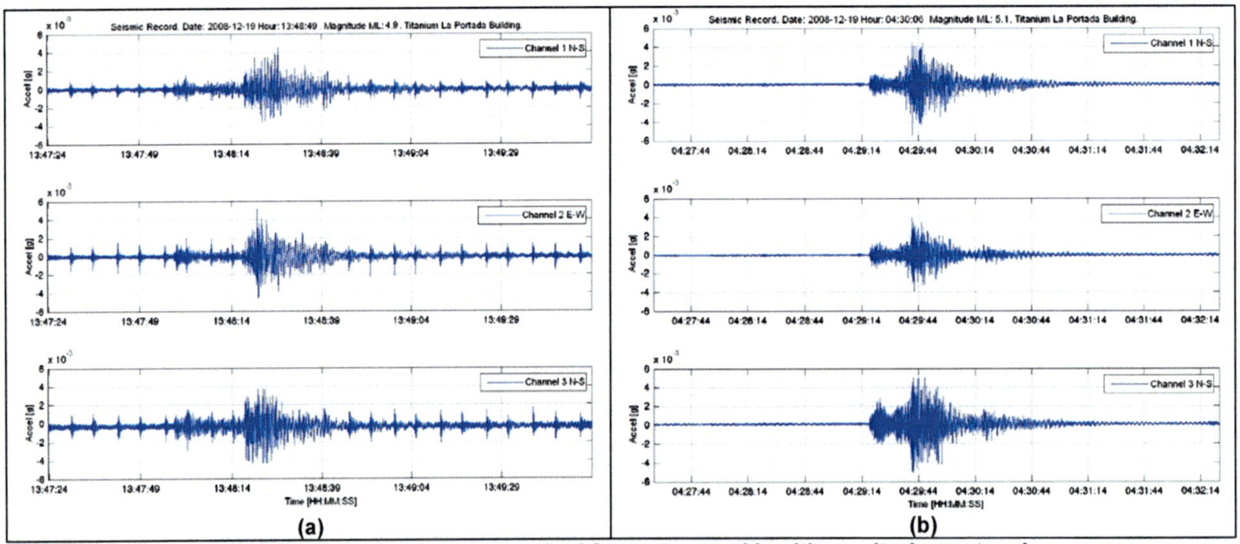

Figure 13. Seismic records, acquired from structural health monitoring network.

The maximum acceleration observed was 0.005 [g]. No damage or detention of construction was done. In Figure 13 concrete pumping is superimposed by the earthquake.

An important result can observed from the comparison between power spectra before, during and after a low level earthquake. Figure 14 shows that during the earthquake (recorded from Figure 13 (b)) higher modes are excited, while modes under 1 Hz are clearly attenuated.

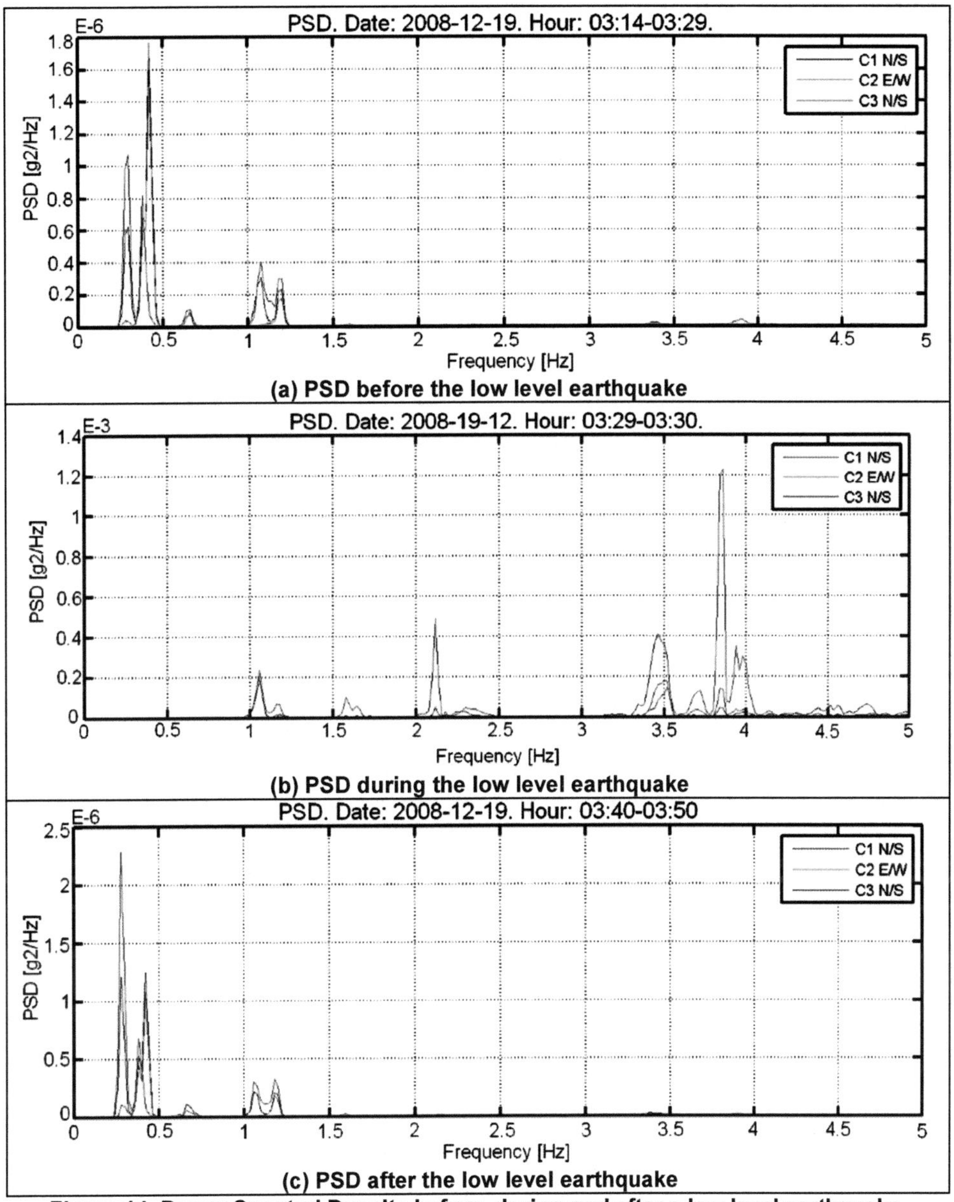

(a) PSD before the low level earthquake

(b) PSD during the low level earthquake

(c) PSD after the low level earthquake

Figure 14. Power Spectral Density before, during and after a low level earthquake.

9 Conclusions

Through continuous monitoring of a high rise building under construction, 13 frequencies and damping ratios were identified. For natural frequencies, weekly and daily variation patterns were determined. Weekly variations of frequency, caused by the construction of additional floors, were found to be in the range between 1.5% and 5.5%, while daily variations, caused by environmental day-night cycles and working routine, reached values of up to 1.6%. The analysis established that the damping ratio for the monitoring period, under ambient vibration and construction process, is in a range between 0.5% and 1.5%. The damping ratio exhibited a strong dependence on the amplitude of vibration.

10 Acknowledgements

The Civil Engineering Department of the University of Chile and the Chilean Council for Research and Technology, CONICYT Fondecyt Project # 1070319 supported this research paper.

ASL Sencorp and Alfonso Larrain Vial. ALV & Associated, Structural Engineers for the valuable information about the building and structural considerations.

11 References

[1] Boroschek R, Castillo A. "Using the nonparametric frequency method and SSI for identifying modal properties of buildings". IX Seismology and Earthquake Eng. Chilean Congress. Concepcion, Chile. 2005

[2] Brincker R, Zhang L and Andersen P. "Modal Identification of output-only system using frequency domain decomposition". Institute of Physics Publishing. Smart Materials and Structures. 10 441-445. 2001.

[3] Brownjohn J. "Ambient vibration studies for system identification of tall buildings". Earthquake Engineering and Structural Dynamics, 2003; 32: 71-95. 2002.

[4] Carreno R. "Variation of dynamic properties of Chilean Chamber of Construction Building: seismic case". Civil Engineering Thesis. University of Chile. Chile. 2009

[5] Clinton J, Bradford S, Heaton T, and Favela J. "The Observed Wander of the Natural Frequencies in a Structure". Bulletin of the Seismological Society of America, Vol. 96, No. 1, pp. 237–257. 2006.

[6] Hernandez F. "Analysis of Methods for Identification Dynamic Properties Variation". MSc. Earthquake Engineering Thesis. University of Chile. Chile. 2009.

[7] Johnson E, Lam H, Katafygiotis L and Beck J. "Phase I IASC-ASCE structural health monitoring benchmark problem using simulated data". Journal of Eng. Mechanics, vol. 130, N° 1, January 2004, pp. 3-15. 2004.

[8] Meirovitch L. "Analytical Methods in Vibrations". Macmillan Publishing CO., Inc. 1967.

[9] Ni Y, Chia Y, Liao W and Ko J. "Technology Innovation in developing the structural health monitoring system for Guangzhou New TV Tower". Structural Control and Health Monitoring. N° 16: pp. 73-98. 2009

[10] Nunez T. "Variation due to changes in mass and stiffness of dynamic properties of a structure in construction process". MSc. Earthquake Engineering Thesis. University of Chile. Chile. 2009

[11] Peeters B. and De Roeck G. "System Identification and Damage Detection in Civil Engineering". PhD. Thesis. Katholieke Universiteit Leuven, Faculteit Toegepaste Wetenschappen. Belgium. 1998.

[12] Ren W. and Zong Z. "Output-only modal parameter identification of civil engineering structures". Structural Engineering and Mechanics, Vol. 17, No 3-4. 2003.

[13] Turek M. "A Method for Implementation of Damage Detection Algorithms for Civil Structural Health Monitoring Systems". PhD Thesis. The Faculty of Graduate Studies (Civil Engineering). University of British Columbia. 2007.

[14] Van Overschee P. and De Moor B. "Subspace algorithms for the stochastic identification problem". Automatica, Vol. 29, no. 3, pp. 649-660. 1993.

[15] Van Overschee P. and De Moor B. "Subspace Identification for Linear Systems: Theory-Implementation-Applications". Kluwer Academic Publishers, Dordrecht, The Netherlands. 1996.

[16] Ventura C, Schuster N. "Structural dynamic properties of a reinforced concrete high-rise building during construction". Canadian Journal of Civil Engineering. Vol. 23, N° 4, pp. 950-972. 1996.

Proceedings of the IMAC-XXVIII
February 1–4, 2010, Jacksonville, Florida USA

Model Updating for a 183m of Reinforced Concrete Chimney

M Garcia-Diéguez [a], K-Y Koo [b], CM Middleton [b], JMW Brownjohn [b], C Goddard [c]

[a] Department of Construction and Engineering Manufacturing, University of Oviedo, Campus de Gijón 7.BC.01 33203 Gijón Spain
[b] Vibration Engineering Section, University of Sheffield Mappin Street, Sheffield S1 3JD, UK
[c] Bierrum International, UK

ABSTRACT
This paper deals with model updating of Rugeley Chimney, a 183m reinforced concrete flue gas stack at a coal-fired power station. Since February 2007 this has been monitored due to safety concerns about vibration levels. A 42-tonne tuned mass damper (TMD) was installed to reduce response levels and the monitoring system was used to track the mode frequencies.

A numerical model has been built and experimental observations have been used to identify the concrete properties and the foundation stiffness of the original chimney through a process of model updating. These updated parameters are used to compare with the actual situation of the chimney and to detect possible changes on the chimney characteristics (stiffness and density) in the present time, comparing with another numerical model with TMD mass and stiffness on the top of the chimney. The updating procedure uses a sensitivity matrix and comparison of the frequencies and mode shapes through MAC values.

1 INTRODUCTION ON EXPERIMENTAL STUDIES OF TALL CONCRETE CHIMNEYS

Performance of tall reinforced chimneys due to both along-wind and cross-wind loading has long been a concern for operators of industrial facilities such as conventional power stations. For example, upstream structures cause interference effects and enhance cross wind response [1], and prediction of response levels requires accurate estimation of modal parameters, in particular damping ratio. Happily new developments of system identification are such that accurate modal parameter estimation requires measurement only of response to random loads such as wind-buffeting so that simple vibration monitoring is all that is required in order to identify modal parameters.

Finite element (FE) model updating [2] is a subject of great importance for mechanical and civil structures and is an optimization problem, in which, the differences between the experimental and numerical modal data have to be minimized by adjusting uncertain model parameters based on modal parameters obtained (for many civil structures) in operational conditions. In addition the measurements of these structures are usually difficult due to the size, localization and usage of the structures. In last decades, these techniques were applied by several research groups with civil structures like buildings [3], bridges [4], towers [5] or chimneys [6]. Other authors [7] used the model updating to do predictions on the model with structural modifications successfully.

This paper deals with the identification of modal parameters for the chimney and their use in explaining the actual situation of the chimney in several conditions through FEM updating.

2 RUGELEY CHIMNEY AND TUNED MASS DAMPER

The chimney at Rugeley Power Station north of Birmingham was built around 1968 and consists of a 183m high reinforced concrete windshield, internally protected by a sectional, acid resisting brickwork lining. The windshield

T. Proulx (ed.), *Dynamics of Civil Structures, Volume 4*, Conference Proceedings of the Society for Experimental Mechanics Series 13,
DOI 10.1007/978-1-4419-9831-6_11, © The Society for Experimental Mechanics, Inc. 2011

tapers from an external diameter of 9.4m at the top to 15.7m at the bottom. Its outer surface has suffered from environmental actions and external reinforced concrete cladding was added in 1998 over the top 24m.

In 2006 a new flue gas desulphurisation (FGD) chimney was built of the same height at a distance of 110m in the direction of prevailing winds. The two chimneys are shown in Figure 1, and were intended to coexist until demolition of the old chimney. The designers (Bierrum International) commissioned an investigation [8] into cross-wind interference effects on the old chimney, which anticipated observed swaying of the old chimney.

Figure 1: The old chimney without TMD on the left; the FGD chimney is on the right.

Figure 2: Flue ducts flanking chimney on the bottom.

Figure 3: The old chimney with the TMD.

The designers recommended installation of a monitoring system to track possible large amplitude response and if necessary a tuned mass damper to control response. Both the monitoring system and the TMD were designed and installed in early 2007, with early data from the monitoring system being used to fine-tune the TMD and later to update the FEM of the original chimney. The chimney is not quiet symmetric as the flue ducts pass through flanking buttresses shown in Figure 2 and there is an expected stiffer (tangential) direction leading to small differences in the two first natural frequencies observed experimentally. The updated chimney model could be a simulator of the planned demolition work, the first part of which would consist in removing 5m of brickwork at the top of the chimney seen in Figure 3 (black part).

The TMD, provided by Multitech, comprises $4.2x10^4$kg of moving mass (steel ring with concrete elements) and five damper units, designed as the smallest moving mass to generate sufficient damping with maximum relative movement of 450mm. The final configuration is shown in Figure 3.

3 MONITORING SYSTEM DESIGN AND ADQUISITION SIGNAL PROCESSING

A major requirement for the monitoring system was to provide alarm of large amplitude response, directly derived from the expected capacity of the structure. Three alert levels were specified, the largest at 0.3 m displacement.

The system comprised four Honeywell QA750 quartz-flex servo-accelerometers oriented in "tangential" and "radial" directions on the chimney in watertight enclosures at two levels (40m and 180m), and a 4-channel National Instruments data acquisition (DAQ) system [9], with ADSL connection from 15th March 2007. A LabVIEW virtual instrument (VI) was written to collect data saved on the DAQ and transmit over broadband using LabVIEW's datasocket protocol and also analysed in-situ using Stochastic Subspace Identification (SSI) techniques, are considered to be the most powerful class of time domain OMA techniques.

Figure 4 shows the SSI stability plot for 8Hz sampled data recovered for September 2007. Around 0.32Hz is a cluster of modes clearly identified as first mode and with the second mode around 1.4Hz. Moreover the monitoring system allowed observing changes on the modes due to structural effects from TMD, in addition to expected changes due to temperature-dependent changes in concrete modulus. Other influences on the modes are the wind and its directional effects due to the construction of the new chimney close to the old one [10].

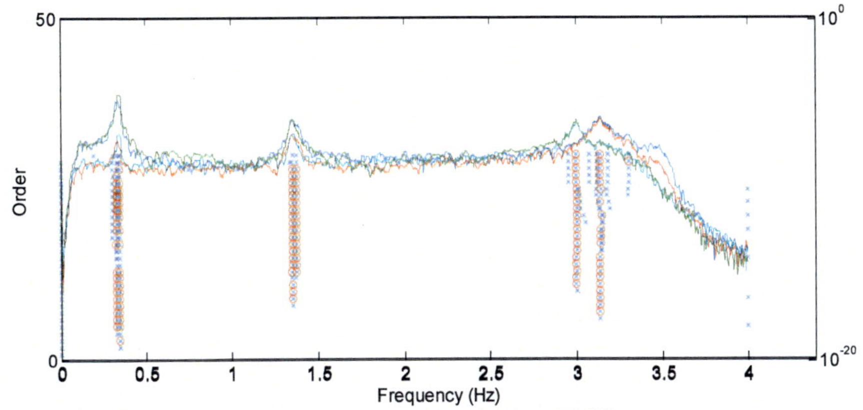

Figure 4 Stability diagram during normal operation (before TMD).

Environmental effects are always present on outdoor civil structures. More usual environmental effects in this type of structures are wind and temperature. These effects have been measured at the same time that the monitoring system was working, to explain their influence on the natural-frequencies. It is difficult to isolate only one effect, however, when looking at data for some low-wind days (i.e. wind speed less than about 5mph), can be seen only temperature effects (Figure 5). The lack of continuity on the curves is due unreliable frequency estimates. On the other hand, can be shown the same frequencies from windy days to check the differences (Figure 6a).

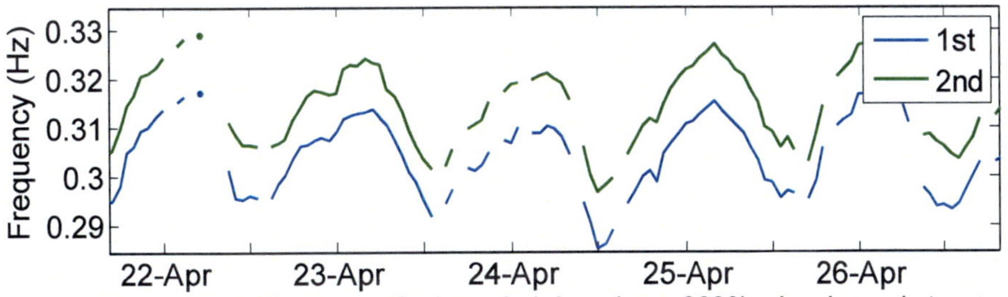

Figure 5: Evolution of the frequency for low wind days (year 2009), showing only temperature effects.

In these pictures (Figure 6a) show the effect of a windy day, besides the temperature effect particularly on 23rd March and 25th March which are windy days, with mean wind speed of almost 25mph (Figure 6b). The frequencies for these days are not so smooth as on any low wind days. Figure 6a shows that the main variation on the frequencies is due to temperature effect. According to TMD aim, it works on first frequency, normally on windy days and its effect is decrease the frequency, so this is the effect shows on the Figure 6a. Thermal and wind effects have not been simulated on the model and these fluctuations in frequencies values should be determined because the results of the model can be included in this experimental range of frequencies.

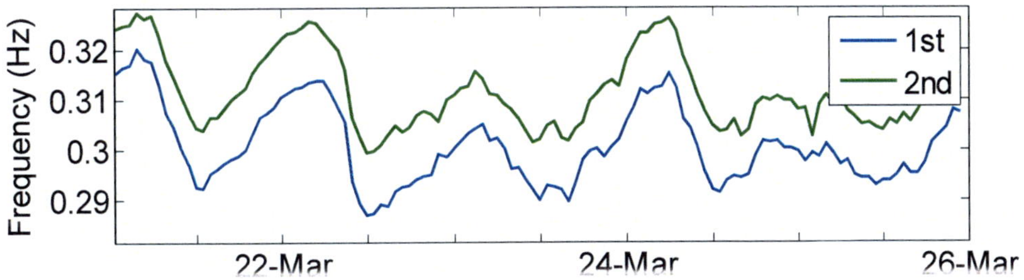

Figure 6a: Evolution of frequencies on windy days (2009)

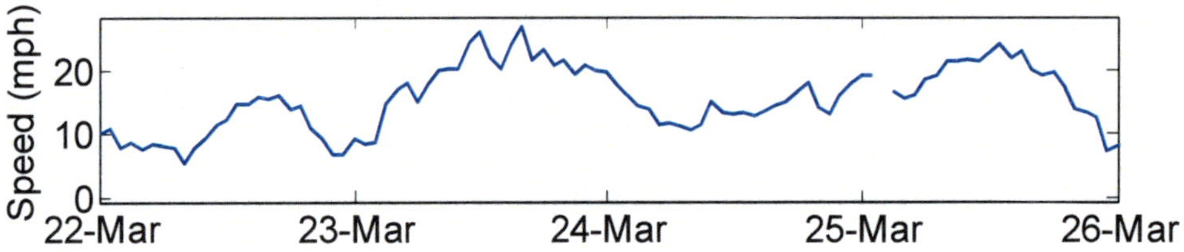

Figures 6b: Evolution of wind speed on several days (year 2009).

4 NUMERICAL MODEL AND EXPERIMENTAL DATA

The main aim of chimney FEM is to explain the behaviour of the chimney during its last years. In particular the variations of frequencies have some peculiar characteristics whose causes are not immediately apparent. The FEM could be a good tool to understand the TMD effect on the frequency changes, to study structural cause effect relationship and to simulate the future demolition of the chimney. The 3D FEM was built in ANSYS.

The FEM needed to represent, in five sections the chimney taper, with different thickness. The mechanical parameters of the concrete were updated, but Poisson coefficient was kept at $\nu = 0.2$. Other parameters to update included the stiffness of the foundation, different in horizontal and vertical directions.

The original model was based on the concrete chimney including the 5m brick section (Figure 7). A second model was built adding TMD mass and stiffness, and was used to compare the evolution of the frequencies with the model and with the experimental data taken on the first days of monitoring system before the installation of the TMD. The models did not account directly for any environmental effects on parameters. The second numerical model was made by adding the TMD near the top of the chimney, modelled as a mass restricted in vertical direction joined to the chimney by several spring-dampers in radial direction. In the Figure 8 is shown the actual situation of the TMD and in the Figure 9 is shown the TMD model.

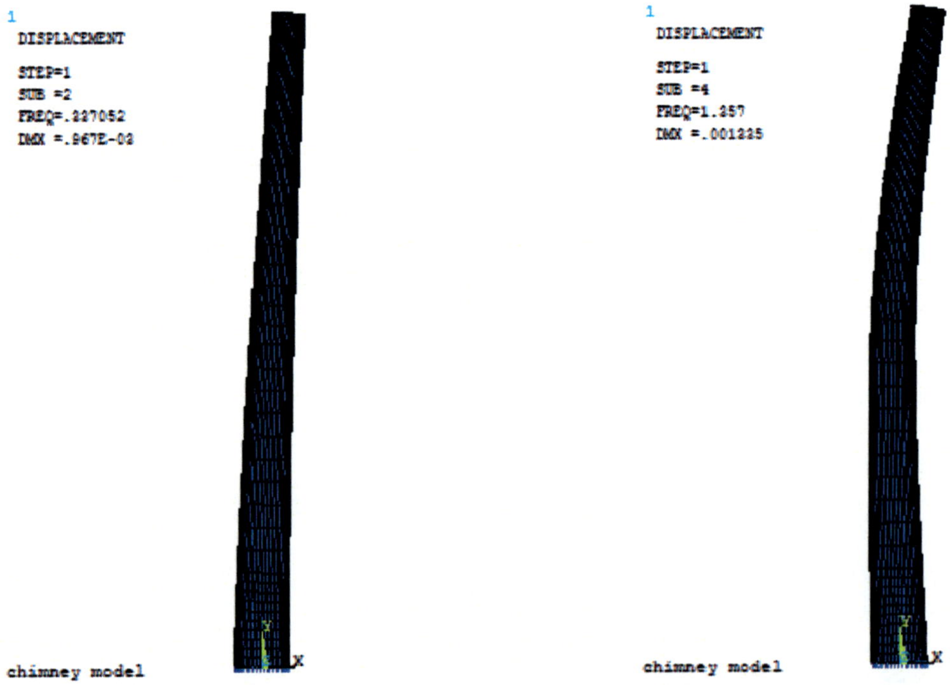

Figure 7: Mode shapes of the original chimney with updated model.

Figure 8: TMD (grey masses)

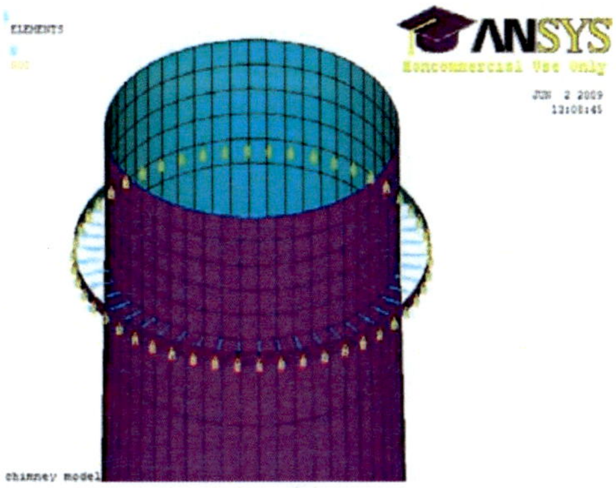

Figure 9: TMD representation.

The real characteristics of the TMD are:

- Mass = 42000 Kg
- Stiffness = 1.8e5 N/m
- Damping coefficient ≈ 2.5%

The updated original model was used with these new TMD parameters. The objective of using this TMD model is to explain the evolution of the frequencies after the installation of the TMD. The experimental data set for this period of time is huge and accompanied by strong variations (Figure 5), due to environmental effects. These lead to varying system properties, for two possible reasons: the amplitude capacity of the damper, which may not always work under its design condition and chimney structure variations, which may include structural deficiencies towards the end of the chimney's life. Hence updating procedure was not used with the TMD model.

5 MODEL UPDATING PROCEDURE

Any model updating procedure includes three basic aspects [11]: First the selection of responses as reference data which are the measured frequencies and mode shapes. Second the selection of uncertain parameters to update, to which changes the selected responses should be sufficiently sensitive. Finally the model tuning which is an iterative process to modify the selected parameters based on the selected reference data. To select the parameters for updating, a sensitivity analysis is required that computes the sensitivity of every parameter to eliminate insensitive parameters from the process. One measure of updating success is the modal assurance criterion (MAC) correlating factor pairs of an analytical and an experimental mode shape, as follows:

$$MAC(\phi_a, \phi_e) = \frac{|\phi_a^T \phi_e|^2}{(\phi_a^T \phi_a)(\phi_e^T \phi_e)}, \quad (1)$$

where, ϕ_a and ϕ_e are the analytical and experimental mode shape vectors, respectively. It is accepted a value in excess of 90% should be attained for correlated modes.

For this exercise, the numerical model and the experimental model obtained from two different software packages are transformed to universal file format for iterative updating using bespoke code developed in MATLAB.

6 RESULTS

6.1 *MODEL UPDATING*

The model updating of four unknown parameters of the model was done as explained. The results for the original model compared with the initial values on the model are shown on the following table.

MODEL PARAMETERS	INITIAL VALUES	UPDATED
Concrete stiffness [N/m^2]	26x10^9	24.72x10^9
Concrete density [kg/m^3]	2400	2436
Horizontal foundation stiffness [N/m^2]	170x10^6	162.2x10^6
Vertical foundation stiffness [N/m^2]	260x10^6	256.2x10^6

The updated frequencies with these parameters and the MAC values are shown in the next table. The experimental frequencies in the principal directions are different due to the non symmetric characteristic of the chimney (Figure 2). The model was built with assumed symmetrical configuration to simplify it and the differences in the data are not significant and are excluded to avoid complicating the model.

FREQUENCIES	NUMERICAL	EXPERIMENTAL	UPDATED	MAC values
First	0.3446	0.335938	0.3370	0.9977
Second	0.3446	0.339844	0.3370	0.9958
Third	1.397	1.35156	1.357	0.9879
Fourth	1.397	1.36328	1.357	0.9268

The MAC values are above 90% and the frequencies are very close, which means than the values are acceptable; consequently the model updating is success, allowing for symmetry effects.

6.2 *RESULTS WITH TMD*

With the updated parameters and the TMD characteristic, the effects of the TMD model can be checked, and the frequencies resulting compared with the actual experimental data. These results can be seen on the pictures (Figure 11). The first frequency of this model decreases, with respect original model, both on the model and in reality. In addition there appears an extra mode, close to the first one, which is the effect of the additional degree of freedom of the TMD. The second frequency (above 1Hz) is unchanged.

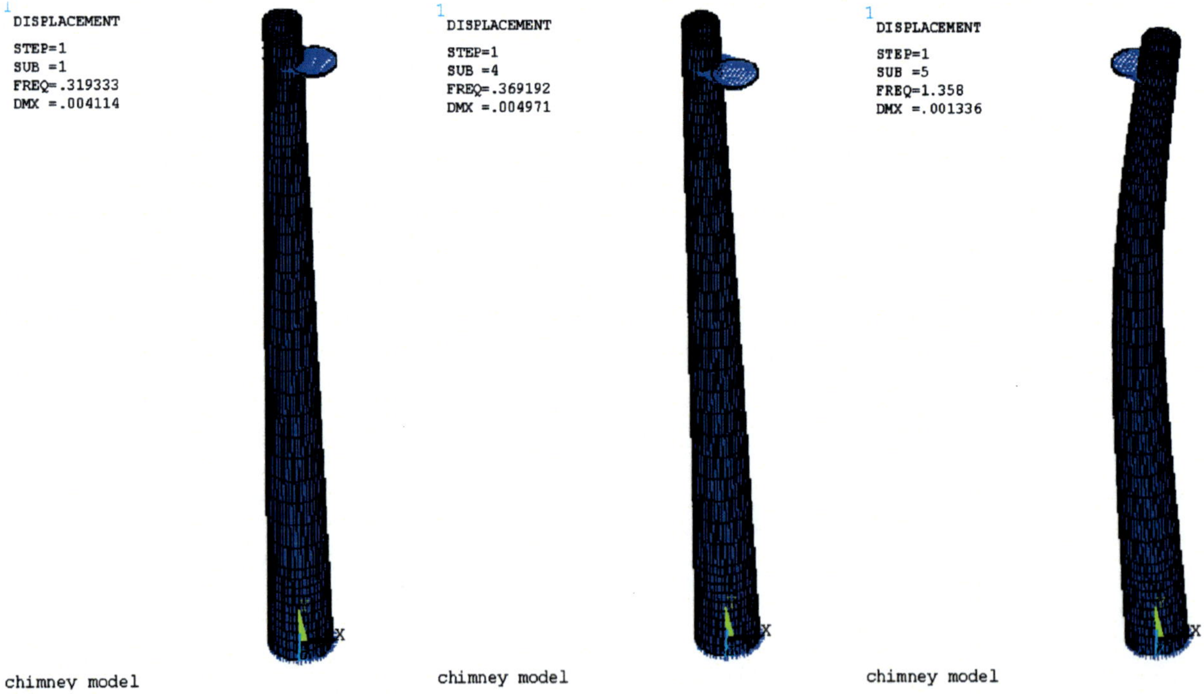

Figure 11: Mode shapes of TMD model

6.3 *RESULTS IN THE FIRST PART OF DEMOLITION*

The first part of the demolition consisted in decreasing the height of the chimney, by first removing the top 5m of brickwork; this is not really damage but a change in its physical characteristics. It is reasonable that the modal frequencies would increase due to this change as a result from the numerical model of removing about 5m from the top of the chimney. The results (Figure 12) show that all the frequencies increase slightly as expected.

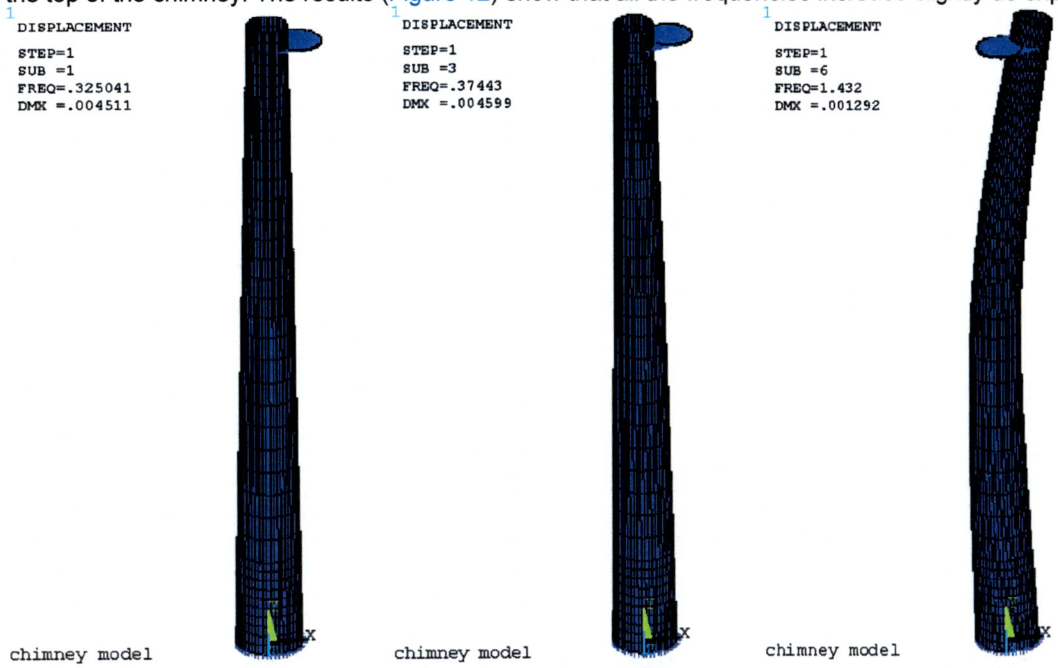

Figure 12: Mode shapes and frequencies in the first part of demolition

To summarise, three models were used to explain the behaviour of the chimney during the last period of its life. The first model, for the original chimney, was used to update the unknown parameters. The second model, with the TMD installed was used to explain the drop in frequencies for the experimental data and the last model was used to justify the increase in the frequencies in the first part of the demolition. The evolution of the first mode frequency for these states and equivalent experimental results are given in Figure 13, the changes show good agreement, despite difficulty matching equivalent environmental conditions for experimental data.
The major influence on dynamic parameters for the period with intact chimney and TMD operation is the thermal effect. According to [12] concrete Young modulus change with the temperature about ±5% for a difference in the temperature from -5°C to 35°C. The reference temperature for the TMD model is 20°C, and with temperature variations affecting Young's modulus alone, the variations in first mode frequency are given in Figure 14.

The range of variation on the model is, much narrower than the experimentally observed variation, which though thermal expansion is believed to lead to changes in the stiffness contributions of parts of the structure.

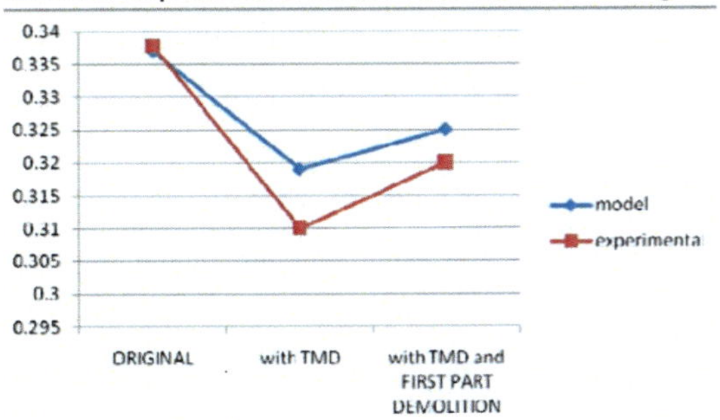

Figure 13: Evolution of the first natural frequency according to the model and the experimental data.

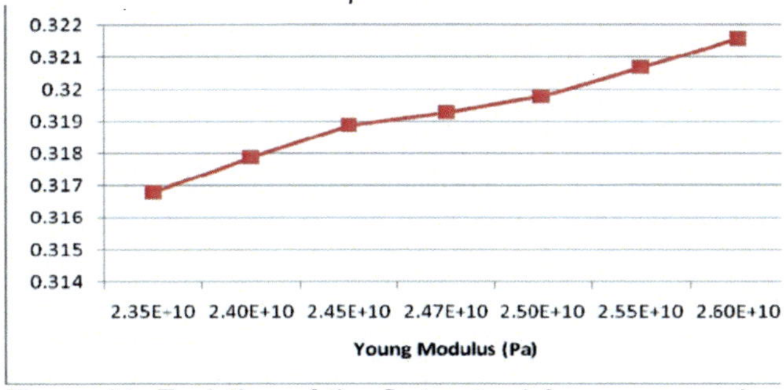

Figure 14: Evolution of the first natural frequency varying Young modulus, according to the model.

7 CONCLUSIONS

The model of the original chimney was successfully developed and the updated parameters were very useful for explaining the behaviour of the chimney afterwards. The TMD model predicted quite well the evolution of the first frequency. There are some discrepancies that could be by several reasons apart from the environmental effects. These other reasons could be foundation stiffness changes or concrete stiffness changes due to the inactivity of the chimney. Anyway the effect of the TMD is clearly shown on the model. In addition this model could be a good simulator for the demolition phase; in fact, the predictions for the first part of demolition match those observed, so that the remainder of the demolition process could also be simulated.

ACKNOWLEDGEMENTS

The economic support given by the Spanish Ministry of Education and Innovation trough the project BIA2006-15266-C02-01 and the European Social Fund, is gratefully appreciated.

8 REFERENCES

[1] Goddard, C. R., "The design of the 183m new chimney, Rugeley, UK -interference effects", in *CICIND*, (2007).

[2] Friswell M.I., Mottershead J.E. “Finite Element Model Updating in Structural Dynamics”, Kluwer Academic Publishers, Dorcdrecht, (1995).

[3] Wu J.R., Li Q.S., “Structural parameter identification and damage detection for a steel structure using a two-stage finite element model updating method.”, Journal of Constructional Steel Research, 62 231–239 (2006).

[4] Teughels A., De Roeck G., “Structural damage identification of the highway bridge Z24 by FE model updating.”, Journal of Sound and Vibration, 278 589–610 (2004).

[5] Gentile C., Saisi A., Gallino N., “Operational Modal Analysis and FE modelling of a masonry tower.”, 3rd International Operational Modal Analysis Conference, (May 2009).

[6] Aoki T., Sabia D., Rivella D., “Influence of experimental data and FE model on updating results of a brick chimney.”, Advances in Engineering Software, 39 327–335 (2008).

[7] Modak S.V., Kundra T.K., Nakra B.C., “Prediction of dynamic characteristics using updated finite element models.”, Journal of Sound and Vibration, 254(3), 447-467 (2002).

[8] Galsworthy, J. K. & Vickery, B. J., “Wind loads and interference effects for new and existing chimneys at the Rugeley power station”, UK, Alan Davenport Wind Engineering Group, Boundary Layer Wind Tunnel Laboratory, London, Ontario, Canada, N6A 5BP (2006).

[9] Brownjohn JMW, Carden EP, Goddard RC, Oudin G, Koo K, “Real-time performance tracking on a 183m concrete chimney and tuned mass damper system.” 3rd International Operational Modal Analysis Conference, (May 2009).

[10] Brownjohn JMW, Carden EP, Goddard RC, Oudin “Real-time performance monitoring of a TMD for a 183m reinforced concrete chimney.” 4th International Conference on Structural Health Monitoring on Intelligent Infrastructure, (July 2009).

[11] Brownjohn J.M.W., Xia P.-Q., Hao H., Xia Y., “Civil structure condition assessment by FE model updating: methodology and case studies”, Finite Elements in Analysis and Design, 37 761-775 (2001).

[12] Morita T. et al., “Residual Mechanical Properties of High Strength Concrete Members Exposed to High Temperature – Part 1. Test on Material Properties”, Summaries of Technical Papers of Annual Meeting, Architectural Institute of Japan, Naigata, (1992).

Proceedings of the IMAC-XXVIII
February 1–4, 2010, Jacksonville, Florida USA
©2010 Society for Experimental Mechanics Inc.

Finite Element Model Updating and Dynamic Design of Spot Welded Structures

Choudhury M.K., Modak S.V, Kundra T.K
Indian Institute of Technology Delhi (Hauz khas, New Delhi-16)

ABSTRACT

Spot welds are used extensively in the automotive industry to join panels and for construction of other sub-assemblies that contains several thousands of spot-welds, so it is not practical in structural analysis to model each and every spot weld joint in detail. So a simple Finite element (FE) model of spot-welds needs to be used. There is not much work reported on dynamic testing, correlation and updating of the spot welded structures. This work explores the use of FE model updating to improve the dynamic characteristics of the spot welded structures. A hat like structure, used in earlier studies, is built and experimental modal analysis done on the spot welded structure. An FE model for the hat structure is built and correlated mode pairs are identified and attempts are made to update the FE model. Structural dynamic modification studies have been carried out to evaluate the effectiveness of the updated FE model of the spot welded structure for dynamic design. It is found that the updated FE model predicts more accurately the changes in dynamic characteristics as compared to the original FE model. Effect of spot welds on natural frequencies and damping of the spot welded structures is also studied.

INTRODUCTION

Spot welding is one of the widely used methods for manufacture of thin sheet components, especially in mass-production industries such as the automobile industry. Spot welds are used extensively in the automobile industry to join panels and car bodies that contain thousands of spot welds. So it is not practical in structure analysis to model each and every spot weld joint in detail. Modeling spot welds is difficult, mainly because there are many local effects such as geometrical irregularities, residual stresses, material in homogeneities and defects due to the welding process that are not taken into account by finite element modeling. Mainly two types of spot weld models exist, those which required require the stress within the weld spot to be calculated and those that do not. In the first case very detailed models are necessary to compute a smooth stress field at the spot weld .A very detailed model produces a detailed and smooth stress field, but it will not necessarily accurately predict the stiffness of real spot welds and their effect on the rest of the structure [4]. Detailed models will produce apparently reliable stress fields, whereas they may poorly estimate the forces that are interchanged between the spot weld and the rest of the structure. In the second case the only requirement from the model is to simulate, as closely as possible, the stiffness (and mass) characteristics of the real spot welds and their influence on the rest of the structure. Simple models that use few elements need to be built. These simplified joint models must be able to produce same results as that of a more refined model [3]. Model updating based on vibration characteristics is an important tool to improve the accuracy of the finite element models. In this paper, dynamic design of a spot welded structure is presented. Following sections deal with finite element modeling, experimental modal analysis, and FE model updating and structural dynamic modifications on spot welded structure.

FE modeling of the spot-welded structure

A Hat structure that is spot-welded from one flat and other folded mild steel plate is fabricated for the experimental studies in this work. This structure has been used in another study [4]. A "single hat" structure (SH) is shown in Figure-1.The structure consists of a hat section made up of mild steel plates (density7800kg/m^3) joined together by spot welds at the flanges, and are designed to represent simplified models of the beams[1] used in the construction of car bodies, for example the roof pillars. Figure-2 shows the section of the hat structure plates that form the SH benchmark. The thickness of the plates is 1 mm and the structure has 12 spot welds 10.5 mm apart (Figure-3).

T. Proulx (ed.), *Dynamics of Civil Structures, Volume 4*, Conference Proceedings of the Society for Experimental Mechanics Series 13,
DOI 10.1007/978-1-4419-9831-6_12, © The Society for Experimental Mechanics, Inc. 2011

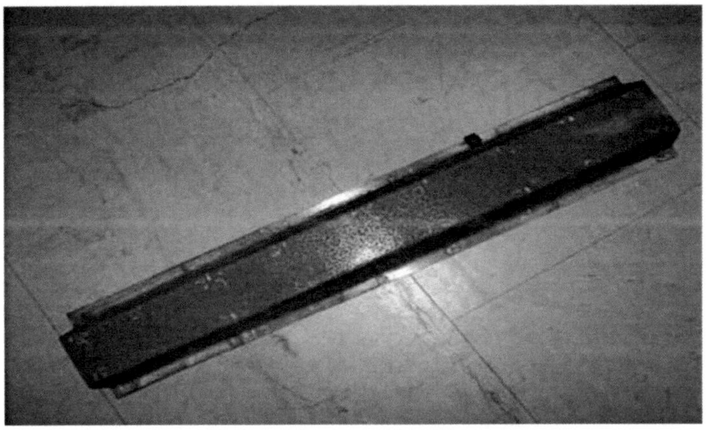

Figure-1 Single hat (SH) structures

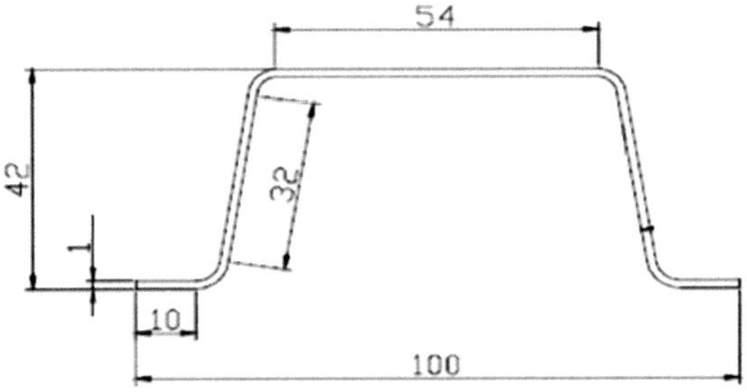

Figure 2- Section view single hat structure

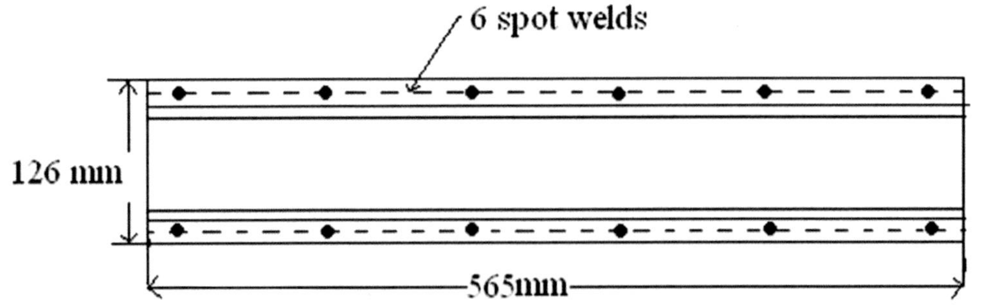

Figure -3 Top view of single hat structure (SH)

Finite element model of the hat structure is built in ANSYS [7] taking four nodded quadrilateral shell elements for the hat surface and beam elements for the spot weld. For applying the beam element at the spot weld location the surface should be concurrently meshed so that the node connecting the two ends of the beam lie perpendicular to each other. Since during actual spot welding the region around the spot weld become rigid. This rigid region is also created in the FE model by using constraint equation. One master node, which is same as the node at which beam element is placed and six slave nodes around the master node, is selected to create six constraint equations. Modal analysis is done in ANSYS to get the natural frequencies and mode shapes through FEA.

Experimental Modal Analysis (EMA) on spot-welded structure.

Spot welded structure is suspended at the two ends with elastic tapes to ensure the free-free condition. Experimental modal analysis (EMA) is done on the SP structure with 36 measurement points. A single axis accelerometer is mounted at location 12 as shown in figure 5 for reference. Excitation is given by impact force using modal hammer. The analysis is done by moving hammer to different locations. The data is recorded using Experimental setup that is shown in Figure-4.

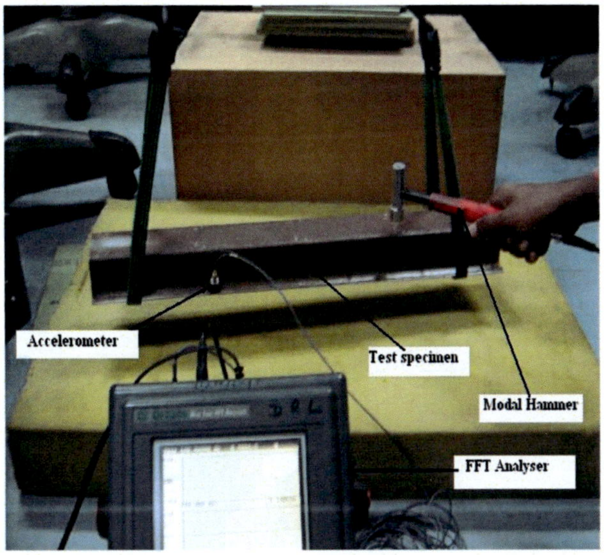

Figure-4 Experimental Set- up

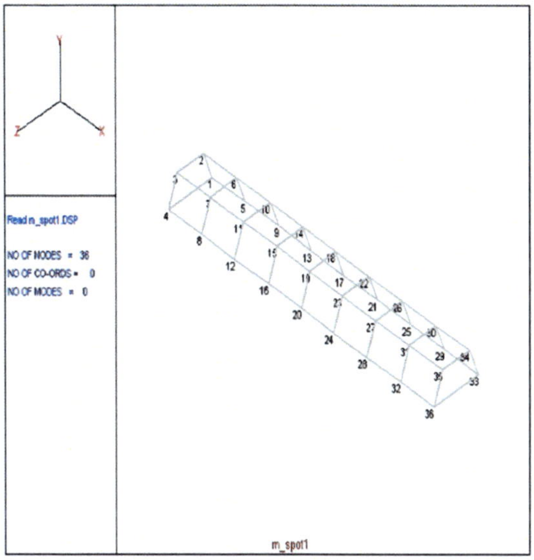

Figure-5 Hat structure showing the measurement points

Table 1 Natural frequencies and damping factor obtained from EMA

Mode no	EMA frequency (Hz)	Damping	Comments
1	380.1	0.0107	1st Twisting
2	397.3	0.0089	2nd Twisting
3	533.5	0.0062	Bending of hat and base mode
4	627.4	0.0043	1st Base mode
5	671	0.0053	2nd Base mode
6	748	0.0086	3rd Base mode
7	860.4	0.0091	Bending mode
8	924.4	0.0073	4th Base mode
9	994	0.0078	Twisting of hat and base mode

Comparison of experimental and numerical results

The results obtained from FE model of spot welded structure are compared with EMA results to find out the correlated mode pairs. A considerable difference between the FEA and EMA frequencies is found for higher modes. Figure-6 shows the comparison of finite element mode shapes obtained from ANSYS [7] and the experimental mode shapes obtained by analysis of measured FRFs in ICATS [5]

Mode shapes obtained by FE analysis

Mode shape obtained by experiment

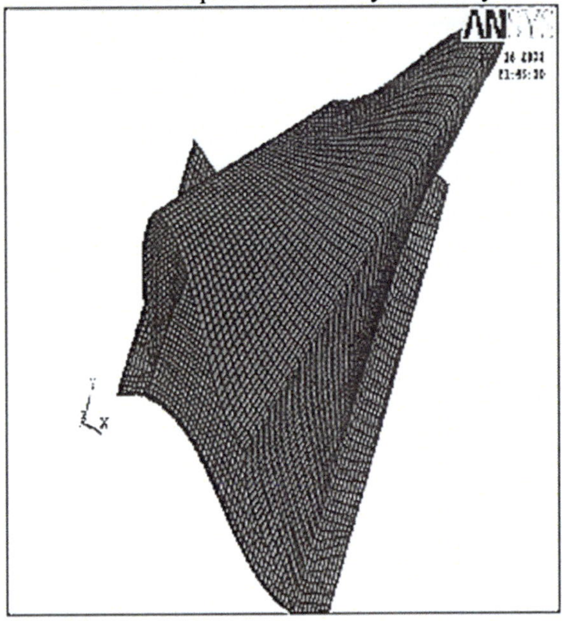

Mode 1 at 386.7Hz (twisting)

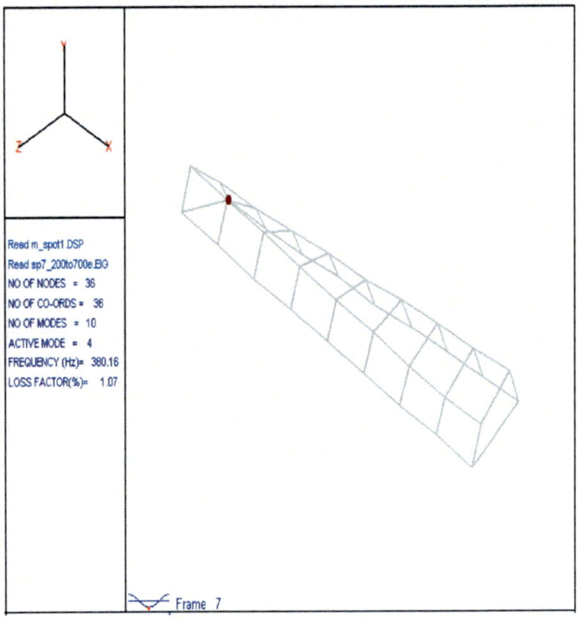

Mode 1 at 380.1Hz (twisting)

Mode 2 at 409.4Hz (twisting)

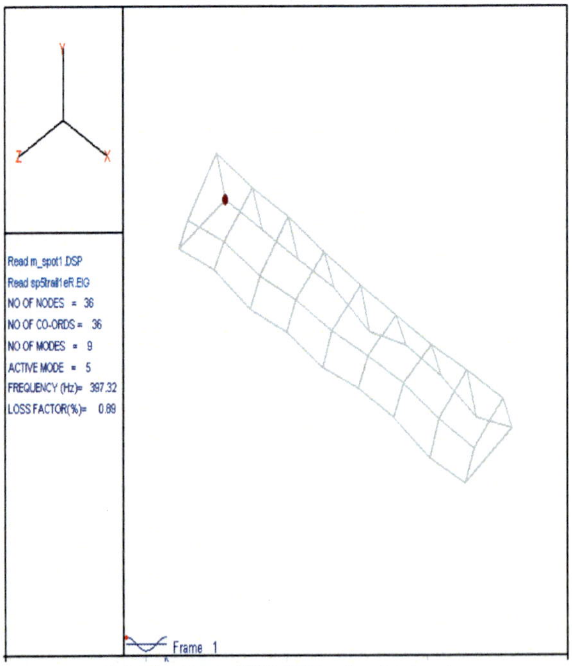

Mode 2 at 497.3Hz (twisting)

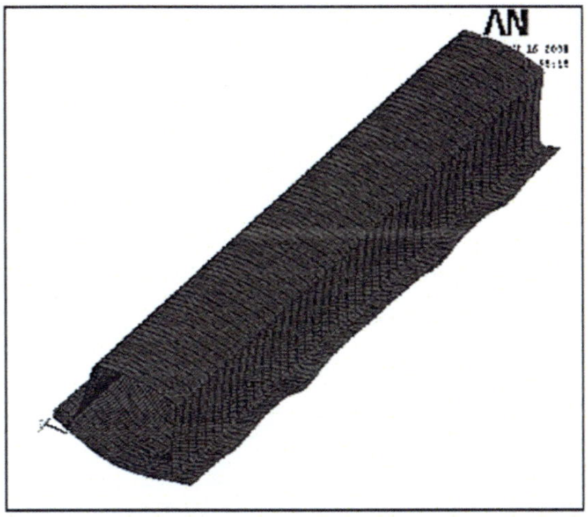

Mode 3 at 516.6 Hz (bending and base mode)

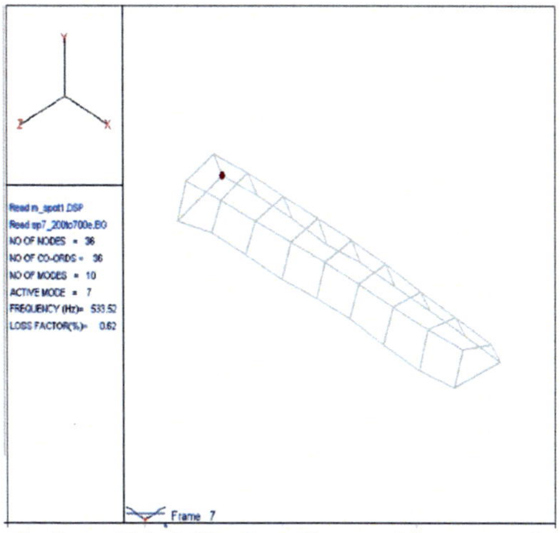

Mode 3 at 533.5 Hz (bending and base mode)

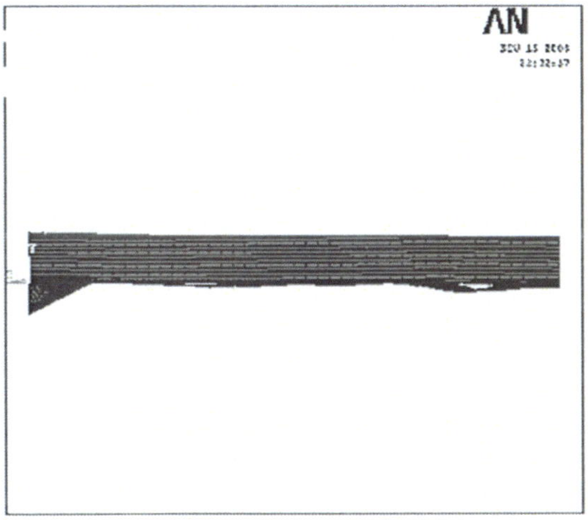

Mode 4 at 648.8 Hz (base mode)

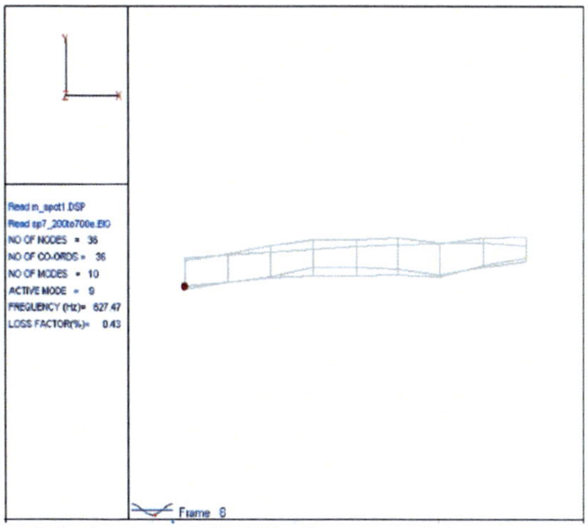

Mode 4 at 627.4Hz (base mode)

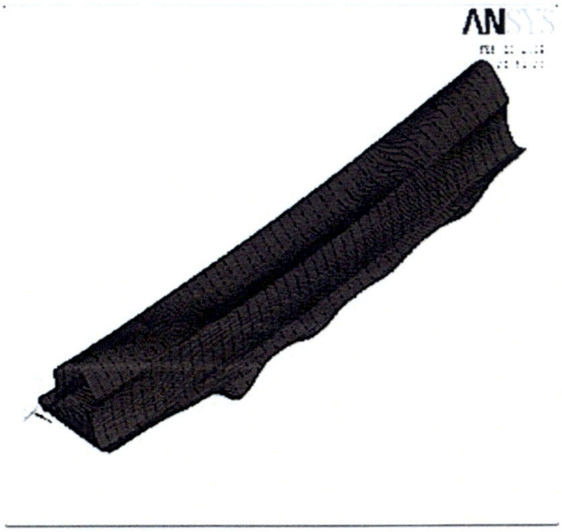

Mode 5 at 873.3 Hz (bending)

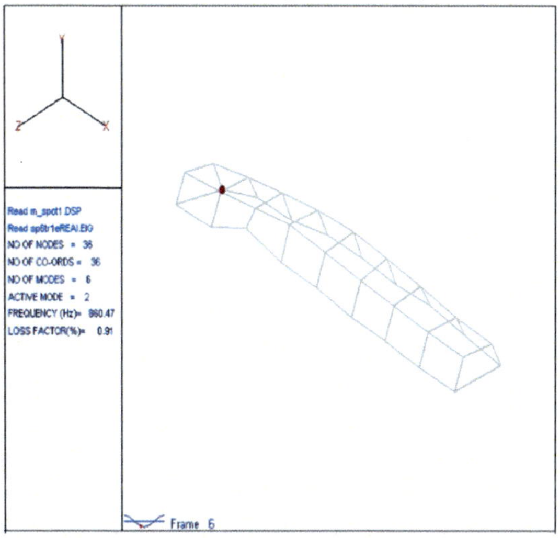

Mode 5 at 860.4Hz (bending)

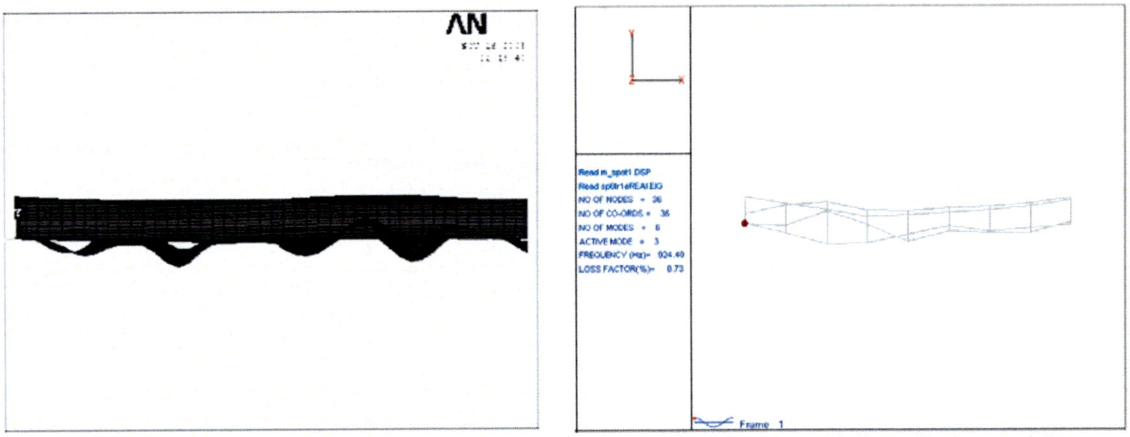

Mode 6 at 1008.9 Hz (base mode) Mode 6 at 924.4Hz (base mode)

Figure 6 Comparison of FE and experimental modes

The updating of FE model is carried out in FEM-Tools software [6]. Young's modulus and moment of inertia of the beam elements are taken as the updating parameters. Percentage change in young's modulus of the beam element after updating is found to be 90%.The large change in young's modulus represents an equivalent reduction in the stiffness of the spot welded structure.

Table 2 shows the comparison of natural frequencies, obtained through EMA, initial FE model and Updated FE model. Here both the parameters are chosen for updating.

Table 2 comparison of original and updated model

Mode	Natural Freq. (Hz)			% Error	
	EMA	FEA (original)	FEA (Updated)	FEA (original)	FEA (Updated)
1	380.1	386.7	386.2	1.73	1.59
2	397.3	409.4	409	3.06	2.96
3	533.5	516.6	512	-3.16	-4.02
4	627.4	648.8	639.8	3.40	1.98
5	860.4	873.3	867.2	1.49	0.78
6	924.4	1008.9	967.7	9.15	4.68
Average (% error)				2.61	0.78

Table 3 shows the comparison of average and maximum error in natural frequency of updated and original FE model. Moment of inertia has not much effect on the updating. So diameter can be varying in the spot welded structure. So, for the further analysis young's modulus has been taken as the updating parameter.

Table 3 Average and maximum error in the updated model

Percentage Error (%)	Young's modulus (%)		Moment of inertia (%)		Both young's modulus and Moment of inertia (%)	
	updated	original	updated	original	updated	original
Average error	1.14	2.61	1.91	2.61	0.78	2.61
Maximum error	7.12	9.15	6.9	9.15	4.68	9.15

Structural Dynamic Modification (SDM) of the spot-welded structure

Model updating techniques are used to update a finite element model of a structure so that an updated model predicts more accurately the dynamics of a structure. The dynamic design using an updated model requires that the model predict the change in the dynamic characteristics due to potential modification with reasonable accuracy [2]. The updated model obtained by matching the natural frequencies is then used for predicting the effects of potential design modifications made to the structure. With this purpose the structural modification in the form of mass and change in number of spot welds are considered.

Mass modification

A mass modification is introduced by attaching a mass of 70 gram at a distance of three-fourth of the length on the top surface as shown in Figure-7.The mass modification is also introduced analytically in the updated model. A comparison of the modified modal data as predicted by the updated model and that of the experimental data is done.

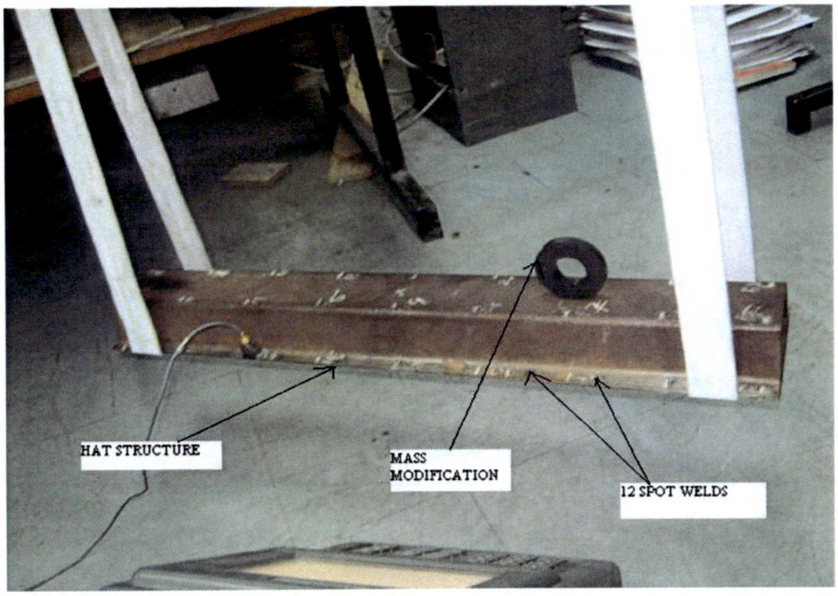

Figure 7 EMA set up with mass modification

Comparison of modified natural frequency of the updated model and that of the experimental data is shown in Table 4.It is observed that the predicted dynamic characteristics (natural frequency) on the basis of updated model are reasonably closer to the measured characteristics for the modified structure. The average error in the original FE model is 4.17% while in the updated model it is reduced to 2.25%. It is seen that in the third mode the results are not satisfactory as the original FE model behaves good in this case. Also, in the fourth mode there is very little change or no change in the natural frequency due to addition of lumped mass.

Table 4 Percentage Error predicted in FE model and updated model

Mode	EMA freq. Before SDM (Hz)	EMA freq. After SDM (Hz)	Original FE freq. After SDM (Hz)	(%) error	Updated FE freq. After SDM (Hz)	(%) error
1	380.1	331.4	379.2	14.42	350.6	5.79
2	397.3	388.2	398.7	2.7	390.9	0.69
3	533.5	505.6	508.1	0.49	498.5	1.4
4	627.4	625.4	648.6	3.79	634.1	1.39
5	860.4	853.6	827	-3.1	854.8	0.14
6	924.4	939	1002.5	6.76	977.5	4.1
Avg(% error)				4.17		2.25

Structural dynamic modification due to variation in number of spot weld

Number of spot weld increased to 22

In order to predict the change in dynamic characteristics of the spot welded structure variation in number of spot welds is done. 10 more number of spot welds is created on the hat structure, 5 on each side of the flanges .Now total no of spot welds become 22 i.e., 11 on each side of the single hat structure. The increase in number of spot weld is also introduced analytically in the updated model. A comparison of the modified modal data as predicted by the updated model and that of the experimental data is done. It is observed that the predicted dynamic characteristics (here it is only natural frequency) on the basis of the updated model are reasonably closer to the measured characteristics for the modified structure.

Comparison of modified natural frequency of the updated model and that of the experimental data in case of 22 spot welds is shown in Table 5 When no .of spot welds are increased to 22 there is increase in the natural frequency of the structure due to increase in stiffness of the structure. The increase in natural frequency is well predicted by the updated FE model .The percentage error in the first twisting mode is reduced to .027% from 1.18 as predicted by original FE model. In first bending mode the error predicted by original FE model is 3.69% which are reduced to 0.67% in case of updated FE model. The average error in original FE model is 3.02% which is reduced to 1.59% in updated FE model.

Table 5 Percentage Error predicted in FE model and updated model with 22 spot welds

Mode no	EMA freq. Before SDM (Hz)	EMA freq. After SDM (Hz)	Original FE freq. After SDM (Hz)	(%) error	Updated Freq SDM (Hz)	Error (%) updated
1	380.1	400.1	404.9	1.18	400.2	0.027
2	397.3	443.7	434.5	-2.07	431.8	-2.66
3	627.4	631.1	655	3.79	639.8	1.39
4	860.4	860.2	892	3.69	866	0.67
5	924.4	963.9	1046.9	8.53	1046.2	8.53
Avg (error)				3.02		1.59

Number of spot weld increased to 42

Comparison of modified natural frequency of the updated model and that of the experimental data in case of 42 spot welds is shown in table 6. When no .of spot welds are increased to 42 there is increase in the natural frequency of the structure due to increase in stiffness of the structure. The increase in natural frequency is well predicted by the updated FE model .The percentage error in the first twisting mode is reduced to 1.02 from 3.7 as predicted by original FE model. The average error in original FE model is 11.15% which are reduced to 6.6% in updated FE model. Maximum error is reduced to 5.27% from 15.49%.

Table 6 Percentage Error predicted in FE model and updated model with 42 spot welds.

Mode no	EMA freq. Before SDM (Hz)	EMA freq. After SDM (Hz)	Original FE freq. After SDM (Hz)	% error	Updated Freq SDM (Hz)	Error (%) updated
1	380.1	404.4	419.4	3.7	400.27	1.02
2	627.4	622.2	718.6	15.49	655	5.27
3	860.4	863.8	987.1	14.27	980.5	3.51
4	924.4	966	1092.7	13	1090.4	12.8
Avg (%error)				11.15		6.6

Table 7 shows the change in damping due to change in number of spot welds. Due to increase in number of spot welds the damping of the spot welded structure has increased in some modes. Increase in damping due to increase in number of spot-welds is useful as then the overall vibration response and the also the radiated noise inside and outside the vehicle will be lower.

Table 7 Damping comparison of the spot welded structure

EMA12spot weld	Damping	EMA 22spot weld	Damping	EMA 42spot weld	Damping
380.1	0.00107	400.16	0.0111	404.4	0.0097
533.5	0.0062	521.4	0.0135	521.2	0.0127
627.4	0.0043	631.1	0.0108	622.2	0.0323
860.4	0.0091	860.2	0.0133	863.8	0.0125
924.4	0.0073	963.9	0.0040	966	0.0030

Conclusion

In this paper correlation, updating and structural modification studies on a spot-welded structure are carried out. Based on studies of model updating of spot welded structure, it is concluded that the updating of spot welded model has the potential to improve the accuracy of spot welded structure. Model updating is attempted through several updating parameters choices. The updated spot weld model can be used reliably for predicting the effect of structural modification. It is observed that as the number of spot welds is increased the natural frequency of the structure is increased. The damping is also seen to increase for most of the mode.

References

1. Galtier A, 1998 “Fatigue life prediction model for spot welded structures”. Fatigue design symp, vol. 1.
2. Modak S.V. ,.Kundra T.K. , Nakra B.C. ,2001,”Prediction of dynamic design characteristics using updated finite element models” Journal of sound and vibration 254(3),Pg447-467
3. Xu S, Deng X., 2004 “An evaluation of simplified finite element models for spot-welded joints” finite element analysis and design 40,pg1175-1194.
4. Friswell M. I., Palmonella M., Mottershead J.E., Lees A.W, 2005 “Finite Element Model s of spot weld in structural dynamics: review and updating”, computers and structures 83,pg 648-661
5. ICATS, 2006, Imperial College, London.
6. FEM Tools version 3.0.1. Dynamic Design Solutions.
7. ANSYS 11

Proceedings of the IMAC-XXVIII
February 1–4, 2010, Jacksonville, Florida USA
©2010 Society for Experimental Mechanics Inc.

Experimental Evaluation of Mass Change Approaches for Scaling Factors Estimation

Pelayo Fernández[1], Paul Reynolds[2] and Manuel López Aenlle[1].

[1] University of Oviedo. Department of Construction and Manufacturing Engineering. Campus de Gijón s/n. Ed. Oeste. Bloque 7. 33203 Gijón. Spain.

[2] University of Sheffield. Department of Civil and Structural Engineering. Sir Frederick Mappin Building. Mappin Street, Sheffield. S1 3JD. UK.

fernadezpelayo.uo@uniovi.es

ABSTRACT

In operational modal analysis the input forces are unknown so the modes shapes are obtained with an arbitrary normalization. In some applications the mass normalized modes shapes are required and the scaling factors have to be estimated. A way of obtaining these factors is to modify the modal properties of the structure by introducing changes in the mass or stiffness; mass change is usually the most convenient way to modify the structural properties in a controlled manner. Different strategies and approaches to perform the mass change method have been proposed in recent years. In this work the scaling factors of a 15 tonnes concrete slab strip are obtained using several equations proposed by different authors and the results are compared.

The results show that the scaling factors can be estimated with reasonable accuracy using the different equations if a good mass change strategy is applied.

Nomenclature

Φ	Mass normalized mode shape
Ψ	Arbitrary normalized mode shape
α	Scaling factors
$[K]$	Stiffness Matrix
$[M_0]$	Mass Matrix
$[M_1]$	Spatial distribution of added masses
β_1	Ratio of added mass
ω_0	Natural frequencies
ω_1	Modified natural frequencies
k	Terms of the diagonal of the projection matrix

T. Proulx (ed.), *Dynamics of Civil Structures, Volume 4*, Conference Proceedings of the Society for Experimental Mechanics Series 13,
DOI 10.1007/978-1-4419-9831-6_13, © The Society for Experimental Mechanics, Inc. 2011

1. Introduction

In operational modal analysis only arbitrary normalized modes shapes can be obtained [1], e.g. maximum component of unity or arbitrary length. If the modal masses are required the scaling factors must be estimated. The scaling factors are constants that relate each mass normalized mode shape with the correspond arbitrarily normalized one. That is:

$$\Phi = \alpha \Psi \tag{1}$$

where Φ is the mass normalized modal shape, Ψ is the correspond arbitrary normalized modal shape and α is the scaling factor.

A way to estimate the scaling factors is to modify the dynamic behavior of the structure changing the stiffness, the mass or both and perform a new operational modal testing and analysis. These methods based on dynamic modification use both the modal parameters of the unmodified and modified structure, so if the modal parameters are related the scaling factors can be obtained.

In recent years, some different approaches [2, 3, 4, 5, 6, 7] and strategies [8, 9, 10] in the mass change method have been proposed. The method consists of modifying the dynamic behavior of the structure by attaching masses to several points on the structure where the mode shapes are known. The accuracy obtained in the scaling factor estimation depends on both the accuracy obtained in the modal parameter identification and the mass change strategy used to modify the dynamic behavior of the structure. The mass change strategy refers to the definition of the magnitude, the location and the number of masses to be attached to the structure.

In this paper, the scaling factors of a 15 tonnes concrete slab strip by means of Operational Modal Analysis (OMA) are estimated with the mass change method using different proposed approaches and strategies and the results are compared. The results are also compared with the estimated scaling factors using classical modal analysis (CMA).

2. Mass Change Method - Theory

The mass change method consists of performing operational modal analysis on both the original and the modified structure. The modification is carried out by attaching masses to several points of the structure where the mode shapes of the unmodified structure are known. The user selects the number, the magnitude and the location of the masses. To facilitate the mass modification and the calculation of the scaling factors, lumped masses are often used, so that, the mass change matrix becomes, in general, diagonal.

Different approaches to determine the scaling factors can be derived mainly from the eigenvalue equations of both the original and the modified structure. The classical eigenvalue equation in case of no damping or proportional damping is given by:

$$[M_0]\Phi_0 \cdot \omega_0^2 = [K]\cdot\Phi_0 \tag{2}$$

where Φ_0 is the mode shape, ω_0 the natural frequency, $[M_0]$ the mass matrix and $[K]$ the stiffness matrix. If a mass change is made so that the new mass matrix is given by $[M_0 + \beta\ M_1]$, then the eigenvalue equation becomes:

$$[M_0 + \beta M_1]\Phi_1 \cdot \omega_1^2 = [K]\cdot\Phi_1 \tag{3}$$

where Φ_1 and ω_1 are the mode shape and the natural frequency of the modified problem, respectively, M_1 is a matrix of arbitrary rank describing the spatial distribution of the mass modification and β is a scaling parameter with respect to which the mass change magnitudes are computed. Combining equations (1), (2) and (3) and after several assumptions a number of different expressions can be derived to estimate the scaling factors. A general form of these expressions can be summarized in the following formula [7]:

$$\alpha_0^2 \cong \frac{\omega_0 - \omega_1}{a_1} \frac{1}{\beta_1} \frac{a_3}{(\Psi_0^T [M_1] a_2)} \quad (4)$$

where the group of parameters (a_1, a_2, a_3) are different in each expression, e.g.:

($\omega_0/2$, Ψ_0, 1)	in the expression from Parloo et al. [2],	(5)
(ω_1, Ψ_0, 1)	in the expression from Brincker et al. [5]	(6)
(ω_1, Ψ_1, 1)	in the expression from Aenlle et al. [6], and	(7)
(ω_1, Ψ_1, κ)	in the Bernal projection expression [7].	(8)

The parameter β_1 is taken so that the [ΔM] mass change matrix that it is used in eqs. (5), (6) and (7), can be obtained by means of the relation [ΔM] = β_1 [M_1]. In the projection approach, eq. (8), the main idea is to express the mode shapes of the modified state on the basis of the unmodified modes shapes [7]. To introduce this projection the vector of constants κ is used, which is obtained from the diagonal of the matrix [q]= $[\Psi_0]^{-1}$ $[\Psi_1]$.

For the above described methods the following notation will hereafter be used: α_{00P} for eq. (5), α_{00} for eq. (6), α_{01} for eq. (7) and α_{01B} for eq. (8).

If only the modified modes shapes are used in eq. (7), that is, the parameters (a_1, a_2, a_3) are given by ($\Psi_1^T M_1 \Psi_1$), the α_{11} scaling factor is obtained and a new approach can be used combining α_{00} and α_{11}.

$$\alpha^* = (\alpha_{00} + \alpha_{11})/2 \quad (9)$$

It can be observed from eqs. (5) to (9) that different results from the scaling factors would be expected in each expression since different parameters are used in each one. For example, in eqs. (5) and (6) the unmodified modes shapes of the structure are used, so if the mode shapes change significantly, when adding the masses to the structure, the results would be expected to be less accurate. On the other hand, the eqs. (7), (8) and (9) take into account both the modified and unmodified modes shapes. Eq. (7) is expected to be better for small changes in the modes shapes and eqs. (8) and (9) are expected to be better for large changes in the mode shapes.

To obtain good results with the aforementioned equations it is important to take into account that different strategies can be used. These strategies are related to the mass magnitude and the number and location of the masses in the experiment [8, 9, 10].

3. Experimental Program

In this work, several OMA measurements using different mass strategies are done in a concrete slab. In each mass configuration the natural frequencies and mode shapes are estimated and then, the scaling factors are estimated with the mass change method using the different expressions. For comparing the estimated scaling factors obtained with OMA, CMA measurements are also done in the slab and the scaling factors are calculated from the FRF's.

3.1. Structure and Test Configurations

The structure used for the experiments is a simply supported in-situ cast post-tensioned slab strip of span 10.8 m. Its total length is 11.2 m, which includes 0.2 m overhangs over the knife edge supports. It has a width of 2 m, depth of 0.275 m and weighs approximately 15 tonnes. There is a known non-linearity in the slab at one of the supports, where exist a gap between the knife-edge support and the slab. There is also a potential for friction at the supports. Due to these effects the frequencies can change by 2-3% depending on the amplitude of excitation.

In both modal analysis, OMA and CMA, 21 Honeywell QA750 accelerometers were used. They were located at degrees of freedom (DOFs) 4 to 24, as indicated in Figure 1, and were installed on leveled Perspex base plates. For digital data acquisition, a Data Physics DP730 spectrum analyzer (Mobilyzer II) with 24 input channels and 4 output channels (24-bit resolution) was used. The Mobilyzer was controlled by a laptop PC connected to it via an Ethernet connection. Figure 2 shows images of the slab strip during the measurements, both with and without additional mass blocks present. The sampling rate and acquisition time used were 128 Hz and 15 minutes, respectively.

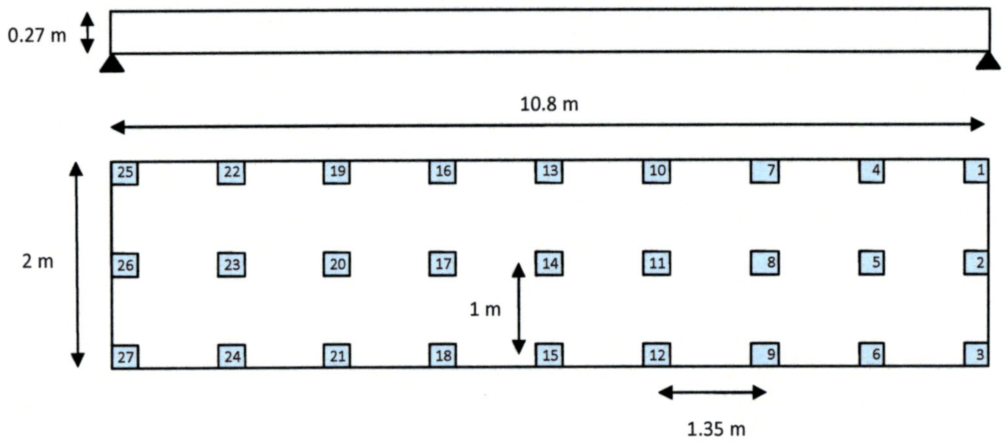

Figure 1: Slab dimensions and test grid

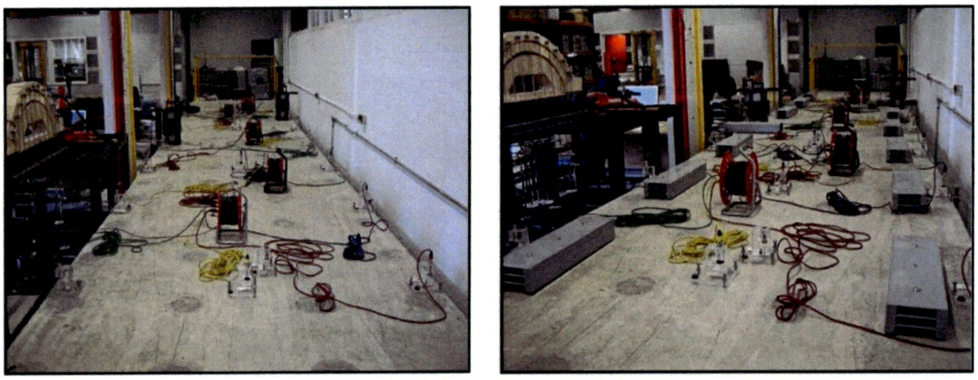

Figure 2: Concrete slab strip

To obtain the modal properties of the concrete slab from OMA, the Enhanced Frequency Domain Decomposition (EFDD) and Stochastic Subspace Iteration (SSI) techniques were used, both implemented in the ARTeMIS Extractor software. Since the natural excitation level is too low (the slab is in a very quiet lab) and due to the non-linearities, artificial excitation is used in the tests to avoid variations in frequencies that cannot be related with the mass change configurations. Two APS Dynamics Model 113 shakers were used as source of excitation. The slab was driven with uncorrelated random signals from both shakers with the same excitation level for all the experiments. The shakers were located at DOFs 15 and 19.

In CMA, the modal properties were estimated from Frequency Response Function (FRF) data calculated using both the excitation and response time histories with the same configuration of shakers and accelerometers than the OMA analysis, that is, shakers located at DOFs at 15 and 19 and accelerometers located at DOFs 4 to 24. To obtain the modal parameters from FRF data, the ME'scopeVES software was used in a MIMO modal analysis with two inputs and 21 outputs.

The first four significant natural frequencies of the concrete slab, obtained from OMA and CMA, are presented in Table 1.

Table 1: First four natural frequencies of the slab

	Mode	1: Bending	2: Bending	3: Torsion	4: Bending
OMA	Freq. (EFDD) [Hz]	4.497	16.844	25.717	37.600
	Freq. (SSI) [Hz]	4.499	16.835	25.767	37.612
CMA	Freq. (FRFs) [Hz]	4.499	16.845	25.750	37.612

3.1. Mass configuration and strategies

The quantity of mass reduces the aforementioned gap between the slab and its support and therefore affects the non-linearity introduced in the dynamic properties of the slab. To minimize this effect the same quantity of mass, 1050 kg, is used in all the experiments. There is one exception, where a mass of 600 kg was used to examine the difference between different changes in mass magnitude. The different mass configurations used in the measurements are presented in Table 2, which were selected to compare the different expressions described in paragraph 2 as well as different mass change strategies [10]. As the mass magnitude is always the same, except for case 3b, the strategies are related to the location and number of masses used in the experiments. The results are expected to be of decreasing accuracy from case 1 to case 5, since a more uniform distribution of masses would be expected to perform better over a range of modes. Cases 1 and 2 try to optimize the first four modes of the structure, cases 3 and 3b try to optimize modes 1 and 4 of the structure whereas cases 4 and 5 are strategies for which a significant mode shape modification is expected, hence allowing the comparison of the different expressions that take into account the modified mode shapes.

Table 2: Mass configurations

Case	№ of Masses	Mass per DOF	Mass change		DOF's with Mass
1	21	50	1050 Kg	7 %	All except [1 2 3 25 26 27]
2	14	75	1050 Kg	7 %	[4 6 7 9 10 12 13 15 16 18 19 21 22 24]
3	6	175	1050 Kg	7 %	[7 9 13 15 19 21]
3b	6	100	600 Kg	4.3 %	[7 9 13 15 19 21]
4	3	350	1050 Kg	7 %	[7 13 19]
5	4	---	1050 Kg	7 %	[7 (425 kg) 13 (150 kg) 19 (200 kg) 21 (275 kg)]

4. Results

The FRF's of some channels (MIMO analysis) and the EFDD singular value decomposition (OMA) are presented in Figure 3. The first four natural frequencies of the slab were presented in Table 1.

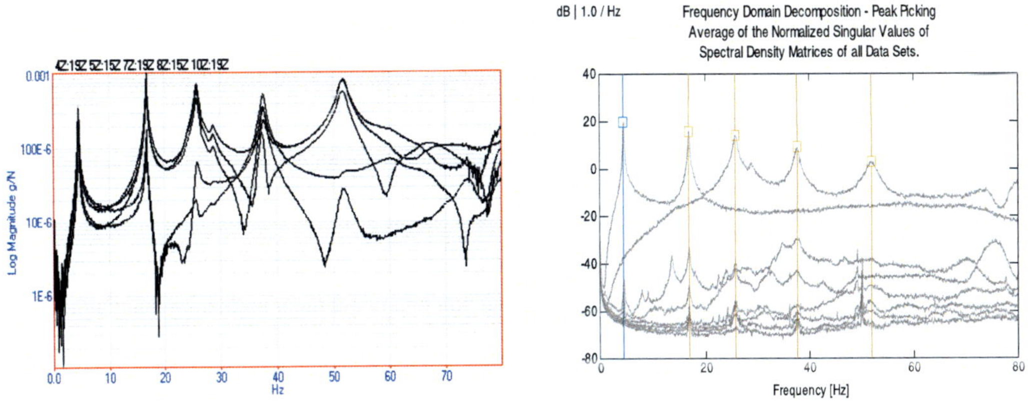

Figure 3: FRFs and SVD

The scaling factors obtained with CMA are presented in Table 3. These are obtained from two different data sets, m00 and m00b, without masses on the slab. In both data sets the inputs are uncorrelated random and the shakers are in the same location. As two reference inputs are used the strongest reference is used for each mode for the residue estimates.

It is important to remark that although the modal masses, and hence the scaling factors, were obtained directly from the FRF's, there are several factors such as the location of the shakers, lines used in the FRF's or the fit model used in the analysis that can result in variations of 10-20% in the results. In order to compare the results with those from OMA through the mass change method, the average of the estimated scaling factors provided by CMA is used for each mode, see Table 3.

Table 3: Scaling factors from CMA analysis

	Data Set						
	m00			m00b			
	Number of Lines in the FRFs						
MODE	1024	2048	4096	1024	2048	4096	Average
1	0.01236	0.01242	0.01260	0.01320	0.01240	0.01315	0.01269
2	0.01250	0.01255	0.01290	0.01288	0.01270	0.01310	0.01277
3	0.01440	0.01370	0.01430	0.01440	0.01360	0.01565	0.01434
4	0.01322	0.01317	0.01325	0.01370	0.01366	0.01373	0.01345

4.1. Symmetric mass change configurations

The scaling factors, corresponding to cases 1, 2 3, and 3b, obtained from EFDD analysis and using the expressions described in paragraph 2, are presented in Table 4. The column $\Delta\omega$ is the frequency shift that is obtained for each mode in the different configurations and the MAC is the modal assurance criteria between the unmodified and modified mode shapes.

It can be seen that in cases 1 and 2, where the attached masses are uniformly distributed across the slab, the scaling factors are estimated with a good accuracy and the different expressions provide similar results. As was expected, the changes in mode shapes are very small.

There is a significant difference in the third mode shape between cases 1 and 2. This mode is the first torsion mode of the slab at 25.72 Hz. Due to the gap in one of the slab supports, there is a rocking motion at 28.5 Hz that, depending of the mass configuration, can interfere in the stability of the torsion mode. In Figures 3a and 3b are presented the SSI stabilization diagram for the configurations with 14 and 21 masses respectively. In these figures can be seen that the first singular value and how the magnitude of the rocking motion is higher in the configuration with 21 masses. This effect can reduce the accuracy in the estimation for this particular mode. On the other hand, the configuration of 14 masses is a better strategy for the torsion mode (masses on both sides of the slab). In table 4 it can be observed that high frequency shift and MAC are obtained with this mass distribution, so that better results are expected.

Table 4. EFDD scaling factors

EFDD	Case 1: 21 MASSES 50 kg Each DOF's [All except supports]							
Mode	α_{00}	α_{01}	α^*	α_{00P}	α_{01B}	αFRF's	Δω [%]	MAC
1	0.01244	0.01241	0.01241	0.01206	0.01244	0.01269	4.11328	0.99998
2	0.01278	0.01278	0.01278	0.01241	0.01278	0.01277	3.87354	0.99993
3	0.01249	0.01251	0.01250	0.01218	0.01247	0.01434	3.35588	0.99687
4	0.01354	0.01328	0.01354	0.01307	0.01354	0.01345	4.61459	0.99928
EFDD	Case 2: 14 MASSES 75 kg Each DOF's [4 6 7 9 10 12 13 15 16 18 19 21 22 24]							
Mode	α_{00}	α_{01}	α^*	α_{00P}	α_{01B}	αFRF's	Δω [%]	MAC
1	0.01263	0.01259	0.01259	0.01222	0.01263	0.01269	4.25529	0.99998
2	0.01312	0.01313	0.01313	0.01271	0.01312	0.01277	4.18566	0.99991
3	0.01424	0.01430	0.01430	0.01356	0.01423	0.01434	6.33416	0.99937
4	0.01377	0.01354	0.01378	0.01325	0.01352	0.01345	4.98406	0.99965
EFDD	Case 3: 6 MASSES 175 kg Each DOF's [7 9 13 15 19 21]							
Mode	α_{00}	α_{01}	α^*	α_{00P}	α_{01B}	αFRF's	Δω [%]	MAC
1	0.01228	0.01220	0.01220	0.01185	0.01228	0.01269	4.66487	0.99987
2	0.01358	0.01353	0.01353	0.01305	0.01358	0.01277	5.10335	0.99993
3	0.01683	0.01679	0.01678	0.01562	0.01677	0.01434	9.39348	0.99783
4	0.01875	0.01839	0.01863	0.01741	0.01857	0.01345	9.34624	0.98862
EFDD	Case 3b: 6 MASSES - 100 kg Each DOF's [7 9 13 15 19 21]							
Mode	α_{00}	α_{01}	α^*	α_{00P}	α_{01B}	αFRF's	Δω [%]	MAC
1	0.01251	0.01270	0.01260	0.01213	0.01270	0.01269	3.04230	0.99986
2	0.01231	0.01255	0.01246	0.01120	0.01255	0.01277	2.85250	0.99968
3	0.01269	0.01290	0.01280	0.01187	0.01290	0.01434	3.52880	0.99836
4	0.01400	0.01413	0.01401	0.01346	0.01412	0.01345	3.26300	0.99866

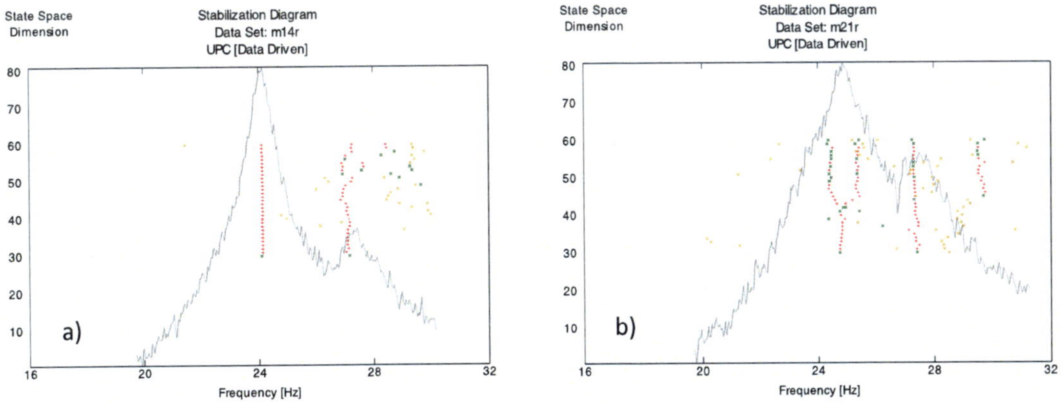

Figure 3a and 3b: SSI stabilization diagram for 21 and 14 masses configurations.

In cases 3 and 3b 6 masses are located at the same position but with different mass magnitude. The results present quite significant differences. In case 3 the scaling factor corresponding to the first mode are acceptable but in the values corresponding to the other three modes are too high. In case 3b, which has the same mass locations but less mass magnitude, the accuracy in the results improves. On one hand, it is not a good strategy to optimize many modes simultaneously with only a few heavy masses. On the other hand, the gap and the friction in the supports change depending on the location and the magnitude of the masses, so that, the non linear effects are mass dependent. So, although better results are expected when the ratio ω_0/ω_1 is higher [10], in this particular case, the dynamic behavior change of the slab is too much and the results don't present good accuracy. That is, although the mode shapes do not change significantly the dynamic behavior of the structure is quite different, as shown in Figure 4. Anyway, in both cases the results obtained for the different mass change methods are similar, indicating that none of them improve the results when a poor strategy is used.

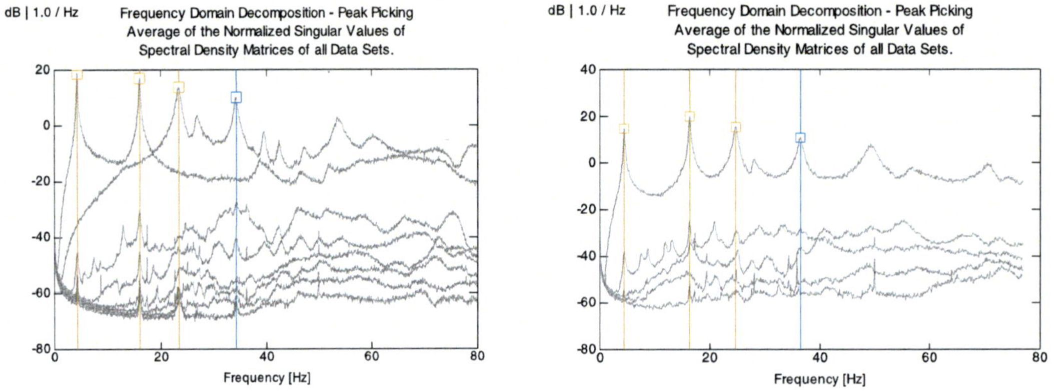

Figure 4: SVD of cases 3 (left) and 3b (right)

4.2. Non-symmetric mass configurations

Cases 4 and 5 correspond to two non-symmetric mass configurations. The results are presented in Table 5 for both cases with EFDD analysis. It can be seen that some mode shapes change significantly. As in case 3, there is a lot of concentrated mass on the slab and therefore there are important changes in the dynamic behavior of the structure due to the non linear effects. As expected, the results from α_{00P} and α_{00P} are the most inaccurate and α_{00P}, in particular, gives lower values than the other approaches. Although expressions α_{01}, α^* and α_{01B} use the modified mode shapes, the results are still not particularly good. In α_{01B} (the projection approach) the κ factor that is introduced to improve the results is not enough due to large changes in mode shapes, where the new modified mode shape can be a projection of more than one un-modified mode shape. Nevertheless, α_{01B} improves the results compared with α_{01} and α^*.

Table 5: Non-symmetric mass configurations (EFDD)

EFDD	Case 4: 3 MASSES - 350 kg Each DOF [7 13 19]							
Mode	α_{00}	α_{01}	α^*	α_{00P}	α_{01B}	αFRF's	$\Delta\omega$ [%]	MAC
1	0.01219	0.01216	0.01216	0.01176	0.01216	0.01269	3.69273	0.99992
2	0.01455	0.01430	0.01430	0.01393	0.01440	0.01277	8.30466	0.97431
3	0.01862	0.02190	0.02192	0.01703	0.02027	0.01434	8.72393	0.99300
EFDD	Case 5: 4 MASSES - DOF's [7 (425 kg) 13 (150 kg) 19 (200 kg) 21 (275 kg)]							
Mode	α_{00}	α_{01}	α^*	α_{00P}	α_{01B}	αFRF's	$\Delta\omega$ [%]	MAC
1	0.01452	0.01452	0.01451	0.01378	0.01452	0.01269	6.51488	0.99957
2	0.01382	0.01421	0.01415	0.01291	0.01417	0.01277	8.23031	0.99104
3	0.01592	0.01751	0.01731	0.01487	0.01737	0.01434	8.81142	0.96573
4	0.01781	0.01832	0.01724	0.01651	0.01749	0.01345	9.12996	0.86864

5. Conclusions

- Different mass change expressions have been applied to estimate the scaling factors of a 15 tonnes concrete slab. The results show that all the methods estimate the scaling factors with good accuracy when a good mass change strategy is used.
- The approach of Parloo et al. [2] is the one that estimates the scaling factors with the largest errors for all the mass change configurations.
- The scaling factors obtained directly from the FRF's present the same or even higher uncertainties that the scaling factors obtained by means of the mass change method.
- The Bernal projection approach [7] provides better results when mode shapes change significantly.

Acknowledgements

The economic support given by the Spanish Ministry of Education through the project BIA2005-07802-C02-02 and the European Social Fund, is gratefully appreciates. Special thanks to Dr. Donald Nyawako for the support in the experimental part of this work.

References

[1] R. Brincker, C. Ventura and P. Andersen. "Why Output-Only Modal Testing is a Desirable Tool for a Wide Range of Practical Applications". In Proc. Of the International Modal Analysis Conference (IMAC) XXI, paper 265, February, 2003.

[2] Parloo, E., Verboven, P., Cuillame, P., and Overmeire, M. V..“Sensitivity-based mass normalization of mode shape estimates from output-only data.” *Proc., Int. Conf. on Structural System Identification.* 2001. pp. 627–636.

[3] D. Bernal and B. Gunes. "Damage Localization in Output-Only Systems: A Flexibility Based Approach". In Proc. Of the International Modal Analysis Conference (IMAC) XX, pp. 1185-1191, February 2002.

[4] M. López Aenlle, R. Brincker, A. Fernández Canteli. L. M. Villa. "Scaling Factor Estimation by the Mass Change Method". Proc. Of the International Operational Modal Analysis Conferencee (IOMAC), Copenhagen, 2005.

[5] R. Brincker and P. Andersen. "A Way of Getting Scaled Mode Shapes in Output Only Modal Analysis". In Proc. Of the International Modal Analysis Conference (IMAC) XXI, paper 141, February 2003.

[6] M. López Aenlle, R. Brincker, A. Fernández Canteli. "Some Methods to Determine Scaled Mode Shapes in Natural Input Modal Analysis". In Proc. Of the International Modal Analysis Conference (IMAC) XXIII, paper, February 2005.

[7] Dionisio Bernal. Modal Scaling from Known Mass Perturbation. "Journal of Engineering Mechanics", Vol. 130, No. 9, September 2004, pp. 1083-1088.

[8] M. López Aenlle,P. Fernández, R. Brincker, A. F. Canteli. Scaling Factors Estimation using an Optimized Mass Change Strategy, Part 1: Theory. In Proc. Of the International Operational Modal Analysis Conference (IOMAC) II. 2007

[9] P. Fernández, M. López Aenlle, L. M. Villa García. and R. Brincker. Scaling Factors Estimation using an Optimized Mass Change Strategy, Part 2: Experimental Results. In Proc. Of the International Operational Modal Analysis Conference
(IOMAC) II. 2007.

[10] M. López-Aenlle,et al., Scaling factor estimation using an optimized mass-change strategy, Mechanical Systems and Signal Processing (2009), doi:10.1016/j.ymssp.2009.06.011.

Proceedings of the IMAC-XXVIII
February 1–4, 2010, Jacksonville, Florida USA

Structural Dynamics of a Frame Including Axial Load Effects

L.N. Virgin, Professor, Department of Mechanical Engineering and Material Science, Duke University, Durham NC, 27708

T.C. Lyman, Graduate Student, Department of Mechanical Engineering and Material Science, Duke University, Durham NC, 27708

Nomenclature

ω_i	Natural Frequency
$[M]$	Mass Matrix
$[K]$	Stiffness Matrix
$[K_G]$	Geometric Stiffness Matrix
E	Young's Modulus
I	Area Moment of Inertia
A	Cross Sectional Area
L	Length of Beam Element
ρ	Density
y_i	Transverse Displacement
ϕ_i	End Rotation
m	Mass
J	Moment of Inertia
P	Axial Load
q_i	Displacement Vector
$\ddot{q}_i$	Acceleration Vector

ABSTRACT

This paper considers the free vibration of a plane, rectangular, portal frame consisting of slender members. Natural frequencies and mode shapes are influenced by the addition of mass at the corners of the frame. The members are sufficiently slender that axial effects occur, and may ultimately lead to buckling. The results from both theoretical and experimental studies are presented.

Introduction

A recurring theme in the design of structures is to consistently push for them to be thinner and lighter. This is especially true in the case of the aerospace industry, where any weight savings translates into better performance and efficiency. However, with any design goal, there are trade-offs that need to be taken into consideration. As structures become lighter and more slender, loadings can start to have an effect on dynamic behavior as well as static behavior. For frame structures, axial loads on the members can affect the stiffness, and in turn, the natural

T. Proulx (ed.), *Dynamics of Civil Structures, Volume 4*, Conference Proceedings of the Society for Experimental Mechanics Series 13,
DOI 10.1007/978-1-4419-9831-6_14, © The Society for Experimental Mechanics, Inc. 2011

frequencies of the frame. Several studies have investigated the effects of applied axial loads to simple frame structures. Lieven and Greening [1] conducted modal tests on a frame with an induced forced due to the shortening of a turnbuckle in one of the members. In their study they also utilized numerical matrix methods to predict mode shapes and natural frequencies. Similarly, Mead [2] analytically investigated the variation of natural frequencies and mode shapes of a planar frame structure that was subjected to externally applied axial loads.

The experimental setup considered in this paper is a simple portal frame structure with slender members. Instead of an external load being directly applied to the structure itself, axial loads are induced by adding masses to the corners of the frame. The masses themselves have an effect on the dynamics of the structure in addition to degrading the stiffness by adding a compressive axial load to the columns. Modal tests were conducted to find the variation in the natural frequencies as mass was added to the frame. Numerical simulations were then ran in ANSYS and FRAME3DD finite element packages to model and gain further insight from the behavior of the structure [3][4].

Theory

Frames are essentially an assembly of beams connected at the ends, and can be modeled by solving a system of partial differential beam equations whose boundary conditions are determined by the joints of the frame. These systems of equations generally produce very large transcendental equations that are only tractable when the structure has a very simple geometry and several other simplifying assumptions can be made [5]. In order to analyze more complex structures, matrix methods have been developed that discretize frame structures into a set of beam elements and lumped masses [6]. When the mass and stiffness properties are discretized, a matrix set of equations describing the behavior of the system can be written as

$$[\mathbf{M}]\ddot{q}_i + [\mathbf{K}]q_i = 0, \tag{1}$$

where the mass and stiffness matrices are the global mass and stiffness matrices for the entire structure, with the q_i vector consisting of displacement and rotational degrees of freedom.

The global mass and stiffness equations are assembled from local beam mass and stiffness matrices, where $q_i = [y_k \ \phi_k \ y_m \ \phi_m]^T$ is a vector of end displacements and rotations. The displacements and rotations for a single beam element are shown in Fig. 1.

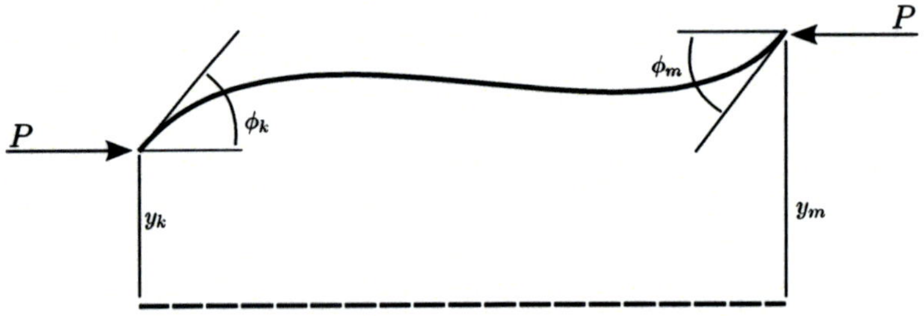

Figure 1: Original and deformed beam with local coordinates. *After* [7].

The local mass, $[\mathbf{M}]$, and stiffness, $[\mathbf{K}]$, matrices for a beam element of a frame are

$$[\mathbf{M}] = \frac{\rho AL}{420}\begin{bmatrix} 156 & 22L & 54 & -13L \\ 22L & 4L^2 & 13L & -3L^2 \\ 54 & 13L & 156 & -22L \\ -13L & -3L^2 & -22L & 4L^2 \end{bmatrix}, \tag{2}$$

and

$$[\mathbf{K}] = \frac{EI}{L^2}\begin{bmatrix} 12 & 6L & -12 & 6L \\ 6L & 4L^2 & -6L & 2L^2 \\ -12 & -6L & 12 & -6L \\ 6L & 2L^2 & -6L & 4L^2 \end{bmatrix}. \tag{3}$$

When compressive axial loads are applied to beams their stiffness is degraded by the load. This is called a 'geometric stiffness' effect and, in the static sense, is the mechanism that is responsible for buckling. In order to account for this in the matrix stiffness method, a separate geometric stiffness matrix,

$$[\mathbf{K_G}] = \frac{P}{30L}\begin{bmatrix} 36 & 3L & -36 & 3L \\ 3L & 4L^2 & -3L & -L^2 \\ -36 & -3L & 36 & -3L \\ 3L & -L^2 & -3L & 4L^2 \end{bmatrix}, \tag{4}$$

is factored into the traditional beam stiffness formulation [8].

By adding in the axial load effects on the stiffness of the system, a geometric stiffness matrix augments the traditional stiffness matrix. Equation (1) now becomes

$$[\mathbf{M}]\ddot{q}_i + [\mathbf{K} - \mathbf{K_G}]q_i = 0. \tag{5}$$

Equation (5) can either be used in a time marching algorithm to produce dynamic simulations, or if q_i is assumed to be simple harmonic motion, $\ddot{q}_i = -\omega^2 q_i$, the equation

$$([\mathbf{K} - \mathbf{K_G}] - \omega_i^2[\mathbf{M}])q_i = 0 \tag{6}$$

can be solved as a general eigenvalue problem to attain the natural frequencies and mode shapes of the structure.

Experimental Setup

The experimental laboratory setup consisted of a simple portal frame comprised of polycarbonate beam members. The polycarbonate beams had a cross section of dimensions, 2.554 cm x 0.154 cm, and a density of $\rho = 1157$ kg/m^3. A schematic and picture of the setup is shown in Fig. 2. For the experiments conducted, the bottom supports of the columns were clamped and the corners were connected at 90 degrees using angled aluminum fixtures. The base fixtures for the columns could also be converted to pinned connections in order to allow the bottom of the columns to freely rotate. The column length, L_1, was 15.24 cm and the length of the cross beam of the frame, L_2, measured 45.72 cm. Threaded rods were attached to the aluminum brackets such that masses could be added to the two corners.

Modal data were taken using an Ometron VH300+ laser doppler vibrometer, and an Endevco 2302-50 modal impact hammer. The data were recorded with a Dell laptop computer with PULSE data acquisition software and a Brüel & Kjær type 3109 input module. All of the hammer impact hits and vibration measurements took place in the plane of the frame.

Figure 2: Schematic of experimental frame setup and picture of physical laboratory model

Simulations

Simulations of the frame were ran in both the commercial finite element software package ANSYS, and FRAME3DD, an open source frame analysis software package. The frame was modeled using 2-D beam elements, with fixed end supports for the columns. The experimentally added mass was modeled as a lumped mass added to the nodes at the top two corners of the frame. The simulations also allowed for the option of adding a lumped inertia term to the corners nodes. The geometric stiffness effects from the masses were modeled as a force in FRAME3DD, in addition to the lumped mass. The geometric stiffness was accounted for in ANSYS by applying an acceleration to the frame equal to that of gravity, creating an equivalent force on the columns from the lumped masses.

Results

The experimental frequency response functions were loaded into and analyzed using ME'scopeVES [9]. Figure 3 shows the experimental mode shapes for the lowest four modes of the frame. Figure 4 shows the four lowest calculated mode shapes from ANSYS.

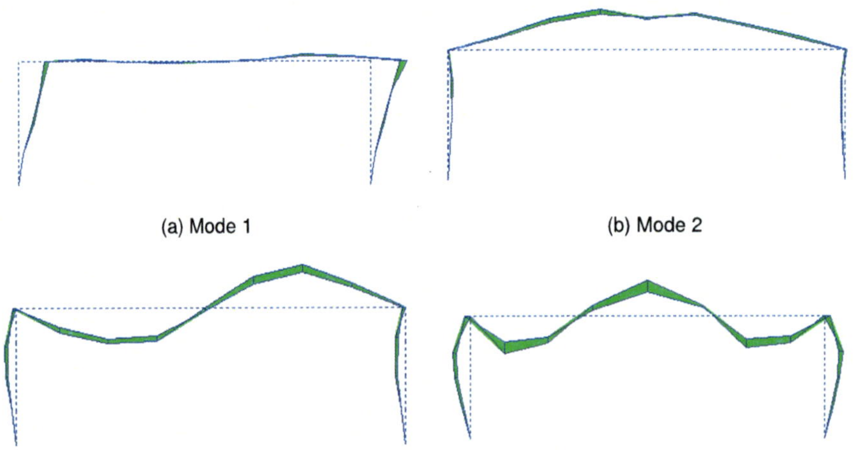

(a) Mode 1 (b) Mode 2

(c) Mode 3 (d) Mode 4

Figure 3: Experimentally determined mode shapes

The experimentally determined mode shapes all corresponded with those that were predicted by the simulations from ANSYS and FRAME3DD. The mode shapes also retained their ordering, both experimentally and in simulation, as mass was added to the frame.

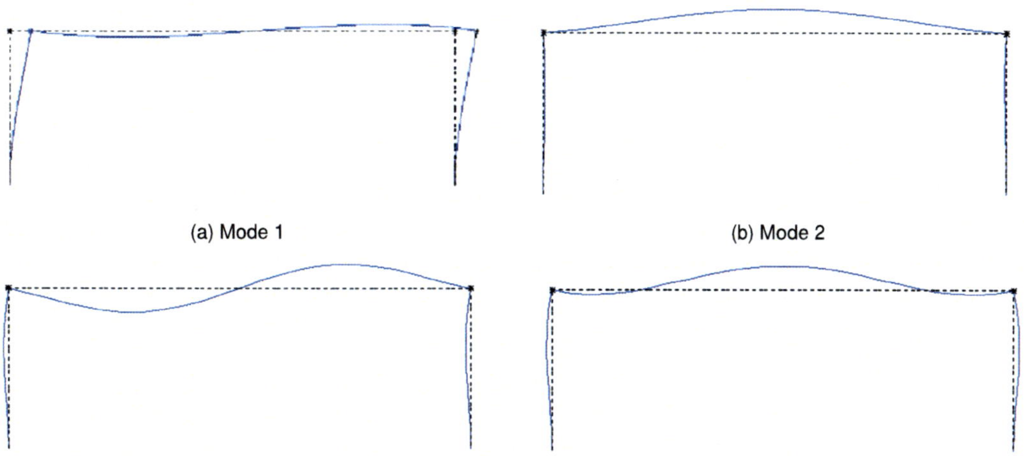

(a) Mode 1

(b) Mode 2

(c) Mode 3

(d) Mode 4

Figure 4: Numerically determined mode shapes from ANSYS

Figure 5 shows the variation of the natural frequencies of the first four modes of the frame with increasing axial load due to added mass. Both ANSYS and FRAME3DD programs produce very similar results for all four sets of natural frequencies. Experimentally the structure was found to be very lightly damped, and damping was not considered in the numerical results. Additionally, initial geometric imperfections that are bound to be present in experimental systems were also not included in the numerical results.

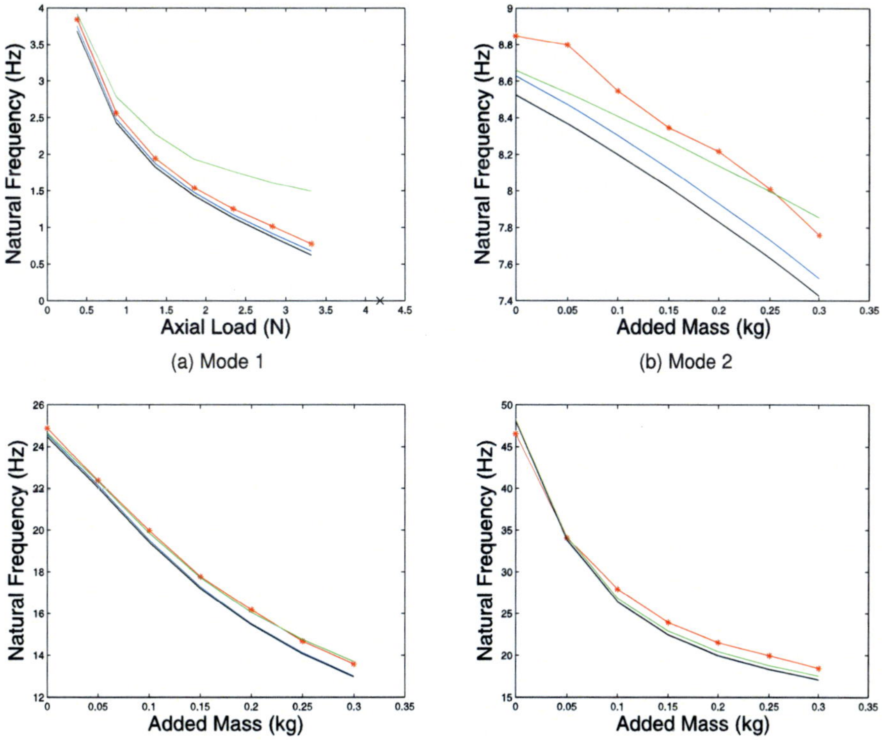

(a) Mode 1

(b) Mode 2

(c) Mode 3

(d) Mode 4

Figure 5: Natural frequency variations with added mass. Red - experimental data. Blue - Frame3DD simulation results. Black - ANSYS simulation results. Green - ANSYS simulation results without geometric stiffness.

An interesting point to investigate with this structure is what factors contribute to this change in natural frequency. Unlike several of the other studies on geometric stiffness effects, mass and rotational inertia were being added to the frame structure which can also have an effect on the natural frequencies of the system. Using numerical simulations, the effects of the axial loading can be separated from those of mass and rotational inertia. The green trace in Fig. 5 shows the results of ANSYS simulations where the natural frequencies are calculated without any geometric stiffness effects, i.e. assuming negligible axial loading effects on the lateral stiffness of the columns. Mode 1, which is very similar to the static buckling mode, was the most affected by neglecting the axial loading. Figure 5a shows the natural frequency plotted versus the axial load on each column, which is caused by gravity acting on the masses added to the corners, $P = mg$. The independently calculated static buckling load from ANSYS is shown with an 'X' at an axial load value of 4.185 N. Modes 2-4 are primarily affected by the mass and show little change in their behavior from neglecting the axial loading. Figures 5b-5d show the variation of the natural frequency of modes 2-4 with the amount of mass, m, added to each corner of the frame.

Conclusions

A planar frame with slender, flexible members was investigated both experimentally and numerically to attempt to characterize a change in modal behavior with added mass and increased axial load. Natural frequency was shown to vary due to not only geometric stiffness effects but to added mass and rotational inertia present in the experimental system as well. Additionally, the change in dynamic behavior due to axial loading did not change all of the modes of vibration in the same way. The loading did cause a significant difference in the prediction of the fundamental frequency of the frame as mass was added, but did not have much of an effect on the higher modes.

References

[1] N.A.J. Lieven and P.D. Greening. Effect of Experimental Pre-stress and Residual Stress on Modal Behavior. *Philosophical Transactions of the Royal Society of London*, 359:97-111, 2001.

[2] D.J. Mead. Free Vibrations of Self-Strained Assemblies of Beams. *Journal of Sound and Vibration*, 249:101-127, 2002.

[3] ANSYS, Software Package, Ver. 11.0, ANSYS, Inc. Canonsburg, PA, 2007

[4] FRAME3DD, Software Package. http://frame3dd.sourceforge.net, 2009

[5] W. Nowacki. *Dynamics of Elastic Systems*, John Wiley & Sons. Inc., 1963.

[6] J.S. Przemieniecki. *Theory of Matrix Structural Analysis*, McGraw-Hill, 1968.

[7] A. Kassimali. *Matrix Analysis of Structures*. Brooks/Cole Publishing Company, Pacific Grove, California, 1999.

[8] L.N. Virgin. *Vibration of Axially Loaded Structures*, Cambridge University Press, 2007.

[9] ME'scopeVES, Software Package, Ver. 3.0, Vibrant Technology, Inc., Jamestown, CA, 2001

Proceedings of the IMAC-XXVIII
February 1–4, 2010, Jacksonville, Florida USA
©2010 Society for Experimental Mechanics Inc.

Experimental investigation of the pre-tension effects on the modal parameters of a slender pre-tensioned concrete beam

Alfredo Cigada, Alessandro Caprioli, Marcello Vanali
Politecnico di Milano, Dipartimento di Meccanica, via La Masa 34 20156 Milano Italy

ABSTRACT

Slender pre-tensioned concrete structures are nowadays of common use, due to their unique features. Slender beam can be used to sustain large structures and important loads leaving the architects a lot possibilities in the structure design. The health of such structures is strictly related to the conditions of the tendon that, in many cases, are not accessible. Visual inspection is the most used technique to assess the tendon condition even if in many cases access to the inspection points is very difficult or impossible. This paper investigates the possibility of detecting a pre-stress loss or a tendon failure by means of modal analysis. A pre-stressed concrete beam has been built and tested under laboratory conditions, varying the applied pre-stress and the number of active tendon. Modal analysis has been performed in each of the experimental condition and attention has been focussed on the variation of all identified modal parameters, frequency, damping and mode shapes. Furthermore a set of damage indexes have been computed to highlight the most sensible magnitude able to identify a change on the structure. Obtained results showed that an accurate analysis is needed in order to identify a change in modal parameters due to variations in the pre-stress, while tendon failure leads to more important changes in the identification results. The performed study are the starting point in order to properly tune a numerical model of the beam useful to accurately interpreter structural changes.

INTRODUCTION

A large amount of slender concrete structures are nowadays built using prestress techniques. This technique allows to build slender structures sustaining large weights, and thus giving the architects a number of additional degrees of freedom in the design process. Another important achievement is the amount of space left to be used due to the reduced thickness of floors and walls.

The prestress condition is obtained by means of steel tendon which provide the compression that allow to increase the structural performances reducing the positive (elongation) strain and stress. The health of prestressed concrete structures is thus strictly related to the conditions of the tendon that, in many cases, are often included in the concrete and there is no way to directly inspect them.

One of the main problem of prestressed concrete is therefore the tendon conditions which may deteriorate due to corrosion, or exceptional loads. To this aim many non-destructive diagnostic techniques have been proposed [1,2] but it has to be considered that these techniques are always difficult to apply and their results are often affected by a high level of uncertainty. In the last decades a lot of papers have been written concerning structural health monitoring via modal analysis, the structure condition is assessed by its modal parameters that are considered representative of the actual structure dynamic behaviour. A change in these parameters identify an evolution in the dynamic behaviour and a possible ongoing damage [3,4].

The latter approach pushed forward to the exploitation of the possibility to consider the relationship between prestress conditions and dynamical behaviour of a structure in assessing the tendon conditions and therefore the health of prestress concrete.

The tendon presence and the prestress load influence the structure dynamical behaviour, therefore a reliable identification of the modal parameter of a structure could offer the opportunity to monitor the prestress condition. This sort of monitoring is really attractive: it is non destructive and can be performed continuously during normal life of the structure if operational modal analysis is considered.

Different researches have been developed in this direction but the connection between damage and modal parameter changes has not been completely cleared up yet[5,6,7,8]. One of the main aspects is the entity of the modal parameter change due to tendon damage: in many cases it can be of the same order of magnitude

T. Proulx (ed.), *Dynamics of Civil Structures, Volume 4*, Conference Proceedings of the Society for Experimental Mechanics Series 13,
DOI 10.1007/978-1-4419-9831-6_15, © The Society for Experimental Mechanics, Inc. 2011

of the modal parameter estimation uncertainty. A deeper investigation in this field is therefore necessary to pursue damage detection by means of modal parameter analysis.

In a previous paper [2] a first approach has been developed: other researches results have been analyzed and a numerical analysis (FEM Modelling) has been performed to investigate the possibility to detect prestress changes by means of dynamic behaviour identifications. The obtained results were encouraging and therefore an experimental campaign has been designed and a test specimen has been built to perform extensive tests in laboratory controlled conditions.

As one of the goal of this research is the investigation of the relationship between damage and dynamical behaviour modifications the test structure has been designed to allow the simulation of different sort of known tendon damages. Furthermore the possibility to apply different damage entities allows to test the performances of different techniques: in this paper the first results will be presented: they have been obtained analyzing the relationship between natural frequencies, Mode Shape Area and Flexibility Matrix changes with different damage levels. At first the test specimen and the measurement campaign are illustrated putting into evidence the simulated damage conditions, then the main results are shown and finally some concluding remarks are given.

In the next paragraphs the test specimen, the measurement set-up and the performed experiments are described.

THE TESTED BEAM AND THE MEASUREMENT CAMPAIGN

As the aim of the experimental campaign is to investigate only the prestress effects on the beam dynamic behaviour the measurement set-up has been kept simple. Tests have been carried out in laboratory controlled conditions on the post-tensioned concrete beam depicted in Figure 1. The specimen is a single span (6 m long) beam with a constant rectangular section (20 cm x 30 cm). The concrete used to build the specimen has been prepared according to the European standard [9,10]. The beam is post-tensioned by means of a single straight unbounded tendon, being in the simplest framework of ungrouted post-tensioning. The tendon, located in the middle of the beam section, at 6 cm from the lower side of the beam, is made of three strands whose characteristics are assumed to be in agreement with the prEN10138 norm. The adopted diameter is about 15 mm (0,6"), the elastic modulus Es=209000 N/mm2. The steel structure connect the strands and the beam, those structures are equipped with 3 ring load cells that allow to measure the load of every strands (Figure 2). The beam is simply supported by 2 steel round beams which diameter is 5 cm (Figure 3).

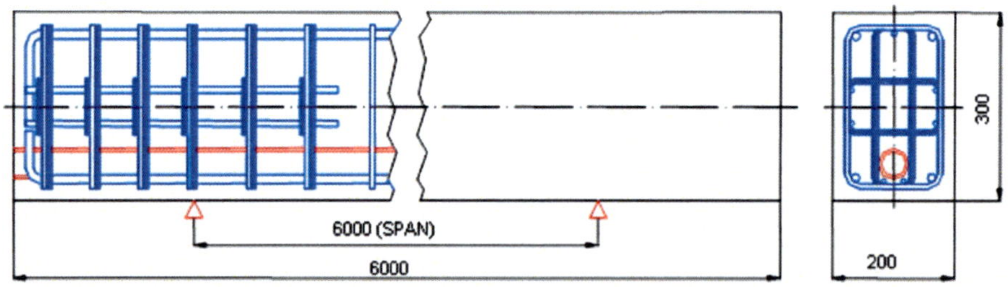

Figure 1 Layout of the concrete beam object of the analysis

Figure 2 The concrete beam in the laboratory with the accelerometers and the impact hummer.

Figure 3 Detail of the constraints

The choice to build the single tendon with 3 strands has been taken to grant the possibility to simulate different damage kinds: it is possible to apply different loads with the same number of strands, the stiffness of the strands remain the same, and, on the other side, apply the same load with different number of strands. The first case is used to study the effect of a prestress loss while in the second case a damage of the tendon can be studied. In Figure 4 the test matrix resuming all the possible damages scenarios is shown.

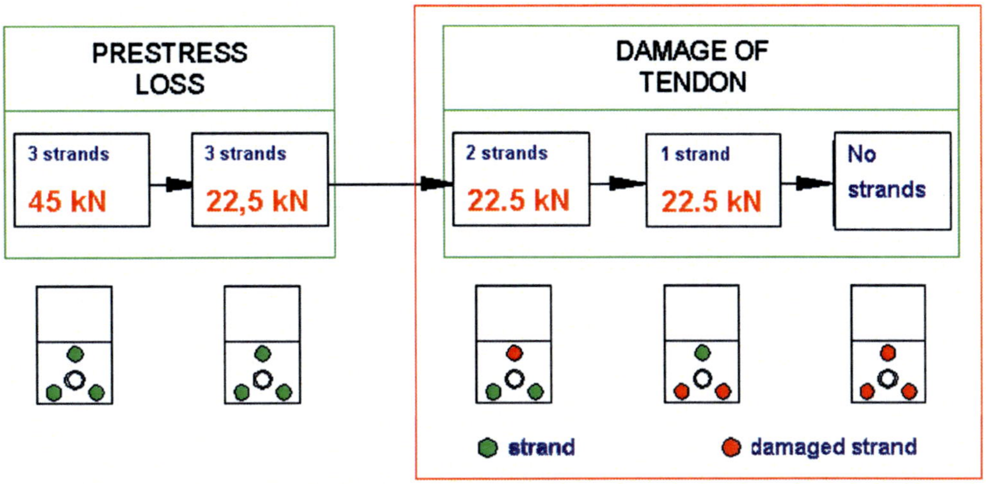

Figure 4 Different configurations of the 3 tendon allow to simulate different sort of damages

As the idea of this paper is to detect prestress changes by means of dynamic behaviour analysis a series of indexes have to be chosen in order to give a quantitative index of the beam conditions starting from the identified modal properties, frequencies, damping and mode shapes. Due reasons that are well explained in literature [11] the damping value is the index affected by the highest uncertainty in the identification and therefore is not considered in the study.
Starting from the other modal parameters, frequency and mode shapes, 3 different indexes have been related to the damage level [4]: natural frequencies themselves, mode shape area and the flexibility matrix.
The first index is self explaining, while the other two are defined as:

$$\text{MSA}_i = \int_{X=0}^{X=L} \left| X(\omega_i) dx \right|$$

where x is the position on the beam, i is the considered vibration mode and $X(\omega_i)$ is the modal shape of the considered mode.

$$\text{FlexMatr} = [F] = [\Phi]\left[diag\left(\frac{1}{\omega_i^2} \right) \right][\Phi]^T$$

where $[\Phi]$ is the matrix of the modal shapes and ω_i are the eigenfrequencies.

All these properties are calculated starting from the identified modal parameters, the latter have been estimated using the Polyreference least square frequency domain identification algorithm [12,13]. It has to be pointed out that natural frequencies and mode shape area are properties of the whole structure, on the other hand every element of the flexibility matrix gives local information. As the simulated damages are related to the whole structure (a prestress load loss interest all the beam) a synthetic index related to changes in the flexibility matrix has been considered: in particular the index that has been chosen is the standard deviation between the matrix evaluated when the beam is in nominal preload condition and the one evaluated when the beam is in damaged condition.
In the next section the measurement set-up is described.

THE MEASUREMENT SET-UP AND FIRST RESULTS

In order to provide a deep knowledge of the beam dynamic behaviour the measurement set-up was made of 20 piezo-accelerometers placed on the symmetry plane on the upper surface of the beam; Measurement points have been equally spaced along the upper surface.
Excitation has been given by means of an impact hammer and the response has been measured in the 0 – 200 Hz range. A first set of measurements has been taken to ensure that the impact had the needed energy

in the considered frequency range. All subsequent analyses have been carried out averaging ten impacts and always checking coherence information.
The first condition that has been tested is the simple beam without any preload, results are shown in Figure 5 (blue line). The first four in plane vibration modes are clearly visible in the frequency response function at 12,5 Hz, 51,1 Hz, 109,1 Hz and 175,1 Hz. Then preload has been applied and the same analysis have been repeated.

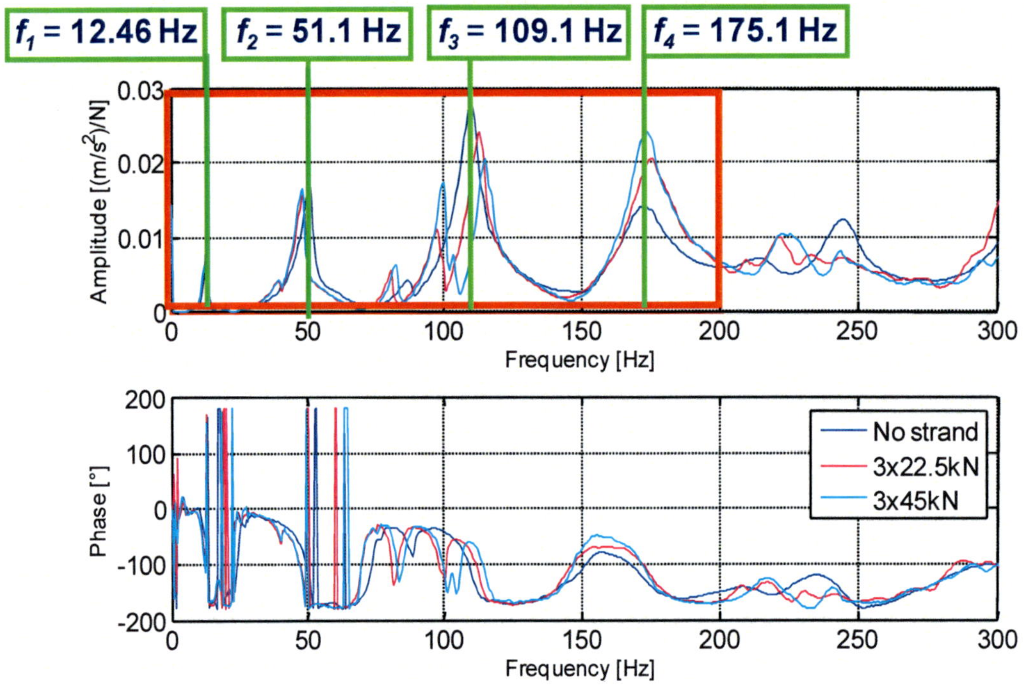

Figure 5 Frequency responses measured at 1/3 of the beam span with different tendon configurations

The analysis of the accelerometer signals showed that other natural frequencies appear when preload is applied (Figure 5, light blue and violet lines). A new series of tests has been carried out measuring also the out of plane vibrations. This analysis showed that the new natural frequencies that appear when a preload is applied are related to torsional and lateral mode shapes, probably due to a non perfect symmetry of the applied prestress load.
In the following analyses only the "in plane" mode shapes will be considered postponing the other results to a further step of the research. The results obtained simulating a prestress loss will be presented in the next section.

PRESTRESS LOSS

This section considers the results obtained applying different preload levels by means of the same number of strands. The test is carried out to investigate only the effect of the preload force on the modal properties of the beam not taking into account any other damage: as the tendon section remains the same his stiffness does not change.
The first set of experiments has been carried out with 3 strands and different preloads: 45 kN, 22,5 kN and 0 kN on every strand. The results, in terms of natural frequencies and mode shapes (1st and 2nd natural frequency), are shown in Figure 6.
Increasing the preload level from 0 to maximum leads to a decrease in the identified frequency values as shown in the left graph of Figure 6.

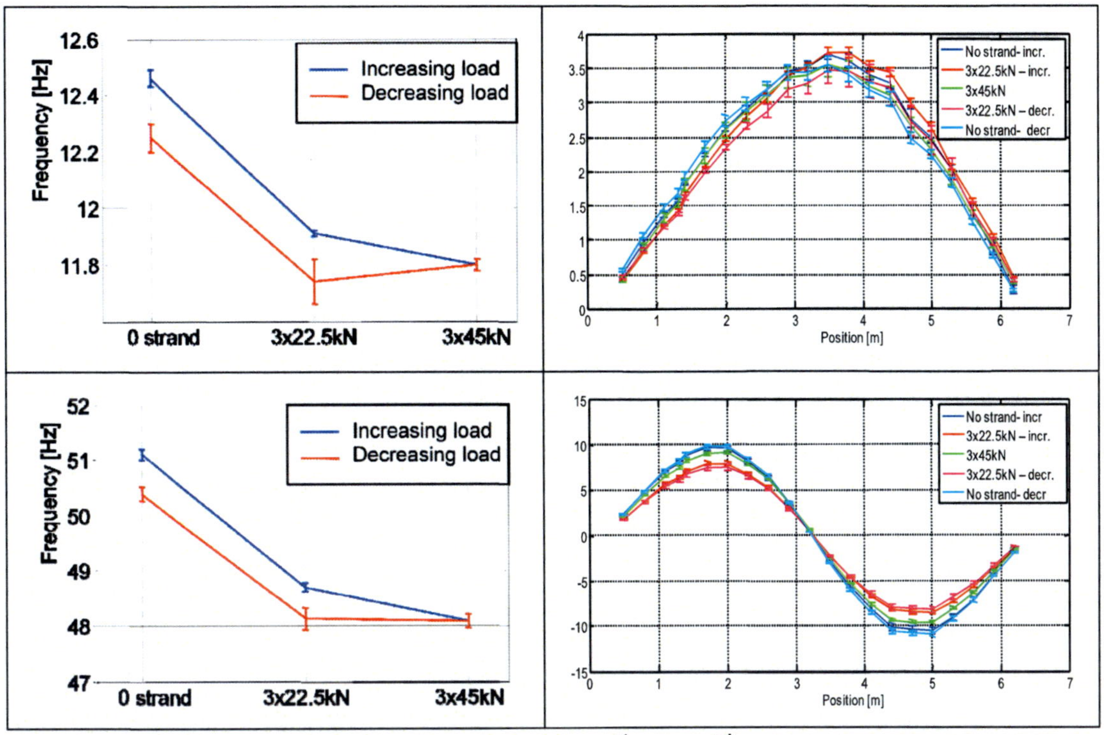

Figure 6 Natural frequencies and modal shapes (1st and 2nd mode) measured increasing and decreasing the preload (1st application of the load)

Looking at mode shapes in the right side of Figure 6 a unique trend in shape changes due to the pre-stress value cannot be detected. Even worse than this, if the results pertaining to the prestress decrease are looked at, it can be seen that the same frequency value and shape modification can be attributed to different pre-stress levels. At the end of the test the obtained values showed a difference if compared to the ones obtained at the test start, and this had to be investigated deeper.

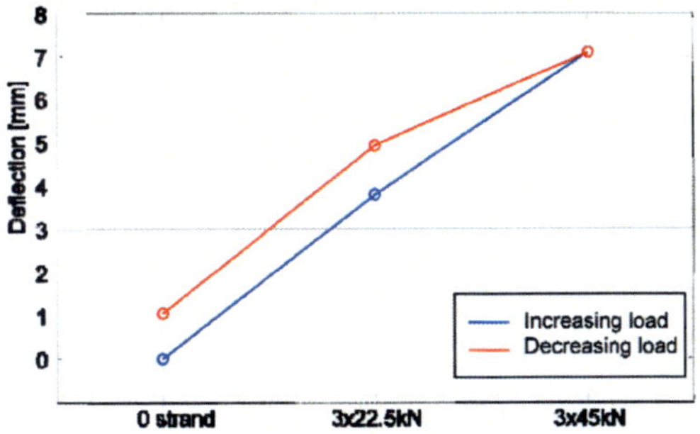

Figure 7 Deflection measured on the beam at the span centre

One of the reason could be identified in the different static equilibrium conditions measured on the beam, Figure 7 shows the centre span deflection measured in the different tested conditions, as can be seen a residual deflection is evidenced at the end of the first test session.

It has then been decided to repeat the cycle using only one tendon, to assure that the preload was applied symmetrically. This new test, highlighted a different beam behaviour (Figure 8). After having removed the preload stress the beam returned to the same static equilibrium position of the beginning. As can be noticed the deflection at full load is in the order of magnitude of 1/3rd of the previously obtained as the total preload is given by 1/3rd of the strands. No residual deformations have been evidenced as the preload has been removed. In this situation too natural frequencies and mode shapes have been extracted and results are shown in Figure 9.

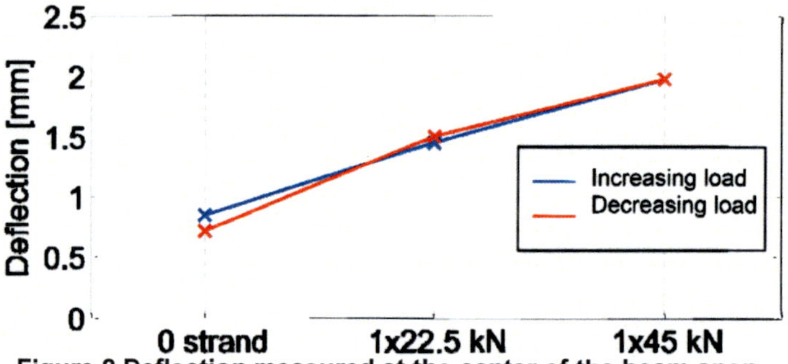

Figure 8 Deflection measured at the center of the beam span

Looking at the graph in Figure 9 it is possible to notice that a decrease in the estimated frequency values is still present as the prestress load increases. Moreover the magnitude of this decrease is in the same order as the one observed in the first test, where the applied load was three times bigger. Looking at mode shapes it is still not possible to identify a clear dependence between mode shape changes and prestress value.

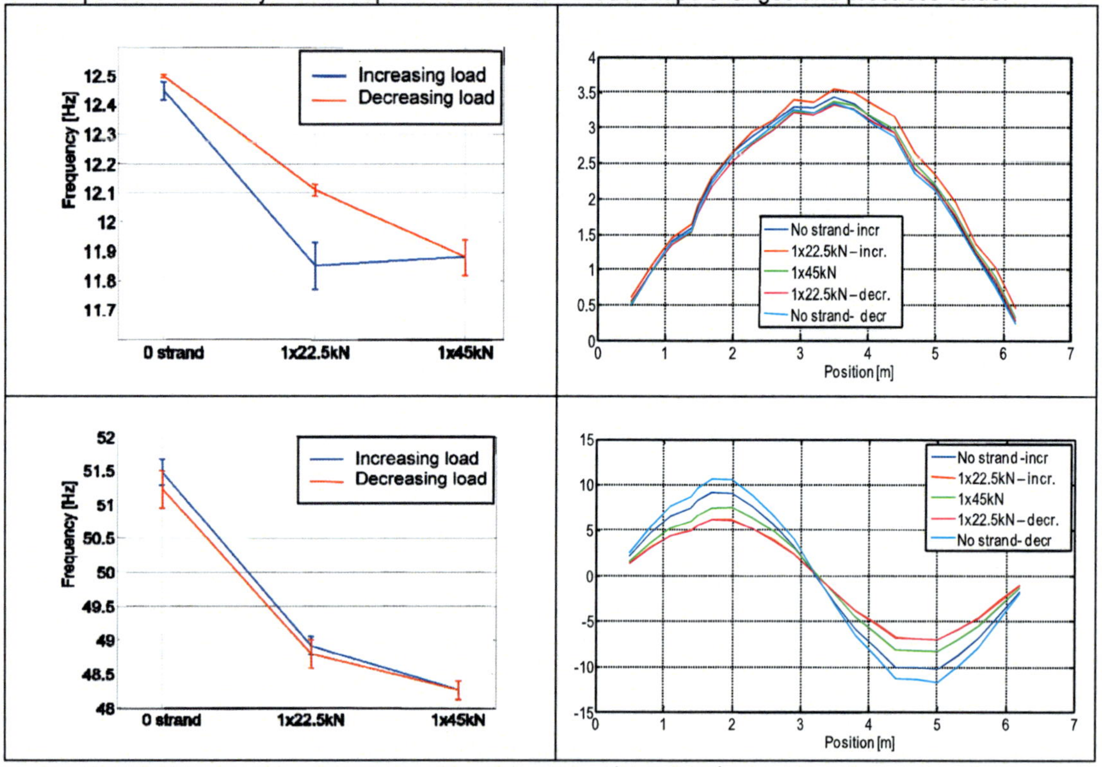

Figure 9 Natural frequencies and modal shapes (1st and 2nd mode) measured increasing and decreasing the preload (2nd application of the load)

Results given form this first test series did not put into evidence a straight relationship between the beam dynamic behaviour and the prestress value. Moreover some of the obtained results deserve a deeper investigation, a totally different load values lead to the same variations in the identified beam eigenfrequencies. A series of hypothesis have been made, but these are still to be verified. A reason could be that friction can affect the prestress distribution on the beam, or that due to imperfections on the beam surface the load is applied differently from one test case to the other. All these possibilities will be further investigated and a numerical model will be developed to better understand the results.

The next investigated kind of damage is the loss of a tendon and the results for this case are shown in the next section.

DAMAGE OF THE TENDON

Another series of tests have been performed to investigate a different sort of scenario, while in the previous section a loss of prestress has been studied, in this section the attention will be pointed on a loss of stiffness that could be due to tendon corrosion. Figure 10 shows how natural frequency and modal shapes (1st and 2nd) change as a function of the number of strand while the preload on each strand remains the same.

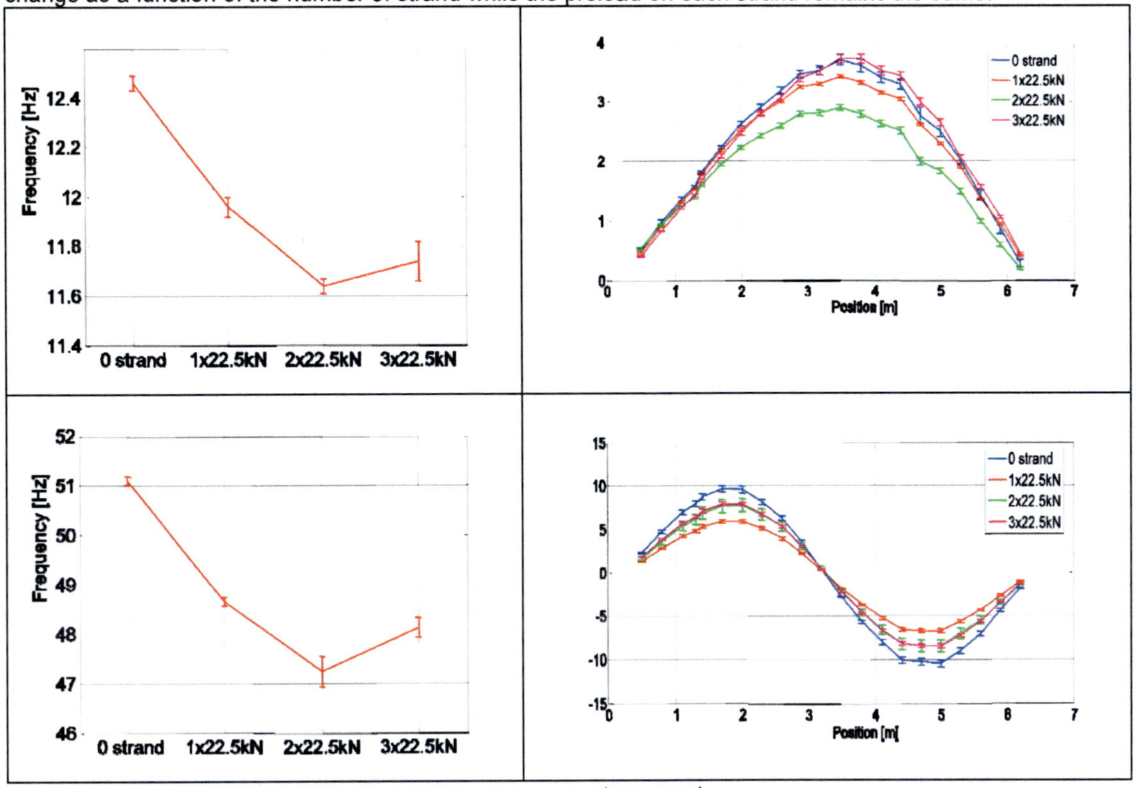

Figure 10 Natural frequencies and modal shapes (1st and 2nd mode) measured in different preload conditions (2nd application of the load)

Looking at the results a clear relation between natural frequencies and number of strands installed cannot be detected: while a meaningful decrease of the natural frequencies has been measured increasing the number of strands from 0 to 2, a further increment to 3 strands involved an increment of the identified natural frequencies. The same behaviour has been detected for the 1st and 2nd mode. This situation is in good agreement with the previous section results, in which the same frequency value was identified with substantially different prestress values as shown in Figure 6 and Figure 9.

If mode shapes are considered some meaningful differences are identified as the number of strand varies but a clear trend is not yet identified. Different simulated damages leads to similar mode shapes, as is the case of 2/3 strands for mode 2 (green and pink curves in Figure 10).

This idea is confirmed by the analysis of the Mode Shape Area. Figure 11 summarizes the results obtained for this damage index: the mode shapes measured with different damage scenarios (different preload conditions and different number of strands are considered) and the correspondent calculated mode shape area are shown.

As can be seen in the left side of Figure 11 different pre-stress conditions lead to changes in the mode shapes, but again an uniform trend is not detected. This is confirmed if the graph on the right side are analyzed. The computed Mode Shape Area is different as prestress and number of strands changes but a straight law connecting its value to the tension value is not detectable.

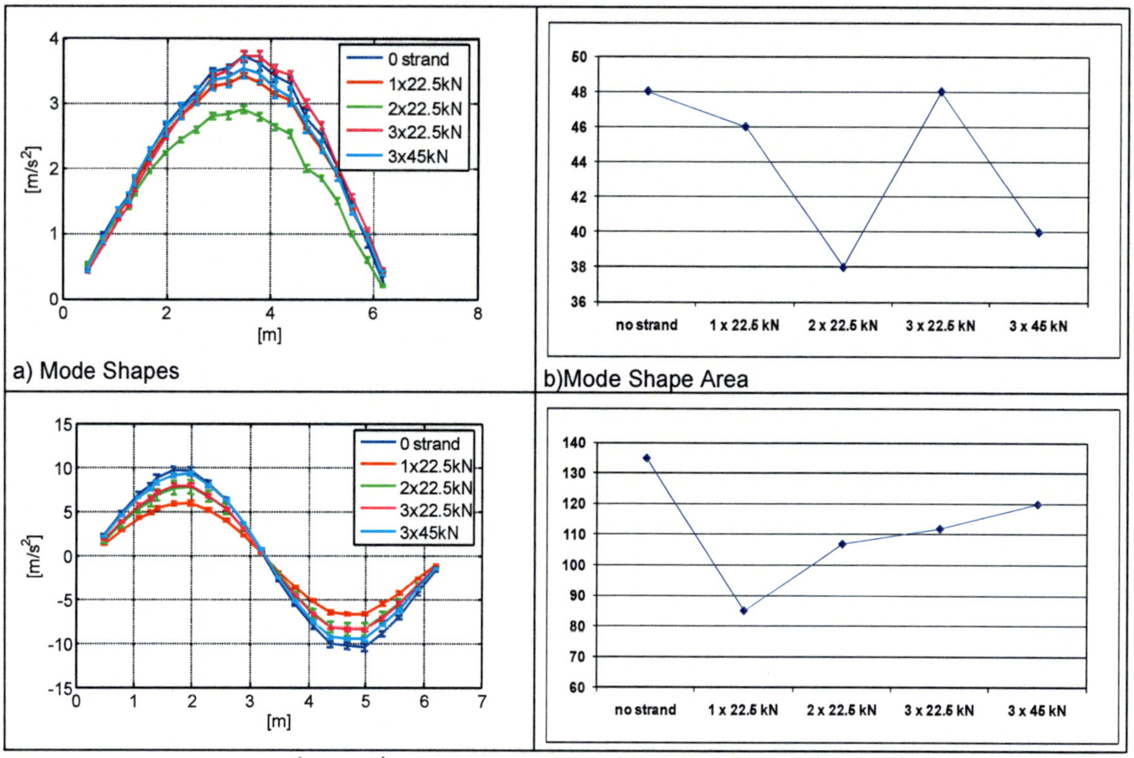

Figure 11 Modal shapes (1st and 2nd modes) and mode shape area index (1st and 2nd modes)

Considering what has been found so far a straight relation between the beam dynamic behaviour and the prestress condition has not been found yet, even if the beam modal parameters are affected by meaningful changes as the prestress is varied. This situation pushes forward to consider a damage index that is related to all frequencies/mode shapes, so that all the information can be exploited at one time. This will be shown in the next section.

FLEXIBILITY MATRIX

As the results obtained considering Natural Frequencies and Mode Shapes alone did not identify a straight relationship between prestress value and parameter evolution, here the possibility to resume all indexes into changes in the Flexibility Matrix have been investigated.

Recalling the definition of flexibility matrix [F]:

$$\text{FlexMatr=}[F]=[\Phi]\left[diag\left(\frac{1}{\omega_i^2}\right)\right][\Phi]^T$$

It can be seen that all modal residues and eigenfrequencies contribute to the final value. It has to be considered that every element of the Flexibility Matrix gives local information while the damages considered in this research are global.

In Figure 12 the squared difference between the matrix in reference conditions and with no preload is presented. Changes are evident and moreover all matrix elements are affected indicating that changes are generalized to the whole structure.

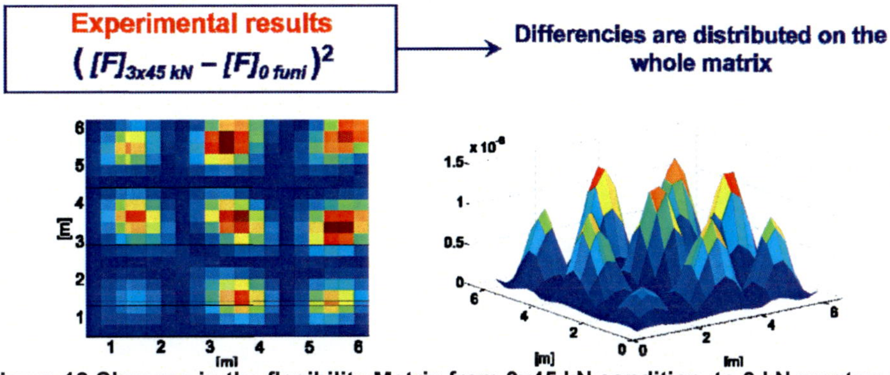

Figure 12 Changes in the flexibility Matrix from 3x45 kN condition, to 0 kN prestress

An index resuming all changes has been therefore adopted: the flexibility matrix in undamaged condition (3 strands installed with the a preload of 45 kN) has been evaluated and used as reference, the same matrix has been evaluated in damaged conditions and the standard deviation between the two matrix has been computed. In Figure 13 the behaviour of the computed Standard Deviation is shown versus the preload condition.

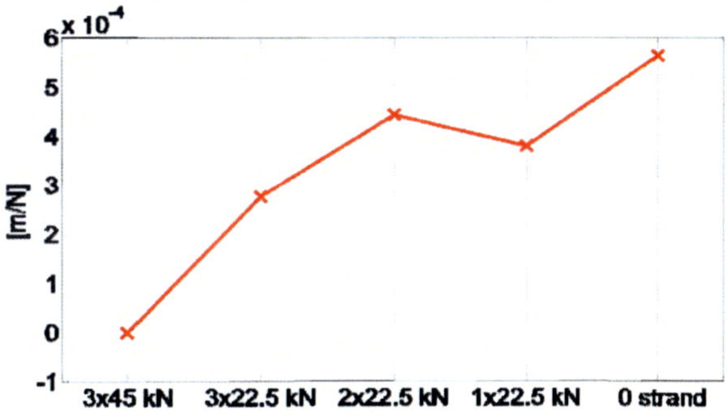

Figure 13 SQM of the flexibility matrix in different conditions

Looking at the graph in Figure 13 an uniform trend in the SQM evolution is evident as the preload situation evolves from nominal to 0. From this index it seems that there is a possibility to detect changes in the beam prestress conditions starting from the modal parameters identification.

CONCLUDING REMARKS

This paper presents a resume of the first results obtained analyzing the prestress influence on the dynamic behaviour of a concrete beam. The idea was to exploit modal identification to assess the prestress condition of the beam. A simple test specimen has been built and analyzed in laboratory controlled condition, so that all the other possible influencing magnitudes have been kept under control. A series of different experiments have been carried out to simulate different damage kinds, at first a loss of tension in the strands and then the total loss of one or more strands have been reproduced. In each of the tested conditions the beam modal parameters have been estimated. The obtained results showed that is really difficult to extrapolate a straight rule linking the prestress condition to single modal parameters, while the use of a global index, considering all modal parameters together seemed more promising. These results are being used to tune a numerical model that will help to interpreter changes in the modal properties.

ACKNOWLEDGMENTS

Authors would like to acknowledge Eng. Mario Scrosati for the help given in building the test specimen.

REFERENCES

1. M. G. Ali, A. R. Maddocks, “Evaluation of corrosion of prestressing steel in concrete using non-desctructive techniques” Corrosion and Materials; Vol. 28, No 5, 42-48, 2003.
2. M. Bruggi, A. Caprioli, M. Vanali, P. Venini, “Investigation of the Pre-Stress Loss in the Dynamical Behaviour of Concrete Beams”, in Proceedings of the 26th International Modal Analysis Conference – IMAC, 2008.

3. W. Doebling, C. R. Farrar, M. B. Prime, D. W. Shevitz "Damage Identification and Health monitoring of Structural Mechanical Systems from Changes in Their Vibration Characteristics: a Literature Review" Los Alamos National Laboratory, Los Alamos, NM, USA, Tech. Rep. LA-13070-MS, May 1996.
4. G. De Roeck "The State-of-the-Art of Damage Detection by Vibration Monitoring: the SIMCES Experiences on a Prestressed Concrete Bridge" Journal of Structural Engineering, Vol. 131, pp. 1898-1910, 2005.
5. M. Saiidi, B. Douglas, S. Feng, "Prestress force effect on vibration frequency on concrete bridges" J. Structural Engineering, Vol 120, No 7, 2233-2241, 1994.
6. J.T. Kim, C.B Yun, Y.S. Ryu, H.M. Cho, " Identification of prestress-loss in PSC beams using modal data", Structural Engineering and Mechanics, Vol. 17, No 3-4, 467-482, 2004.
7. G. De Roeck, "The state-of-the-art of damage detection by vibration monitoring: the SIMCES experience," Journal of Structural Control, vol. 10, pp. 127, 2003.
8. J.F. Unger, A. Teughels, G. De Roeck, "System Identification and Damage Detection of a Prestressed Concrete Beam" Journal of Structural Engineering, 132, 1691-1698, 2006 ASCE..
9. UNI EC 1-2006 UNI EN 1992-1-1:2005 "Eurocode 2 – Design of Concrete Structures Part 1-1: General rules and rules for buildings"
10. UNI EN 206-1:2006 "Concrete – Part 1: Specification, performance, production and conformity"
11. A. Cigada, A. Caprioli, M. Redaelli, M. Vanali "Vibration testing at Meazza Stadium: reliability of operational modal analysis to health monitoring purposes" Journal of Performance of Constructed Facilities, Vol. 22, No 4, 228-337, 2008.
12. Peeters, B. "The PolyMAX frequency-domain method: a new standard for modal parameter estimation?" Shock and Vibration, 11(3-4), 395-409, 2004.
13. Caprioli, A., Cigada, A., and Vanali, M. "Comparison between different operational modal analysis techniques for the identification of large civil structure modal parameters" in Proceedings of the 24th International modal analysis conference, St. Louis, Mo, USA, 2006.

Proceedings of the IMAC-XXVIII
February 1–4, 2010, Jacksonville, Florida USA
©2010 Society for Experimental Mechanics Inc.

System Identification of a Three-Story Infilled RC Frame Tested on the UCSD-NEES Shake Table

Babak Moaveni[1], Andreas Stavridis[2], and P. Benson Shing[2]

[1]Department of Civil & Environmental Engineering, Tufts University, Medford
[2]Department of Structural Engineering, University of California, San Diego

ABSTRACT

A 2/3-scale, three-story, two-bay, infilled RC frame was dynamically tested on a shake-table. The test objectives were to assess the seismic performance of existing, non-ductile, infilled RC frames and provide data for the evaluation of newly developed analytical methods predicting the behavior of such structures. The shake-table tests were designed to induce damage on the structure progressively through scaled earthquake records of increasing intensity. Between the earthquake records, the response of the frame to low-amplitude ambient vibration and white-noise base excitation tests was measured. At these low levels of excitations, the structure can be considered as a quasi-linear system with parameters depending on the damage state. The deterministic-stochastic subspace identification method based on system input and output signals has been used to estimate the modal parameters of the test structure at its various damage states. The identification is conducted considering the white-noise base excitation and the resulting structural response measured by accelerometers. The study has quantified the decrease of natural frequencies and the increase of structural damping at progressive damage states. The identified modal parameters have been used for damage identification of the infilled frame in a companion paper.

Introduction

Changes in dynamic characteristics of structures have been used extensively for the purpose of condition assessment and damage identification of structures. In recent years, vibration-based structural health monitoring based on modal parameters has attracted increasing attention in the civil engineering research community as a potential tool to develop methods through which structural damage can be identified at the earliest possible stage. Extensive literature reviews on damage identification based on changes in vibration characteristics have been provided by Doebling et al. [1, 2] and Sohn et al. [3]. The study presented in this paper focuses on the experimental modal analysis of a three-story infilled reinforced concrete (RC) frame at different damage levels due to a sequence of shake-table tests.

The frame was a 2/3-scale model of a prototype structure designed to represent non-ductile RC frames built in California in the 1920s. It was tested at the University of California San Diego (UCSD) in November 2008, using the Large High Performance Outdoor Shake-table, which is part of the George E. Brown, Jr. Network for Earthquake Engineering Simulation (NEES). The testing sequence is part of a collaboratory project involving researchers from Stanford University, the University of Colorado at Boulder and UCSD. The goals of the research team are to evaluate the seismic performance of older infilled RC frame buildings and to develop analytical methods and retrofit techniques for this type of structures. This two-bay, three-story frame was the largest structure of this type ever tested on a shake-table. The shake-table tests were designed to induce damage on the structure progressively through a sequence of scaled earthquake records. Between the earthquake excitations, the RC frame was subjected to low-amplitude ambient vibration and white-noise base excitation tests. At these low levels of excitations, the structure can be considered as a quasi-linear system with its parameters depending on the damage level. The deterministic-stochastic subspace identification (DSI) method, based on system input and output signals, has been used to estimate the modal parameters (natural frequencies, damping ratios, and mode shapes) of the frame at its various damage states as well as those of the undamaged structure. The identification method adopted in this study uses the white-noise base excitation and the corresponding structural response measured at different damage states of the structure. Recently, the DSI method was successfully applied by Moaveni et al. [4] for system identification of a seven-story building slice tested on the UCSD-NEES shake-table. It is worth noting that during the application of white-noise base excitations with the shake-table, the

T. Proulx (ed.), *Dynamics of Civil Structures, Volume 4*, Conference Proceedings of the Society for Experimental Mechanics Series 13,
DOI 10.1007/978-1-4419-9831-6_16, © The Society for Experimental Mechanics, Inc. 2011

table response was distorted by the system dynamics and, thereby, deviated significantly from a white noise/broadband excitation. Therefore, the application of output-only system identification methods to the white-noise test data will result in modal parameters with significant estimation error.

Test Setup

Three-Story Infilled RC Frame

The prototype structure was designed by Stavridis and Shing [5] to represent a structure built in California in the 1920s era. The design only considered the gravity loads in accordance with the engineering practice of that era. However, the design was based on contemporary construction materials which were used for the construction of the test specimen. The plan view of the prototype structure and the elevation of an external frame are presented in Figure 1. The structural system consists of a non-ductile RC frame, which is infilled with three-wythe unreinforced masonry walls on the exterior. Such structural system can be found in many existing buildings in the western United States, including pre-1930s buildings in California. Moreover, this type of construction is common in many regions of the world with high seismicity such as the Mediterranean and Latin America.

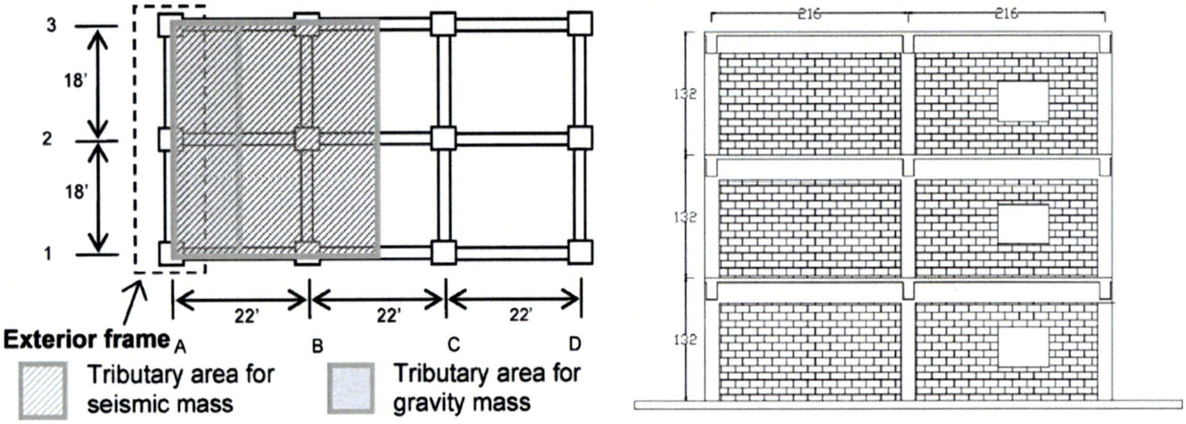

Fig. 1 Plan view of the prototype structure (left) and Elevation view of an exterior frame (right)

The specimen tested on the outdoor shake-table at UCSD is shown in Figure 2. It corresponds to the external frame of the prototype structure with all its dimensions scaled by 2/3 and infill walls consisting of two wythes of brick units. The scaling factors were determined according to the similitude requirement [6] and are summarized in Table 1. From this point on, the ground motion levels mentioned in the paper are with respect to the full-scale prototype structure. As shown in Figure 2, two steel towers were secured on the shake-table on the north and south side of the structure to prevent a potential out-of-plane collapse of the structure during severe shaking. However, they did not interact with the structure during the tests as they were placed with a 0.75 inch gap from the specimen. Further details regarding the design of the specimen and the shake-table tests can be found in [7].

Table 1. Scale factors

Quantity	Scaling Factor
Length	2/3
Elastic modulus	1.00
Mass	0.20
Seismic acceleration	2.27
Force	4/9
Stress	1.00
Strain	1.00
Time	0.542
Frequency	1.85
Gravitational Acceleration	1.00

Fig. 2 Front view (left) and side view of the specimen (right)

Instrumentation Layout

The specimen and the steel towers were instrumented with an extensive array of sensors. They included 135 strain gages placed in the longitudinal and transverse reinforcing steel bars of the first story beams and columns where damage was anticipated, 71 string pots and Linear Variable Differential Transformers (LVDTs) to measure relative deformations of the structural members, and 59 uniaxial accelerometers. Since each accelerometer could only measure the acceleration along one direction of motion, metal cubes were attached on the structure to mount the accelerometers and obtain measurements along X, Y and Z directions, with X being the direction of motion, Y the out-of-plane direction, and Z the vertical direction. Six additional accelerometers were mounted at the middle and at the west end of the foundation slab of the specimen. The recordings from the middle accelerometer are considered as the input excitation for the structure. In this study, measured response data from 9 longitudinal (3 per floor), 9 vertical (3 per floor) and 6 transverse (2 per floor) acceleration channels, as shown in Figure 3, have been used to identify the modal parameters of the test structure. The channels have been named as LF-D, in which L is the location on the floor and can be W for West, C for Center, or E for East, F is equal to 1, 2, or 3 and denotes the floor level the accelerometer is located, and D denotes the direction of acceleration measured and can be X, Y, or Z. It should be mentioned that Channel C1-X (C: center of floor; 1: first floor; X: measuring in X direction) did not function properly during all the tests and Channel E3-X did not function correctly during Tests 5 and 9 (see Table 2). However, there is sufficient redundancy in instrumentation that this does not present a problem.

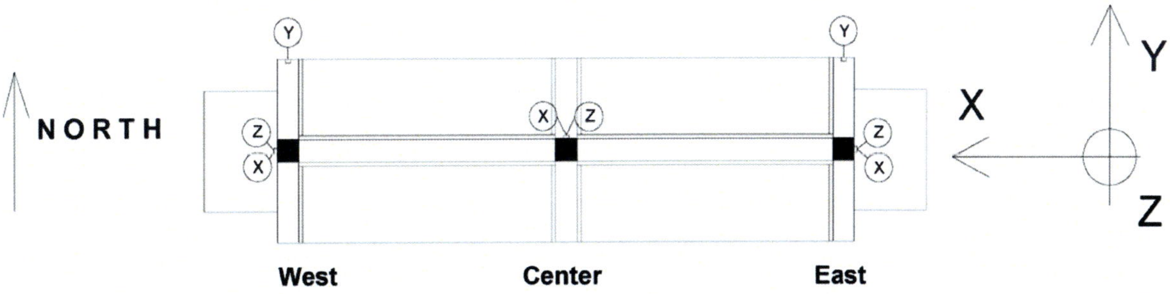

Fig. 3 Locations and directions of accelerometers on each floor which are used in this study

The measured acceleration responses were sampled at 240Hz resulting in a Nyquist frequency of 120Hz, which is higher than the modal frequencies of interest in this study (< 60Hz). Before applying the system identification method to the measured data, all the absolute acceleration time histories were band-pass filtered between 0.5Hz and 70Hz using a high order (1024) Finite Impulse Response (FIR) filter.

Dynamic Tests Performed

A sequence of 44 dynamic tests was applied to the test structure including ambient vibration tests, free vibration tests, and forced vibration tests (white noise and seismic base excitations). The main goal of the testing sequence was to gradually damage the structure by subjecting it to seismic ground motions of increasing amplitude. Between the earthquake records, low-amplitude white noise base excitation tests as well as ambient vibration tests were performed, so that the modal parameters of the structure can be identified at the different damage states. The ground motion accelerograms were obtained by scaling time and amplitude of the acceleration time history recorded at the Gilroy 3 station during the 1989 Loma Prieta earthquake using the scaling factors shown in Table 1. For structures with a natural frequency close to the frequency of the infilled frame studied here, 67% of the Gilroy 3 motion corresponds to a design basis earthquake (DBE) for Seismic Design Category D. Moreover, for this structure, the 100% level of the Gilroy 3 motion corresponds to a maximum considered earthquake (MCE). Between the earthquake tests, recordings were obtained from ambient vibration as well as low-amplitude white-noise base excitation tests for the purpose of modal analysis. In this study, eight damage states in the testing sequence have been analyzed including the initial state of the undamaged structure. Damage state S0 is defined as the undamaged (baseline) state of the structure before its exposure to the first seismic excitation, while damage states S1 to S7 correspond to the state of the structure after exposure to the earthquake ground motions of increasing amplitude. The design basis earthquake (67% of the Gilroy) was applied twice on the shake-table and visible cracks were observed after the structure was exposed to this earthquake for the second time. Table 2 summarizes the dynamic tests used in the present study and the corresponding damage states of the test structure. It should be noted that the ambient vibration data are not used for the system identification of the infilled frame structure because the signal to noise ratio of the measured ambient acceleration response is very low.

Table 2. Dynamic tests used in this study (WN: white noise base excitation; EQ: earthquake base excitation)

Test No.	Test Date	Test Description	Damage State
5	11/3/2008	0.03g RMS WN, 5 min	S0
8	"	20% Gilroy EQ	
9	"	0.03g RMS WN, 5 min	S1
12	11/6/2008	40% Gilroy EQ	
13	"	0.03g RMS WN, 5 min	S2
21	11/10/2008	67% Gilroy EQ (DBE)	
25	11/12/2008	0.04g RMS WN, 5 min	S3
26	"	67% Gilroy EQ (DBE)	
27	"	0.04g RMS WN, 5 min	S4
28	"	83% Gilroy EQ	
29	"	0.04g RMS WN, 5 min	S5
33	11/13/2008	91% Gilroy EQ	
35	"	100% Gilroy EQ (MCE)	
36	"	0.04g RMS WN, 5 min	S6
40	11/18/2008	120% Gilroy EQ	
41	"	0.04g RMS WN, 5 min	S7

System Identification of the Infilled Frame

In the study presented here, the deterministic-stochastic subspace identification (DSI) method, an input-output system identification method, has been used to estimate the modal parameters of the infilled frame at various damage states. Output-only system identification methods have been successfully applied for system identification of linear systems [4, 8]; however these methods are based on the assumption of a broadband (ideal white-noise) input excitation. In this experiment, the white noise base excitation inputs have been significantly modified by the shake-table system dynamics and the table outputs are far from broadband signals. Therefore, the application of output-only system identification methods to these "white noise" base excitation test data will result in large estimation errors in the modal parameters. Figure 5 shows the Fourier Amplitude Spectrum (FAS) of the unfiltered "white noise" base excitation acceleration measured on the shake-table during Test 9 (damage state S1). The largest peak in the FAS of the input acceleration can be due to the effects of the oil column resonance, which is around 11.5Hz. The mechanical characteristics of the shake-table are available in [9].

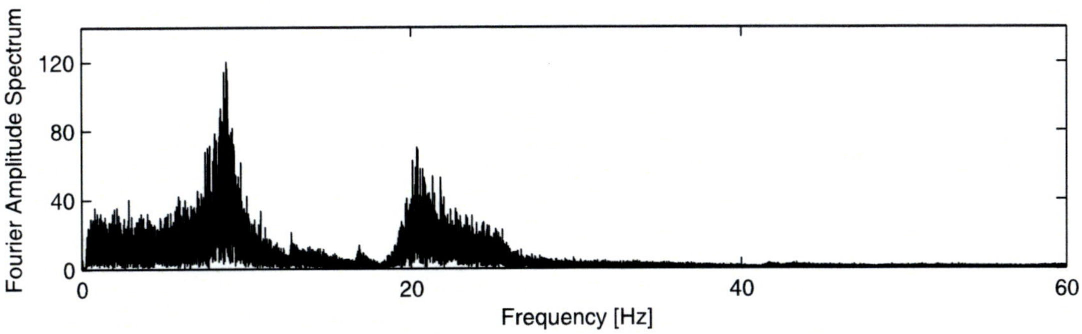

Fig. 5 Fourier Amplitude Spectrum of the base excitation measured on the shake-table during Test 9

Review of Deterministic-Stochastic Subspace Identification Method

The deterministic-stochastic state-space model for linear time-invariant systems can be written as

$$\begin{aligned} \mathbf{x}(k+1) &= \mathbf{A}\mathbf{x}(k) + \mathbf{B}\mathbf{u}(k) + \mathbf{w}(k) \\ \mathbf{y}(k) &= \mathbf{C}\mathbf{x}(k) + \mathbf{D}\mathbf{u}(k) + \mathbf{v}(k) \end{aligned} \quad (1)$$

where **A**, **B**, **C** and **D** are state-space matrices, $\mathbf{u}(k)$ and $\mathbf{y}(k)$ denote the input and output vectors, respectively, and $\mathbf{x}(k)$ is the state vector. In the deterministic-stochastic model, the process noise $\mathbf{w}(k)$ represents disturbances (small unmeasured excitations) and modeling inaccuracies, while the measurement noise $\mathbf{v}(k)$ models the sensor inaccuracies. However, in the data-driven stochastic subspace identification (SSI-DATA) method, an output-only method, both noise terms (**w** and **v**) also implicitly include the input excitation since it is impossible to distinguish the input information from the noise terms. Considering the assumptions that the deterministic input $u(k)$ is uncorrelated with both the process noise $\mathbf{w}(k)$ and the measurement noise $\mathbf{v}(k)$, and that both noise terms are not identically zero, a robust identification algorithm was developed by Van Overschee and de Moore [10] to identify the state-space matrices in the combined deterministic-stochastic system. Similar to SSI-DATA, robust numerical techniques such as QR factorization, SVD, and least squares are used in this method. At each damage state, the DSI method has been applied to 5-minute long filtered input-output data records with a sampling frequency of 240Hz. The measured input (shake-table acceleration) and the absolute longitudinal (7 channels during Tests 5 and 9; 8 channels during other tests) and vertical (9 channels) acceleration response data were filtered between 0.5 to 70Hz using a 1024 order FIR filter. For each dynamic test, an input-output Hankel matrix was formed including 20 block rows with 17 or 18 rows each (1 input and 16 or 17 output channels) and 71,962 columns using the filtered data.

Modal Identification Results

Modal parameters of the test structure were identified using the DSI method outlined above based on the input-output data measured from low-amplitude (0.03g and 0.04g RMS) white-noise base excitation tests at various damage states (S0, S1, S2, S3, S4, S5, S6, and S7). Figure 6 shows in polar plots the complex-valued mode shapes of the four most significantly excited modes of the test structure identified based on data from Test 5 (damage state S0). The four most significant vibration modes identified at this damage state consist of the first two longitudinal (1-L, 2-L), the second torsional (2-T) and the second coupled longitudinal-torsional (2-L-T) modes. The torsional modes were excited probably due to imperfections in the construction of the specimen which could induce eccentricities. An additional source of excitation of torsional modes was the interaction between the specimen and the shake-table which could amplify the imperfection of the shake-table and result in input excitation along the Y direction. The real parts of these mode shapes are illustrated in Figure 7. It should be noted that the six measurements perpendicular to the infill wall plane (W1-Y, W2-Y, W3-Y, E1-Y, E2-Y, and E2-Y) are also used to plot the mode shapes in Figures 6 and 7. However, from Figure 7 it is observed that the longitudinal mode shapes have negligible out-of-plane components. The polar plot representation of a mode shape provides information on the degree of non-classical (or non-proportional) damping [11] characteristics of that mode. If all components of a mode shape (each component being represented by a vector in a polar plot) are collinear, that vibration mode is classically damped. The more scattered the mode shape components are in the complex plane, the more the system is non-classically (non-proportionally) damped in that mode. However, measurement noise (low signal-to-noise ratio), estimation errors, and modeling errors can also cause a truly classically damped vibration mode to be identified as non-classically damped. From Figure 6, it is observed that

1-L, 2-L and 2-L-T modes at damage state S0 are identified as almost perfectly classically damped while some degree of non-proportional damping is identified for 2-T mode.

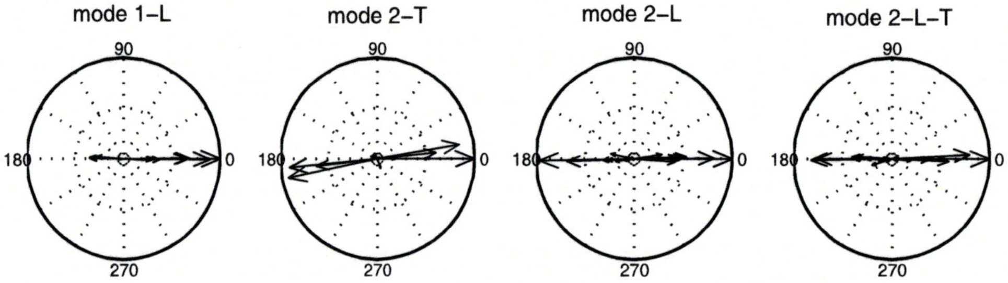

Fig. 6 Polar plot representation of complex mode shapes at damage state S0

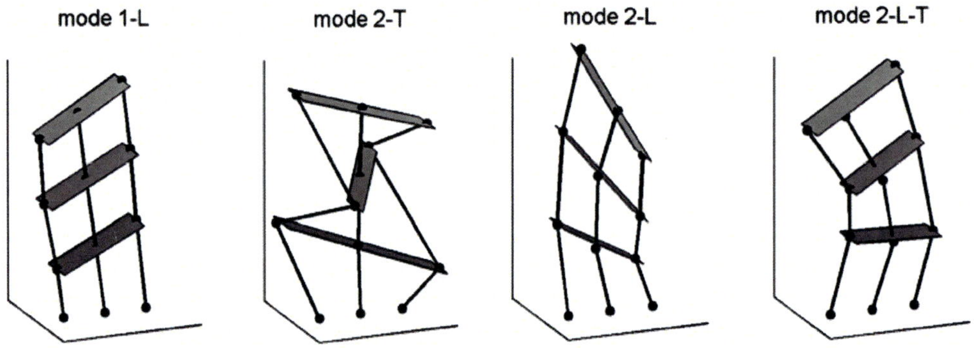

Fig. 7 Vibration mode shapes of the infilled frame at damage state S0

The natural frequencies and damping ratios of these four most significantly excited modes are given in Table 3 for all damage states. Modal Assurance Criterion (MAC) values [12] were also computed to compare each complex-valued mode shapes identified at each damage states with the corresponding mode shape identified at damage state S0. It should be noted the modal parameters of the second torsional mode (2-T) are identified based on the longitudinal (7 channels during Tests 5 and 9; 8 channels during other tests) and transverse (6 channels) measurements while the other three modes are identified based on longitudinal and vertical (9 channels) data. This is due to the fact that the transverse components of longitudinal mode shapes are negligible while these components are small in the 2-L-T mode shape. From Table 3, the following observations can be made. First, the identified natural frequencies decrease consistently with increasing level of damage, while the identified damping ratios increase. Second, the reduction in the natural frequencies of the two longitudinal modes with damage is much more significant than for the 2-T and 2-L-T modes. Similar observation can be made on the level of increase in the identified damping ratios. Third, the MAC values between the identified mode shapes at each damage state with their counterpart identified at S0 also have a monotonically decreasing trend with increasing damage. The high MAC values (close to one) for damage states S1 to S4 indicate that there is little change in the identified mode shapes at these damage states. The identified natural frequencies and mode shapes of the two longitudinal modes have been used in a companion paper for damage identification of the infilled frame at its various damage states.

The modal parameters presented in Table 3 provide useful insight on the structural performance and the damage sustained due to the seismic excitations. The results show that the structure practically remained elastic after the low level earthquakes and the first noticeable deterioration of the structural properties occurred after the 67% of Gilroy, which corresponds to a design basis earthquake (DBE) for this structure. At S3, the inspection of the physical specimen revealed the first visible cracks in the infill panel with a window at the first story, as shown in Figure 8(a). This observation is consistent with the fact that the modal parameters at S3 show the first noticeable differences compared to the modal parameters at S0. The differences are mainly depicted in the drop of the first mode frequency and the increase of the damping ratio for that mode. However, the cracks were insignificant with respect to the structural integrity of the specimen and did not alter the structural properties drastically. Significant change in the structural properties is noted at damage state S6, which occurred after subjecting the structure to a maximum considered earthquake (MCE). At this stage, significant cracks developed in both bays of the first floor, and the columns also cracked as illustrated in Figure 8(b). Although these cracks did not appear to be critical for the structure, they caused a natural frequency reduction of more than 50%, which would be translated to a reduction of the stiffness to approximately one fourth of its initial value. Moreover, the damping value increased from the initial value of 2.0% to 15.7% due to the energy dissipated by the sliding along the fractured mortar joints.

Table 3. Modal parameters of the infilled frame identified at different damage states

Damage State:		S0	S1	S2	S3	S4	S5	S6	S7
Mode 1-L	Frequency [Hz]	18.18	18.11	17.99	16.74	15.93	14.78	8.47	5.34
	Damping [%]	2.0	2.4	1.9	3.3	3.8	6.1	15.7	15.6
	MAC	1.00	1.00	1.00	1.00	1.00	0.98	0.80	0.71
Mode 2-T*	Frequency [Hz]	21.16	21.02	21.32	20.77	20.16	19.69	18.20	17.39
	Damping [%]	1.5	1.5	1.3	1.5	1.8	1.5	1.5	1.6
	MAC	1.00	1.00	1.00	0.99	0.98	0.99	0.95	0.68
Mode 2-L	Frequency [Hz]	41.22	41.09	41.56	40.21	38.56	35.50	27.34	22.57
	Damping [%]	1.1	1.0	1.0	1.4	3.0	4.4	4.8	4.2
	MAC	1.00	1.00	1.00	0.99	0.96	0.92	0.74	0.67
Mode 2-L-T	Frequency [Hz]	57.81	57.35	57.96	56.25	54.64	52.65	45.98	43.33
	Damping [%]	1.1	1.3	1.0	0.7	1.2	2.1	2.1	2.7
	MAC	1.00	1.00	1.00	0.97	0.92	0.87	0.54	0.29

** Mode 2-T is identified based on longitudinal and transverse measurements while the other three modes are identified based on longitudinal and vertical measured data*

The identified first-mode natural frequency at S6 coincides with the frequency corresponding to the peak of the acceleration response spectrum of the scaled Gilroy motion. When the motion amplitude was increased to 120% of the original record, the spectral acceleration at that frequency was 2.87 g. Therefore, significant damage was induced by the earthquake and severe shear cracks developed at mid-height in the middle column. These cracks further reduced the frequency to less than one third of its initial value, although it did not further increase the damping value, which was already high. Figure 9 shows the observed damage at this state.

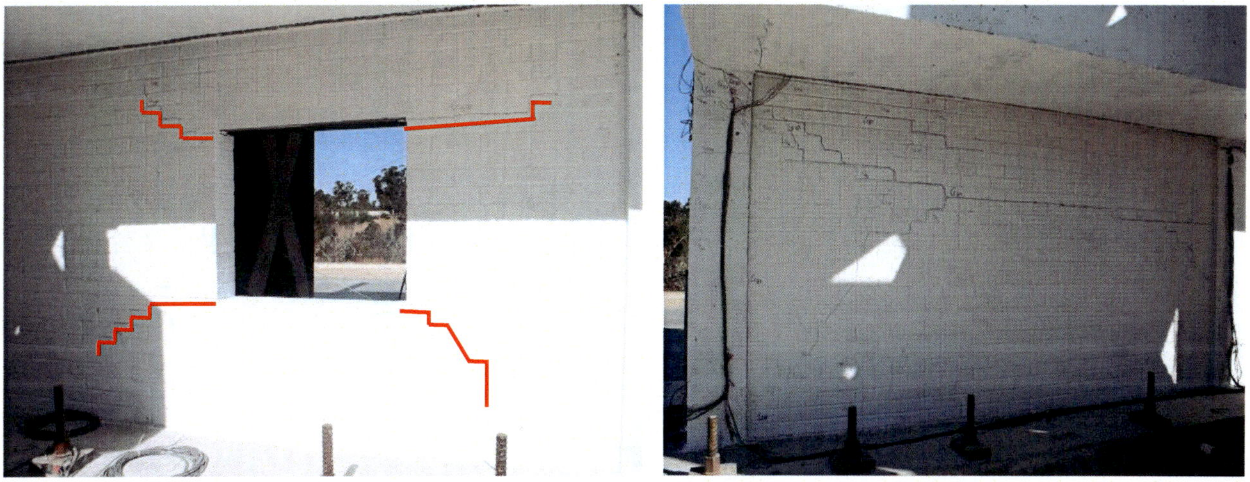

(a) (b)

Fig. 8 Crack pattern at damage states S3 (a) and S6 (b)

Fig. 9 Crack pattern at damage states S6 in the solid infill and the middle column (left), and the infill with opening and the east column (right)

Conclusions

A 2/3-scale, three-story, two-bay, infilled RC frame was tested on the UCSD-NEES shake-table in November 2008. The test objective was to assess the seismic performance of infilled frames and provide data for the evaluation of newly developed analytical methods. The shake-table tests were designed to induce damage on the structure progressively through scaled earthquake records. At various levels of damage, low-amplitude ambient vibration and white-noise base excitations were applied to the frame which responded as a quasi-linear system with parameters evolving as a function of structural damage. The white-noise base excitation tests were used for the system identification of the infilled frame structure in this study. The deterministic-stochastic subspace identification method, an input-output system identification method, has been applied to estimate modal parameters of the frame at various damage states.

The results of this system identification study indicate that: (1) the identified natural frequencies decrease consistently with increasing level of damage indicating a reduction of the stiffness, while the identified damping ratios increase; (2) the reduction in the natural frequencies of the two longitudinal modes with damage is much more significant than that for the 2-T and 2-L-T modes; similar observation can be made on the level of increase in the identified damping ratios; (3) the MAC values that relate the identified mode shapes at each damage state to those of the undamaged state decrease with increasing damage. The modal parameters identified at various damage states in this study provide the input for damage identification of this infilled frame using a sensitivity-based finite element model updating strategy which is the topic of a companion paper [13].

Acknowledgements

The shake-table tests discussed in this paper were supported by the National Science Foundation Grant No. 0530709 awarded under the George E. Brown, Jr. Network for Earthquake Engineering Simulation (NEES) program. Input from other collaborators at Stanford University and the University of Colorado at Boulder, and a Professional Advisory Panel (PAP) throughout this study is gratefully acknowledged. The panel members are David Breiholz, John Kariotis, Gregory Kingsley, Joe Maffei, Ron Mayes, Paul Murray, and Michael Valley. Finally, the authors would like to thank the technical staff at the Englekirk Structural Engineering Center of UCSD and Ioannis Koutromanos for their assistance in the shake-table tests. However, the opinions expressed in this paper are those of the authors and do not necessarily represent those of the NSF, the collaborators, or PAP.

References

[1] Doebling, S.W., Farrar, C.R., Prime, M.B., and Shevitz, D.W. *Damage identification in structures and mechanical systems based on changes in their vibration characteristics: a detailed literature survey.* Los Alamos National Laboratory Report, LA-13070-MS, Los Alamos, New Mexico, USA, (1996).

[2] Doebling, S.W., Farrar, C.R., and Prime, M.B. "A summary review of vibration-based damage identification methods." *The Shock and Vibration Digest,* 30(2), 99-105, (1998).

[3] Sohn, H., Farrar, C.R., Hemez, F.M., Shunk, D.D., Stinemates, D.W., and Nadler, B.R. *A review of structural health monitoring literature: 1996-2001.* Los Alamos National Laboratory Report, LA-13976-MS, Los Alamos, New Mexico, USA, (2003).

[4] Moaveni, B., He, X., Conte, J.P., and Restrepo, J.I. "System identification study of a seven-story full-scale building slice tested on the UCSD-NEES shake table." *Journal of Structural Engineering, ASCE*, under review, (2009).

[5] Stavridis, A. and Shing P.B. "A study on masonry infilled non-ductile RC frames." *2nd NEES/E-Defense Workshop*, Miki, Japan, (2006).

[6] Harris H.G., Sabnis G.M. "Structural modeling and experimental techniques." *CRC Press*, Boca Raton, Florida, (1999).

[7] Stavridis, A. *Analytical and experimental seismic performance assessment of masonry-infilled RC frames.* Ph.D. Dissertation, Department of Structural Engineering, University of California, San Diego, CA, (2009).

[8] He, X., Moaveni, B., Conte, J.P., Elgamal A., and Masri, S.F. "System identification of Alfred Zampa Memorial Bridge using dynamic field test data." *Journal of Structural Engineering, ASCE,* 135(1), 54-66, (2009).

[9] Ozcelik, O. *A Mechanics-based virtual model of NEES-UCSD shake table: theoretical development and experimental validation.* Ph.D. Dissertation, Department of Structural Engineering, University of California, San Diego, CA, (2008).

[10] Van Overschee, P., and de Moore, B. *Subspace identification for linear systems*. Kluwer Academic Publishers, Massachusetts, USA, (1996).

[11] Veletsos, A.S., and Ventura, C.E. "Modal analysis of non-classically damped linear systems." *Earthquake Engineering and Structural Dynamics*, 14(2), 217-243, (1986).

[12] Allemang, R.J., and Brown, D.L. "A correlation coefficient for modal vector analysis." *Proc. of 1st International Modal Analysis Conference*, Bethel, Connecticut, (1982).

[13] Moaveni, B., Lombaert, G., Stavridis, A., Conte, J.P., and Shing, P.B. "Damage identification of a three-story infilled RC frame tested on the UCSD-NEES shake table." *Proc. of 28th International Modal Analysis Conference*, Jacksonville, FL, (2010).

Proceedings of the IMAC-XXVIII
February 1–4, 2010, Jacksonville, Florida USA
©2010 Society for Experimental Mechanics Inc.

Damage Identification of a Three-Story Infilled RC Frame Tested on the UCSD-NEES Shake Table

Babak Moaveni[1], Geert Lombaert[2], Andreas Stavridis[3], Joel P. Conte[3], and P. Benson Shing[3]

[1]Department of Civil & Environmental Engineering, Tufts University, Medford, USA
[2]Department of Civil Engineering, Katholieke Universiteit Leuven, Leuven, Belgium
[3]Department of Structural Engineering, University of California, San Diego, USA

ABSTRACT

Reinforced concrete (RC) frames with masonry infill walls can be frequently found in areas of high seismic risk around the world. Such structures are often designed with older building codes and may experience catastrophic failures during earthquakes. A 2/3-scale, three-story, two-bay, infilled RC frame was tested on the UCSD-NEES shake table to investigate the seismic performance of this type of construction. The frame was designed according to the building practice in California in the 1920s. The shake table tests were designed so as to induce damage in the structure progressively through scaled earthquake records. At various levels of damage, low-amplitude white-noise base excitations were applied to the infilled RC frame, which responded as a quasi-linear system with modal parameters depending on the extent of the structural damage. In this study, the modal parameters identified from the white-noise test data acquired at various levels of damage are used to detect the existing damage. A sensitivity-based finite element model updating strategy is employed to detect, locate, and quantify damage at each damage state considered. This paper presents the results of the damage identification study, which shows that the method can accurately identify the location and severity of damage observed in the tests.

Introduction

In recent years, vibration-based structural health monitoring has attracted increasing attention and is of growing importance in the civil engineering research community. Vibration-based, non-destructive, damage identification uses changes in dynamic characteristics (e.g., modal parameters) of the structure to identify damage. Extensive literature reviews have been provided by Doebling et al. [1, 2] and Sohn et al. [3] on this subject. Sensitivity-based finite element (FE) model updating methods have been applied successfully for condition assessment of structures [4, 5]. These methods update the physical parameters of a FE model of the structure as damage evolves by minimizing an objective function that measures the discrepancy between FE predicted and experimentally identified structural dynamic properties, such as natural frequencies and mode shapes. Optimal solutions of the problem are reached through sensitivity-based constrained optimization algorithms.

A 2/3-scale infilled RC frame, designed according to the engineering practice in California in the 1920s, was tested on the UCSD-NEES shake table in November 2008 to assess the seismic performance of such older buildings as part of a collaboratory project involving researchers from Stanford University, the University of Colorado at Boulder, and the University of California at San Diego. This two-bay, three-story frame was the largest structure of this type ever tested on a shake table. The shake table tests were designed so as to induce damage in the building progressively through scaled earthquake records. Between the earthquake tests, low-amplitude white-noise base excitation tests were applied to the infilled RC frame, which responded as a quasi-linear system with modal parameters evolving as a function of damage. The deterministic-stochastic subspace identification (DSI) method, based on system input and output signals, was used to estimate the modal parameters (natural frequencies, damping ratios, and mode shapes) of the undamaged structure and of the structure in its various damage states [6]. In this study, a FE model updating strategy is applied for damage identification of the structure after it was subjected to earthquake excitations of increasing amplitude. The objective function used here for damage identification is defined as a combination of the discrepancies between

DOI 10.1007/978-1-4419-9831-6_17, © The Society for Experimental Mechanics, Inc. 2011

the FE predicted and experimentally identified natural frequencies and mode shapes. The damage identification results are compared to the damage observed in the physical specimen.

Test Specimen, Test Setup, and Dynamic Experiments

Three-Story Infilled RC Frame

The three-story specimen shown in Figure 1 was tested on the outdoor shake table at the University of California, San Diego in November 2008. The prototype structure was designed by Stavridis and Shing [7] to represent structures built in California in the 1920s era. The specimen corresponds to the external frame of the prototype structure with all its dimensions scaled by 2/3, while the infill walls had two wythes of brick units. The scaling factors were determined according to the similitude requirement and some limitations of the test setup [8]. Time and acceleration scale factors of 0.54 and 2.26, respectively, were applied to the ground motions to satisfy the similitude requirements of the experiments [9]. The ground motion levels mentioned hereafter are with respect to the full-scale prototype structure. As shown in Figure 1, two steel towers were secured on the shake table on the north and south side of the structure to protect the shake table from a potential collapse of the structure during severe shaking. However, they did not interact with the structure during the tests as they were placed with a 0.75 inch gap from the specimen. More details about the test structure can be found in [8] and in a companion paper on system identification of the test specimen [6].

Fig. 1 Front view (left) and side view of the specimen (right)

Instrumentation Layout

The specimen and the steel towers were instrumented with an extensive array of sensors. They included 135 strain gages, 71 string pots and LVDTs, and 59 uniaxial accelerometers. The accelerometers were used to measure the accelerations along directions X, Y and Z, with X being the direction of base excitation (longitudinal), Y the transverse (out-of-plane) direction, and Z the vertical direction. The measured response data from 9 longitudinal (3 per floor), 9 vertical (3 per floor) and 6 transversal (2 per floor) acceleration channels were used to identify the modal parameters of the test structure [6]. The locations of the accelerometers are shown in Figure 2.

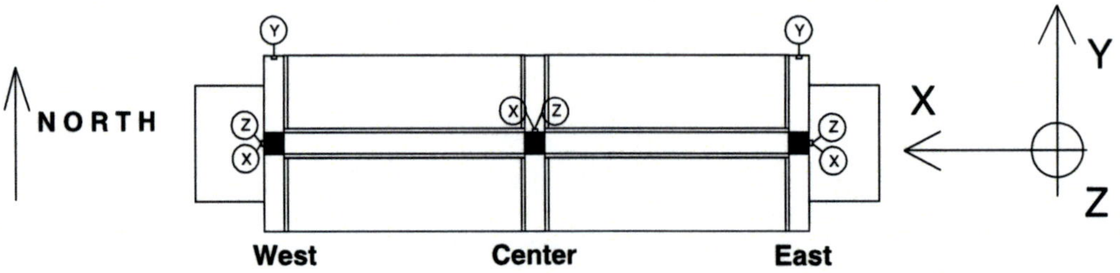

Fig. 2 Locations and directions of accelerometers used in this study on each floor level

Dynamic Tests Performed

A sequence of 44 dynamic tests was applied to the test structure including ambient vibration tests, free vibration tests, and forced vibration tests (white noise and seismic base excitations) using the UCSD-NEES shake table. The testing sequence incorporated earthquake ground motions of increasing amplitude to gradually damage the structure. Between the earthquake records, low amplitude white noise base excitation tests were performed and the modal parameters of the structure were identified at various damage states [6]. The ground motion accelerograms were obtained by compressing the time and scaling in amplitude the acceleration time history recorded at the Gilroy 3 station during the 1989 Loma Prieta earthquake. For structures with a fundamental frequency close to that of the infilled frame studied here, the Gilroy 3 motion scaled at 67% corresponds to a Design Basis Earthquake (DBE) for Seismic Design Category (SDC) D. Moreover, the unscaled Gilroy 3 motion corresponds to a Maximum Considered Earthquake (MCE). From the recorded data, damage was identified at seven damage states of the structure (S0 and S2-S7). Damage state S0 is defined as the undamaged (baseline) state of the structure before its exposure to the first seismic excitation (EQ1), while damage states S1 to S7 correspond to the state of the structure after exposure to the earthquakes of increasing amplitude. Table 1 summarizes the dynamic tests performed and the corresponding damage states of the test structure.

Table 1. Dynamic tests used in this study (WN: white noise base excitation; EQ: earthquake base excitation)

Test No.	Test Date	Test Description	Damage State
5	11/3/2008	0.03g RMS WN, 5 min	S0
8	"	20% Gilroy EQ	
9	"	0.03g RMS WN, 5 min	S1
12	11/6/2008	40% Gilroy EQ	
13	"	0.03g RMS WN, 5 min	S2
21	11/10/2008	67% Gilroy EQ (DBE)	
25	11/12/2008	0.04g RMS WN, 5 min	S3
26	"	67% Gilroy EQ (DBE)	
27	"	0.04g RMS WN, 5 min	S4
28	"	83% Gilroy EQ	
29	"	0.04g RMS WN, 5 min	S5
33	11/13/2008	91% Gilroy EQ	
35	"	100% Gilroy EQ (MCE)	
36	"	0.04g RMS WN, 5 min	S6
40	11/18/2008	120% Gilroy EQ	
41	"	0.04g RMS WN, 5 min	S7

System Identification Results

The modal parameters of the specimen at all the damage states considered in this study were identified using the deterministic-stochastic subspace identification (DSI) method, based on the input-output data collected from low amplitude (0.03g and 0.04g RMS) white noise base excitation tests performed at various damage states (S0, S1, S2, S3, S4, S5, S6, and S7) [6]. The natural frequencies and mode shapes of the first two longitudinal vibration modes (1-L, 2-L) were used in the FE model updating procedure. Figure 3 shows the real part of the longitudinal mode shapes of the test structure identified based on data from Test 5 (damage state S0). The natural frequencies and damping ratios of these modes are reported in Table 2 for the eight damage states considered. Modal Assurance Criterion (MAC) values [10] were also computed to compare the mode shapes identified at each damage state with the corresponding mode shapes identified for the undamaged structure at state S0, which serves as a baseline condition of the structure. From Table 2, it can be observed that the identified natural frequencies and the MAC values between the identified mode shapes and their counterpart identified at S0 decrease as the level of damage increases. The high MAC values (close to one) in damage states S1 to S4 indicate that there is little change in the identified mode shapes between these damage states.

Table 2. Modal parameters of the infilled frame identified at different damage states

Damage State:		S0	S1	S2	S3	S4	S5	S6	S7
Mode 1-L	Frequency [Hz]	18.18	18.11	17.99	16.74	15.93	14.78	8.47	5.34
	Damping [%]	2.0	2.4	1.9	3.3	3.8	6.1	15.7	15.6
	MAC	1.00	1.00	1.00	1.00	1.00	0.98	0.80	0.71
Mode 2-L	Frequency [Hz]	41.22	41.09	41.56	40.21	38.56	35.50	27.34	22.57
	Damping [%]	1.1	1.0	1.0	1.4	3.0	4.4	4.8	4.2
	MAC	1.00	1.00	1.00	0.99	0.96	0.92	0.74	0.67

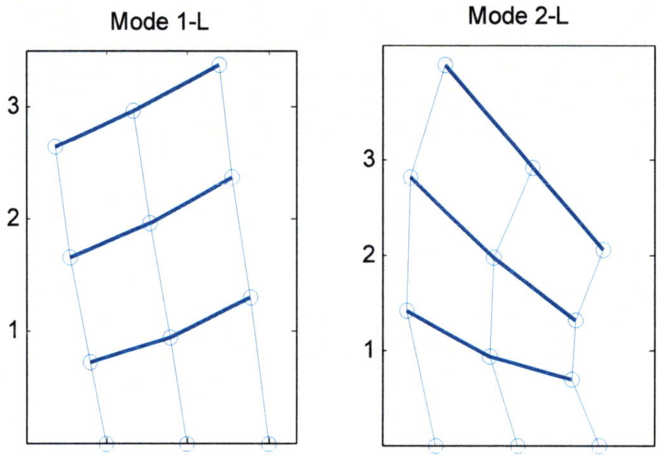

Fig. 3 First two longitudinal mode shapes of the infilled frame at damage state S0

Finite Element Model Updating for Damage Identification

In this study, a sensitivity-based FE model updating strategy is used to identify the damage in the structure after different levels of earthquake base excitation. The residuals used in the updating procedure are based on the identified natural frequencies and mode shapes for the first two longitudinal modes of the specimen. Damage is defined as a relative change in material stiffness (effective modulus of elasticity) of the finite elements in the different substructures of the FE model used for damage identification. For the purpose of damage identification, the effective moduli of elasticity of elements in the various substructures (each assumed to have a uniform value of the effective modulus of elasticity) are updated at each considered damage state of the structure through constrained minimization of an objective function.

Finite Element Model of Test Structure in FEDEASLab

A linear elastic FE model of the structure is developed using a general-purpose structural analysis program, FEDEASLab [11]. This FE model is defined by 35 nodes and 54 linear elastic shell and frame elements, as shown in Figure 4. A four-node linear elastic flat shell element (with four Gauss integration points) available in the literature [12, 13] and implemented in FEDEASLab by He [14] is used to model the infill walls. The members (beams and columns) of the RC frame are modeled with Bernoulli-Euler frame elements. The distributed mass properties of the structure are discretized into lumped translational masses applied at each node of the FE model. The initial FE model of the structure is based on the geometry of the physical specimen and the measured material properties. Figure 4 shows the FE model of the test structure in FEDEASLab with the six infill substructures used in model updating.

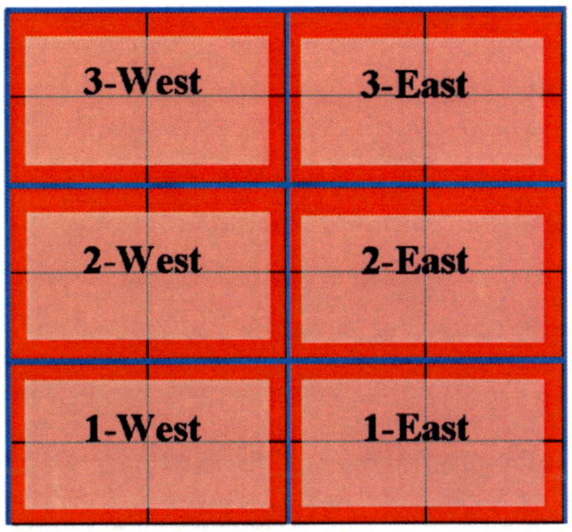

Fig. 4 FE model of the RC frame structure with the six infill substructures used in model updating

Objective Function

The objective function used for damage identification is defined as

$$f(\boldsymbol{\theta}) = \mathbf{r}(\boldsymbol{\theta})^T \mathbf{W}\, \mathbf{r}(\boldsymbol{\theta}) + (\mathbf{a}(\boldsymbol{\theta}) - \mathbf{a}^0)^T \mathbf{W}^a\, (\mathbf{a}(\boldsymbol{\theta}) - \mathbf{a}^0) = \sum_j [w_j r_j(\boldsymbol{\theta})^2] + \sum_k [w_k^a (a_k(\boldsymbol{\theta}) - a_k^0)^2] \quad (1)$$

where $\boldsymbol{\theta}$ = set of physical parameters (i.e., effective moduli of elasticity in the finite elements of the various substructures) which must be adjusted in order to minimize the objective function; $\mathbf{r}(\boldsymbol{\theta})$ = residual vector containing the differences between the FE computed and experimentally identified modal parameters; $\mathbf{a}(\boldsymbol{\theta})$ = vector of dimensionless damage factors representing the level of damage in each of the substructures of the FE model used for damage identification; $\mathbf{a}^0$ = vector of initial damage factors used as starting point in the optimization process. At each considered state of the building, $\mathbf{a}^0$ is selected as the vector of the identified damage factors at the previous state and $\mathbf{a}^0 = \mathbf{0}$ for the first damage state considered in the model updating procedure (which is S2). In Equation (1), $\mathbf{W}$ is a diagonal weighting matrix for modal residuals. Weight factors assigned to mode shape residuals are equal to the weight factor for that mode's natural frequency normalized/divided by the number of mode shape residuals for each mode. Weights of 1 and 0.5 are assigned to the modal residuals corresponding to the natural frequencies of the first and second longitudinal modes, respectively (i.e., $w_1 = 1$, $w_2 = 0.5$). $\mathbf{W}^a$ is a weighting matrix for damage factors, a diagonal matrix with each diagonal coefficient determining the relative cost (or penalty) associated with the change of the corresponding damage factor. The weights for damage factors are used for regularization and can reduce the estimation error of the damage factors in the presence of estimation uncertainty in the modal parameters, especially for the substructures with updating parameters to which the employed residuals are less sensitive. In this study, these weights are set to $w_k^a = 0.02 \times w_1 = 0.02$, with $k = 1, \ldots, n_{sub}$ where n_{sub} denotes the number of substructures used in the FE model updating process. A combination of residuals in natural frequencies and mode shape components is used in the objective function, i.e.,

$$\mathbf{r}(\boldsymbol{\theta}) = \left[\mathbf{r}_f^T(\boldsymbol{\theta}) \quad \mathbf{r}_s^T(\boldsymbol{\theta})\right]^T \quad (2)$$

in which $\mathbf{r}_f(\boldsymbol{\theta})$ and $\mathbf{r}_s(\boldsymbol{\theta})$ represent the eigen-frequency and mode shape residuals, respectively, defined as

$$\mathbf{r}_f(\boldsymbol{\theta}) = \left[\frac{\lambda_j(\boldsymbol{\theta}) - \tilde{\lambda}_j}{\tilde{\lambda}_j}\right], \quad \mathbf{r}_s(\boldsymbol{\theta}) = \left[\frac{\phi_j^l(\boldsymbol{\theta})}{\phi_j^r(\boldsymbol{\theta})} - \frac{\tilde{\phi}_j^l}{\tilde{\phi}_j^r}\right] \qquad (l \neq r),\ j \in \{1 \quad 2 \quad \cdots \quad N_m\} \quad (3)$$

where $\lambda_j(\boldsymbol{\theta})$ and $\tilde{\lambda}_j$ denote the FE predicted and experimentally identified eigenvalues for the jth vibration mode, i.e., $\lambda_j = (2\pi \cdot f_j)^2$, in which f_j is the natural frequency; $\boldsymbol{\phi}_j(\boldsymbol{\theta})$ and $\boldsymbol{\phi}_j$ denote the FE predicted and experimentally identified mode shape vectors, respectively. It should be noted that for each vibration mode, the mode shapes $\boldsymbol{\phi}_j(\boldsymbol{\theta})$ and $\tilde{\boldsymbol{\phi}}_j$ are normalized in a consistent way, i.e., scaled with respect to the same reference component. In

Eq. (3), the superscript r indicates the reference component of a mode shape vector (with respect to which the other components of the mode shape are normalized), the superscript l refers to the mode shape components that are used in the FE model updating process (i.e., at the locations and in the directions of the sensors), and N_m denotes the number of vibration modes considered in the damage identification process. In this study, the natural frequencies and mode shapes of the first two longitudinal vibration modes of the structure are used to form the residual vector $\mathbf{r}(\boldsymbol{\theta})$, which has a total of 34 residual components (based on 17 channels of acceleration response measurements) consisting of 2 natural frequencies and $2\times(17-1)=32$ mode shape component residuals, respectively. One mode shape component cannot be used since it is normalized to one.

In the process of FE model updating, the material stiffness (i.e., effective modulus of elasticity) of each of the damaged substructures is used as an updating parameter in the FE model of the building. Instead of using directly the absolute value of each updating parameter, a dimensionless damage factor is defined as

$$a_k^{Si} = \frac{E_k^{S0} - E_k^{Si}}{E_k^{S0}} \tag{4}$$

where E_k^{Si} is the effective modulus of elasticity of all finite elements in substructure k at damage state Si. Thus, the damage factor a_k^{Si} indicates directly the level of damage (i.e., relative change in effective modulus of elasticity) in substructure k when FE model updating is used for structural damage identification.

Optimization Algorithm

The optimization algorithm used to minimize the objective function defined in Eq. (1) is a standard Trust Region Newton method [15], which is a sensitivity-based iterative method available in the MATLAB optimization Toolbox. In this study, the damage factors were constrained to remain in the selected range [0% - 99%] at all states considered. The optimization process was performed using the "fmincon" MATLAB function, with the Jacobian matrix and a first-order estimate of the Hessian matrix calculated based on the analytical sensitivities of the modal residuals to the updating variables. The use of the analytical Jacobian, rather than the Jacobian estimated through finite difference calculations, increases significantly the efficiency of the computational minimization of the objective function.

Damage Identification Results

In this study, the FE model updating algorithm outlined above was implemented to identify the damage of the test structure. The residuals, $\mathbf{r}(\boldsymbol{\theta})$, used in the objective function were formed using the natural frequencies and mode shapes of the first two longitudinal vibration modes. The first step in the damage identification process consists of deriving a reference/baseline FE model of the undamaged structure (at state S0). In this step, the initial FE model is updated to match as closely as possible the identified modal parameters at the undamaged state of the structure by updating the stiffness (effective moduli of elasticity) of twelve substructures. The twelve substructures used in the calibration of the initial FE model to the reference FE model include six substructures for the infill walls (two per story as shown in Figure 4), and six substructures for the columns (two per story, one for the two outside columns and one for the inside column).The effective modulus of elasticity is assumed to be constant over each substructure.

The effective moduli of elasticity of the different substructures of the initial and the baseline model are reported in Table 3. The values for the initial model correspond to the measured values of the moduli of elasticity of concrete and masonry infill. These were obtained from uniaxial compression tests of concrete cylinders and masonry prisms, respectively, performed on November 10th (See Table 1). The masonry prism test was selected, since this is the closest representation of the masonry assembly. Moreover, a simplifying assumption is made in that the masonry infill is modeled as a homogeneous isotropic material. During the calibration of the initial FE model to the reference FE model, the "damage factors" were constrained in the range [-2, 0.9]. Once the reference model is obtained, the moduli of elasticity of the six infill walls shown in Figure 4 are updated from the reference FE model (at the undamaged/baseline state S0) to states S2, S3, S4, S5, S6 and S7. The stiffness parameters of the elements representing the RC frame members are kept fixed at their values in the reference FE model. This assumes that the concrete frame remains elastic, which is a good approximation for the early damage states based on the inspection of the physical specimen. However, at damage states S5 to S7, this assumption is not accurate. Due to the very small change in the modal parameters from S0 to S1, FE model updating was not performed at state S1.

Table 3. Effective moduli of elasticity of structural components at different substructures for initial and reference FE models

Substructure	Effective Moduli of Elasticity [ksi]	
	Initial FE model	*Reference FE model (S0)*
Infill 1-West	785	687
Infill 1-East	785	761
Infill 2-West	777	936
Infill 2-East	777	816
Infill 3-West	946	963
Infill 3-East	946	908
Column 1-Outside	2380	2424
Column 1-Intside	2380	2389
Column 2-Outside	2530	1987
Column 2-Intside	2530	2539
Column 3-Outside	2463	2251
Column 3-Intside	2463	2465

In updating the FE model at each considered damage state of the infilled frame, the kth dimensionless damage factor was constrained to be in the range $[\mathbf{a}_{Si}^{0} - 0.1, \quad 0.95]$. At each damage state considered, the vector of initial damage factors $\mathbf{a}_{Si}^{0}$, used as starting point in the optimization process, was selected as the damage factors identified at the previous damage state or zero for state S2. The damage factors (relative to the reference FE model or reference state) obtained at different damage states are presented in a bar plot in Figure 5. These results indicate that the severity of structural damage increases as the structure is exposed to stronger earthquake excitations; the extent of damage decreases rapidly along the height of the structure (damage is concentrated in the bottom story) and it is uniform at each story (East and West side infills have similar damage factors). Table 4 presents the natural frequencies computed from the updated FE model at each state considered together with their counterparts identified from white noise base excitation test data as well as the MAC values between FE predicted and experimentally identified mode shapes. To compare the FE predicted and experimental mode shapes, the former were truncated to include only the degrees of freedom corresponding to the locations and directions of the accelerometers. From Table 4, it is observed that the FE predicted natural frequencies and mode shapes for the first two longitudinal modes match well their experimentally identified counterparts. Moreover, the MAC values between FE predicted and experimentally identified mode shapes are very close to unity at all damage states. MAC values are in general higher for the first mode than for the second mode. This can be expected since smaller weight factors are assigned to the residuals corresponding to the second mode shape.

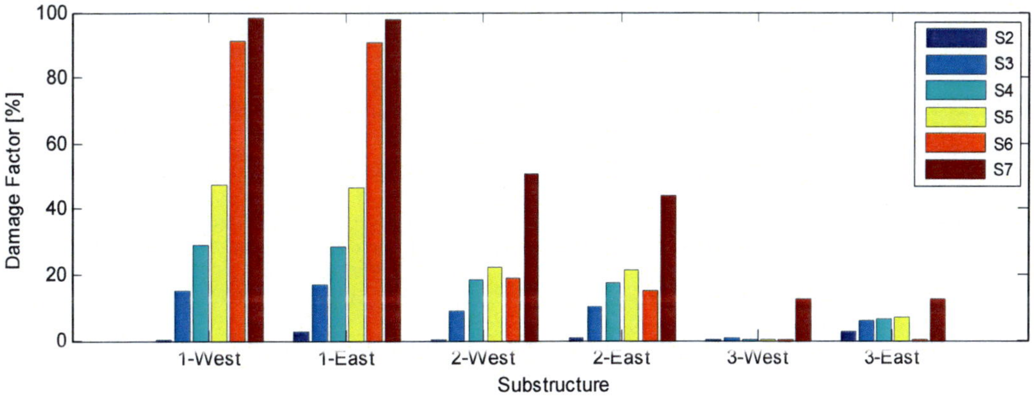

Fig. 5 Identified damage factors at various substructures

Table 4. Comparison of FE computed and experimentally identified modal parameters

Damage State	Experimentally Identified Natural Frequencies [Hz]		FE Computed Natural Frequencies [Hz]		MAC	
	1-L mode	2-L mode	1-L mode	2-L mode	1-L mode	2-L mode
S0 (baseline)	18.18	41.22	17.86	42.59	0.99	0.96
S2	17.99	41.56	17.80	42.41	1.00	0.95
S3	16.74	40.21	16.92	40.25	0.99	0.97
S4	15.93	38.56	16.09	38.29	1.00	0.99
S5	14.78	35.5	14.86	35.37	1.00	0.99
S6	8.47	27.34	8.50	26.40	1.00	0.99
S7	5.34	22.57	5.34	23.48	0.99	0.99

Comparison of Damage Identification Results with Observed Damage

The structure practically remained elastic after the low level earthquakes (20% and 40% Gilroy) and the first noticeable deterioration of the structure occurred during the 67% of Gilroy, which corresponds to a Design Basis Earthquake (DBE) for this structure. At damage state S3, inspection of the physical specimen revealed the first cracks in the structure in the infill panel with the window at the first story, shown in Figure 6(a). However, the cracks were insignificant for the structural integrity of the specimen and did not alter the structural properties significantly. This observation is consistent with the damage identification results according to which no significant damage is identified at damage state S2, while 15-17% loss of stiffness is identified in the first story infill walls at damage state S3 (see Figure 5). Significant change in the structural properties is noted at damage state S6, which occurred after subjecting the structure to a Maximum Considered Earthquake (MCE). At this stage, significant cracks developed in both bays of the first floor and affected the columns which also cracked as illustrated in Figure 6(b). These cracks caused a natural frequency reduction of more than 50%, which translated to a 90% reduction of the stiffness at the first story infills (see Figure 5). The damage in the infill is exaggerated in the finite element model since it represents any damage occurring to non-updating components of RC frame such as the columns. Significant damage was induced in the specimen by the 120% of Gilroy, which included the first dominant shear cracks that developed in the middle column. These cracks further reduced the first mode frequency to less than one third of its initial value. Figure 7 shows the observed damage at this damage state. At this damage state, damage factors of 98% were identified for the two infills at the first story and 44-50% at the second story. These estimates of damage are very high and are not consistent with the experimental evidence. However, it should be noted that any damage to non-updating components of the structure (e.g., columns) will be reflected in the effective moduli of elasticity of infills.

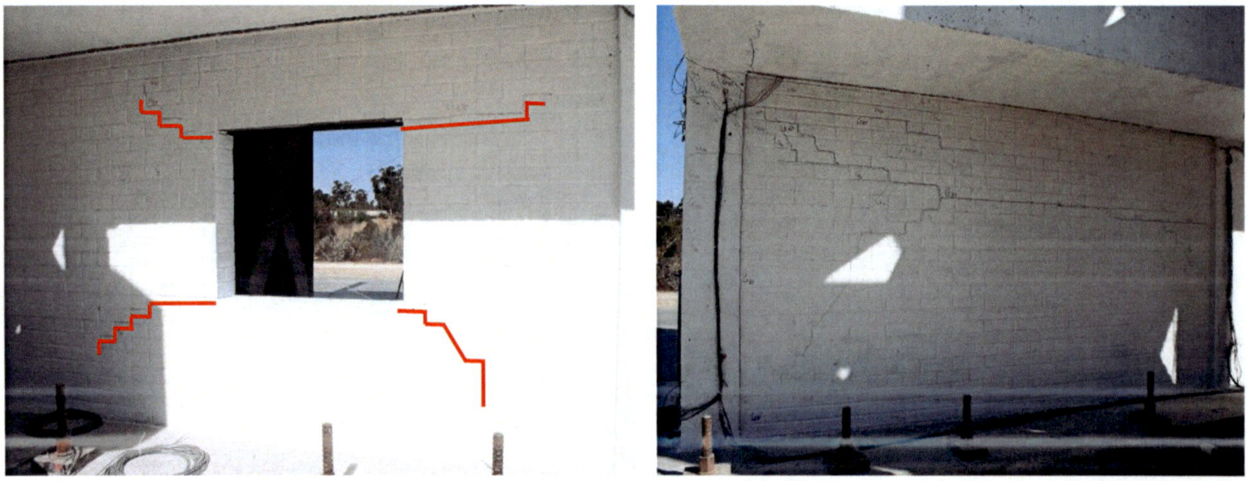

(a) (b)

Fig. 6 Crack pattern at damage states S3 (a) and S6 (b)

Fig. 7 Crack pattern at damage states S6 in the solid infill and the middle column (left), and the infill with opening and the east column (right)

Conclusions

In this study, a FE model updating strategy is applied for vibration-based damage identification of a 2/3-scale, three-story, two-bay, infilled RC frame tested on the UCSD-NEES shake table. The objective function for damage identification is defined as a combination of natural frequency and mode shape residuals measuring the discrepancy between the numerically predicted (using a FE model) and experimentally identified modal parameters. The FE model of the structure is first calibrated through the model updating technique for the reference/baseline (undamaged) state, and then damage identification is performed at six damage states of the structure. These states correspond to the state of the physical specimen after being subjected to earthquake excitations of increasing intensity.

The obtained damage identification results are consistent with the actual damage observed in the structure, which showed a concentration of damage at the bottom story. The analytical modal parameters obtained from the updated FE models are in good agreement with their experimentally identified counterparts, which indicates the accuracy of the updated models. Finally, it should be noted that the success of vibration-based damage identification depends significantly on the accuracy and completeness of the identified modal parameters. Clearly, if estimation uncertainty of the modal parameters is larger than their changes due to damage, it is impossible to resolve/identify the actual damage in the structure.

Acknowledgements

The shake table tests discussed in this paper were supported by the National Science Foundation Grant No. 0530709 awarded under the George E. Brown, Jr. Network for Earthquake Engineering Simulation (NEES) program. Input from other collaborators at Stanford University and the University of Colorado at Boulder, and a Professional Advisory Panel (PAP) throughout this study is gratefully acknowledged. The panel members are David Breiholz, John Kariotis, Gregory Kingsley, Joe Maffei, Ron Mayes, Paul Murray, and Michael Valley. Finally, the writers would like to thank the technical staff at the Englekirk Structural Engineering Center of UCSD and Ioannis Koutromanos for their assistance in the shake table tests. However, the opinions expressed in this paper are those of the authors and do not necessarily represent those of the NSF, the collaborators, or PAP.

References

[1] Doebling, S.W., Farrar, C.R., Prime, M.B., and Shevitz, D.W. *Damage identification in structures and mechanical systems based on changes in their vibration characteristics: a detailed literature survey.* Los Alamos National Laboratory Report, LA-13070-MS, Los Alamos, New Mexico, USA, (1996).

[2] Doebling, S.W., Farrar, C.R., and Prime, M.B. "A summary review of vibration-based damage identification methods." *The Shock and Vibration Digest,* 30(2), 99-105, (1998).

[3] Sohn, H., Farrar, C.R., Hemez, F.M., Shunk, D.D., Stinemates, D.W., and Nadler, B.R. *A review of structural health monitoring literature: 1996-2001.* Los Alamos National Laboratory Report, LA-13976-MS, Los Alamos, New Mexico, USA, (2003).

[4] Teughels, A., De Roeck, G. "Structural damage identification of the highway bridge Z24 by finite element model updating." *Journal of Sound and Vibration,* 278(3), 589–610, (2004).

[5] Moaveni, B., He, X., Conte, J.P., and Restrepo, J.I. "Damage identification study of a seven-story full-scale building slice tested on the UCSD-NEES shake table." *Structural Safety*, under review, (2009).

[6] Moaveni, B., Stavridis, A., and Shing, P.B. "System identification of a three-story infilled RC frame tested on the UCSD-NEES shake table." *Proc. of 28th International Conference on Modal Analysis (IMAC-XXVIII)*, Jacksonville, Florida, USA, (2010).

[7] Stavridis, A. and Shing P.B. "A study on masonry infilled non-ductile RC frames." *2nd NEES/E-Defense Workshop*, Miki, Japan, (2006).

[8] Stavridis, A. *Analytical and experimental seismic performance assessment of masonry-infilled RC frames.* Ph.D. Dissertation, Department of Structural Engineering, University of California, San Diego, CA, (2009).

[9] Harris H.G., and Sabnis G.M. "Structural modeling and experimental techniques." *CRC Press*, Boca Raton, Florida, USA, (1999).

[10] Allemang, R. J., and Brown, D. L. "A correlation coefficient for modal vector analysis." *Proc. of 1st International Modal Analysis Conference*, Bethel, Connecticut, (1982).

[11] Filippou, F.C., and Constantinides, M. *FEDEASLab getting started guide and simulation examples.* Technical Report, NEESgrid-2004-22, http://fedeaslab.berkeley.edu, (2004).

[12] Allman, D.J. "A quadrilateral finite element including vertex rotations for plane elasticity analysis." *International Journal for Numerical Methods in Engineering*, 26(3), 717-730, (1988).

[13] Batoz, J.L., and Tahar, M.B. "Evaluation of a new quadrilateral thin plate bending element." *International Journal for Numerical Methods in Engineering*, 18(11), 1655-1677, (1982).

[14] He, X. *Vibration-based damage identification and health monitoring of civil structures.* Ph.D. thesis, Department of Structural Engineering, University of California, San Diego, (2008).

[15] Coleman, T.F., and Li, Y. "An interior, trust region approach for nonlinear minimization subject to bounds." *SIAM Journal on Optimization*, 6(2), 418-445, (1996).

Proceedings of the IMAC-XXVIII
February 1–4, 2010, Jacksonville, Florida USA

Defect Detection in Concrete Members

Sezer Atamturktur[1]
Clemson University, Civil Engineering Department, Clemson, South Carolina 29634

A rock pocket is a deficient volume within hardened concrete consisting of coarse aggregate and voids that reduce the overall stiffness of the concrete members. The leakage of wet concrete from the form, segregation, or insufficient consolidation during concrete placement may leave rock pockets in concrete construction. This study is concerned with the detection, location and quantification of internal defects, particularly rock pockets, in reinforced concrete members. This is achieved by coupling in situ vibration testing with finite element analysis through Bayesian inference. First, the importance of providing sufficient physical evidence while calibrating the finite element models is illustrated using simulated experiments. With simulated experiments, model calibration successfully detected not only the locations but also the severity of rock pocket defects. Then, the results of impact hammer tests, completed on a concrete beam defected with rock pockets are presented. The finite element model of the test beam is segmented, and the stiffness properties of these segments are independently calibrated with the help of Bayesian calibration techniques using varying amounts of experimental information. The success in detecting defects obtained using simulated experiments, was not observed when the procedure is applied to the scaled concrete beams tested under laboratory conditions. However, the cause(s) of the poor performance with real experiments can be attributed to several factors, each of which requires further evaluation. (Publication approved for unlimited, public release on November-4-2009, Unclassified.)

1. Introduction:

A rock pocket is a local, internal defect, which may be left in concrete members when the mortar paste fails to fill the space around the coarse aggregates. Therefore, the presence of these internal defects typically reduces the load capacity of a concrete member, the severity of this reduction being proportional to the dimensions of the defect. Depending on the size and location of internal defects as well as construction factors (i.e., water-to-cementitious materials ratio, curing, environmental conditions, etc.), degrading effects of these internal defects may either be evident immediately or present themselves after the structural element is in service. For instance, in pre-stressed precast beams, such internal defects, if extensive or if located near the anchorage can cause compression failure during the stressing of tendons. Also, the presence of such inherent construction defects tends to accelerate the degrading effects of aging, operational and environmental conditions.

---

[1] Assistant Professor, Civil Engineering Department. Mailing: Clemson University, Civil Engineering Department, 110 Lowry Hall, Box 340911, Clemson, South Carolina 29634-0911, U.S.A. Phone: 864-656-3000. Fax: 864-656-2670. E-mail: sez@clemson.edu. Currently, long-term visiting faculty staff member at the Los Alamos National Laboratory, Los Alamos, New Mexico 87545, U.S.A.

T. Proulx (ed.), *Dynamics of Civil Structures, Volume 4*, Conference Proceedings of the Society for Experimental Mechanics Series 13,
DOI 10.1007/978-1-4419-9831-6_18, © The Society for Experimental Mechanics, Inc. 2011

Rock pockets, being internal defects, cannot be located through visual inspection, thus traditional techniques of detecting these defects required drilling samples. The traditional techniques were labor intensive, time consuming and semi-invasive. Because the civil engineering community has a pervasive interest in the ability to detect defects in concrete members, a number of nondestructive evaluation techniques have been developed. These techniques are primarily based on the mechanical principle of stress wave propagation; see for instance Mori et al. (2002). Examples of such methods include acoustic impact method (sounding), ultrasonic wave propagation (Acciani et al., 2008) and impact-echo method (Sansalone, 1997). Although these localized nondestructive methods exhibited a level of success in detecting defects near the surface of the member where measurements are taken, they have a drawback of requiring *a priori* knowledge about the vicinity of the defect as well as access to the region with the defect to perform nondestructive testing. In typical applications, the presence of internal defects, as well as their location, severity and kind are not known *a priori*. Therefore, a defect detection method must be capable of detecting defects without prior knowledge of these aspects.

An important step in decision making regarding the maintenance of a concrete element is the classification of the defect- that is the determination of type and severity of the defect. Ultimately, one needs to determine whether the degrading effects of a defect are significant to mandate repair schemes. Therefore, any defect-detection method must be able to determine the type and quantify the severity of defect to be of any use. Moreover, knowing the precise location of the defect would make elaborate and invasive repair schemes possible, which typically include removal of concrete and mortar injections.

The goal of this study is to make a small step towards assessing the feasibility and success of vibration testing coupled with finite element simulation as a rapid assessment tool to detect not only the presence but also the location and severity of internal defects. Although not presented herein, it is the intention of the author to extend the present work to investigate the potential of the method in distinguishing between honeycombing and rock pockets in reinforced concrete beams. Vibration testing based nondestructive method has the potential to provide, in real time, reliable, global information regarding the defects in the concrete member. Therefore, if successful, one output of the proposed approach may be the rapid and inexpensive assessment of the conditions of existing concrete structures. Moreover, the ability to determine the presence of defects, or lack thereof, when made readily available, can also provide venues for quality assessment of prestressed, precast concrete members.

2. Background in Vibration Based Defect & Damage Detection:

The problem of defect detection is not any different than the widely studied problem of damage detection. The assessment of defects in a system can either be calculation based or experimentation based. However, neither of these approaches can yield a complete set of information about internal defects. By complete set of information, we mean the tasks identified by Rytter (1993) for complete damage detection: detection, location, quantification and prognosis. As a result of the natural similarity of the two problems these tasks can easily be transferred to the problem of defect detection.

The detection problem can be handled by using pattern recognition or novelty detection approaches using solely measurements without a strict need for a simulation model (Worden 2003). However for localization, quantification and prognosis, almost certainly one must combine a simulation model with measurements (Friswell, 2008). Therefore, the current trend is in integrating the experimental measurements with FE model calculations in order to obtain a calibrated finite element (FE) model which can be used to make predictions about the state of the structural element unavailable with physical measurements due to a variety of reasons. Detection of damage in structural members through model calibration against vibration measurements has received significant attention from those who are involved in structural health monitoring over the last two decades. A detailed overview of the pertinent literature is deemed out of the scope of this paper, but interested readers are referred to Doebling et al. (1998) and Sohn et al. (2004).

Theoretically, when the FE model is built assuming linearly elastic response, if a sufficient number of vibration modes of a structure were identified experimentally (with minimal or no uncertainty), the calibration exercise would be able to retrieve all parameters necessary to construct a FE model. In other words, when physical evidence is in sufficient quality and quantity, one can effectively calibrate the individual properties of each finite element. Such an approach would also enable the representation of local variability of material properties as well as localized cracks -- an important uncertainty source for civil engineering structures.

Aoki et al. (2005) presented the results of such study. A dynamic identification, FE modeling and FE model calibration campaign was applied to a brick chimney. The FE model of the chimney was built with 20-node isotropic solid elements. A correction factor was assigned for both the mass and stiffness of each finite element. Using a sensitivity based calibration approach, the elemental mass and stiffness correction factors were sought. To alleviate the inevitable incompleteness of the measurements, a weighting function was applied to eliminate the finite elements with low sensitivity. The element stiffness at the base of the chimney was observed to be reduced, while at the corners, the stiffness was observed to be increased. The authors explained the former by chimney soil interaction and the latter by the iron angles at the corners. The way calibration is handled in this study can be considered as a reconstruction of discrete elemental matrices. Such an approach has been demonstrated to be successful for simpler problems. However, for more complex problems with larger numbers of finite elements, a recosntruction approach renders "detection" an undetermined problem.

However, Vestroni (2008) emphasizes that a damage (or defect) detection problem can also be represented as a determined problem. Formulating damage in terms of location and severity instead of the reconstruction of the entirety of the elemental matrices eliminates the need for excessively large amounts of experimental information. Therefore, the mass and stiffness parameters of a structure can be considered to be uniform and equal to the baseline value throughout the structure except at some distinct locations where damage (or defect) is present. Such an approach can effectively be applied to the present problem while aiming at detecting rock pocket defects in concrete structures. The construction defects considered in this study are localized and typically few in number, therefore only a limited number of parameters would be sufficient to describe the defects. However, the absence of knowledge about their location, type and severity makes the evaluation of a large number of candidate scenarios necessary. How this evaluation is completed will be discussed in Section 3.

3. Model Calibration Procedure:

In this study, the FE model parameters that define the internal defects will be sought through calibration of the FE model against experimental measurements obtained from the concrete beam members cast with deliberately located rock pocket defects. To reiterate, the detection problem is defined as a calibration problem. The details of the FE model calibration procedure adapted herein can be found in Higdon et al. (2007–2008) and Kennedy and O'Hagan (2000). In this section, only a brief overview of the method will be provided.

The calibration methodology adapted herein treats both the FE model input and the FE model output probabilistically. Therefore, two types of uncertainty propagation are required. Determining how much uncertainty in the selected calibration parameters causes variability in the output is referred to as forward uncertainty propagation. The inverse uncertainty propagation, in contrast, investigates the sources of uncertainty in the output by focusing on the variability of calibration parameters [Figure 1].

The forward propagation of uncertainty consists of a family of computer runs repeated at the sampled input parameters to observe the variability in the FE model outcomes. As long as a large enough number of samples are generated, this sampling approach converges to the actual distribution of the output parameter. The number of necessary samples depends strictly on the order of complexity of the sampled behavior and on the type of sampling design. On the other hand, inverse propagation of uncertainty is computationally more involved as it conceptually requires the FE model to be inverted. However, for real engineering solutions, the requirement of inverting a FE model is practically prohibitive because of the discretization based approximate approach inherent in the FE analysis.

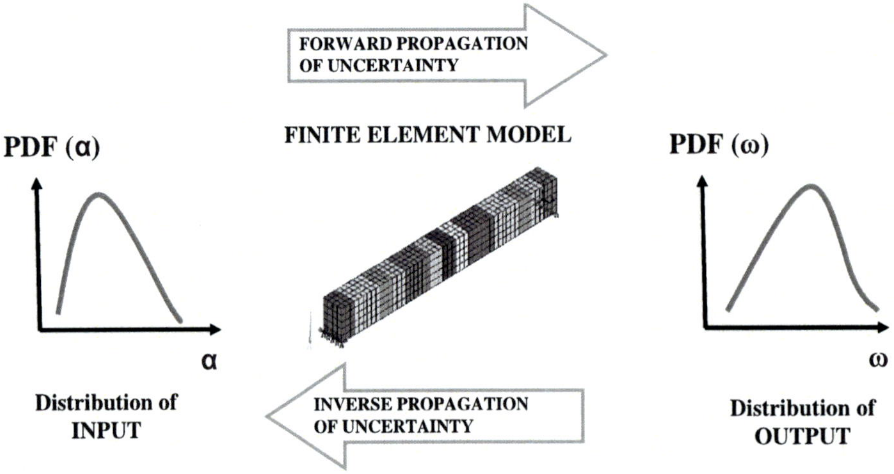

Figure 1: The forward and inverse propagation of uncertainty forms the basis of stochastic calibration.

In stochastic model calibration, the inverse propagation of uncertainty forms the basis of the statistical inference problem. In theory, if the uncertainty in input parameters (particularly those that exercise sensitivity on the output) can be reduced, the uncertainty in the output will also be reduced. Thus, the inverse propagation of uncertainty is usually called upon in model calibration activities to seek the sources of the output uncertainty.

Clearly, reduction in output uncertainty is made possible by the physical experiments. Although experiments constitute our best representation of physical reality, they are subject to random and bias errors. Thus, the calibration process must incorporate experimental errors. Experimental uncertainty is typically categorized in two groups: (1) measurement uncertainty, for instance due to instrumentation and data processing, and (2) natural variability of the structure, for instance in heterogeneous materials. Experimental errors are typically defined as a zero-mean Gaussian random variable. Such an approach is best justified by the central limit theorem. If we assume that the experiments are immune from systematic errors, the experimental error can be considered to be a summation of a large number of independent processes. According to the central limit theorem, these sources would collectively converge to a normal distribution (Hogg and Craig 1978).

In the formulation adapted in this study, the inverse propagation of uncertainty is replaced by a large cohort of forward propagation of uncertainty via Markov Chain Monte Carlo (MCMC) sampling. Because MCMC sampling involves performing multiple runs of a computer code, the computational requirements can rapidly become prohibitive. To remedy this problem, the FE simulation is replaced by a Gaussian Process Model (GPM), a fast running surrogate trained using a limited number of computer runs. A GPM represents the input-output relationship of a numerical model, in our case a FE model, in a purely mathematical way.

MCMC performs a random walk in the domain defined by the calibration parameters defined by $\alpha 1$ and $\alpha 2$ in Figure 2 according to the probability distribution of the calibration parameters: $P(\alpha 1)$ and $P(\alpha 2)$. In the absence of better knowledge, the probability distribution of calibration parameters can be assigned equal probability for all possible values between an upper and lower limit. During each random walk, the model calculates the selected output response according to the sampled parameter values, defined by ω in Figure 2. The acceptance criterion for the sampled parameters is guided by the likelihood estimation $P(\omega \mid \alpha 1, \alpha 2)$. The current sample point (in the domain of the calibration parameters $\alpha 1$ and $\alpha 2$) is rejected if it reduces the likelihood that the set of calibration parameters is correct. If the current sample point is rejected, the random walk returns the last accepted point and the probability distributions of the calibration parameter remains unchanged. However, if the sample point is accepted, the posterior distributions of calibration parameters are obtained. Based on Bayes' theorem, these posterior distributions of the calibration parameters become the prior distributions in the next random walk. According to the current priors, MCMC performs another random walk from the last accepted point to the next point [Figure 2].

The repeated feedback in Bayesian inference progressively characterizes not only the posterior distributions of the calibration parameters, defined by $\alpha 1$ and $\alpha 2$ in Figure 2, but also the hyperparameters of the GPM, the fast running surrogate used to replace the computationally expensive FE simulations. If there are 'n' MCMC iterations, there will be 'n' different sets of estimates for hyperparameters of the GPM and calibration parameters of the FE model that are **accepted** by the likelihood function. By using the hyperparameters of the GPM of the discrepancy term, one can now construct the error model, and by using the calibration parameter values along with the hyperparameters of the simulation emulator, one can construct the GPM surrogate model. Thus, the method defines all possible values of η(x; t) and δ(x) that, when added together, reproduce the experiments.

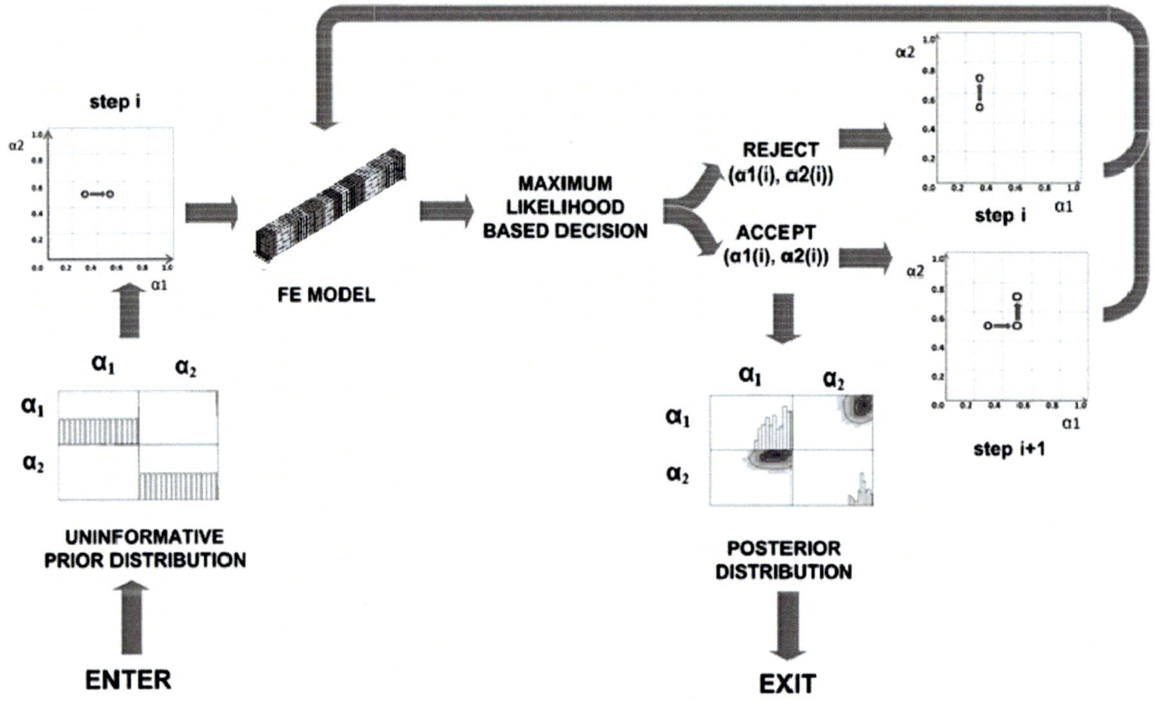

Figure 2: The adapted procedure combines Markov chain Monte Carlo with the context of Bayesian inference.

The mean estimates of the posterior distribution provide the most likely values for the calibration parameters. These values obtained through stochastic calibration, can also be used in a deterministic study. The standard deviation of the posterior distributions captures the remaining uncertainty in the parameter values.

4. Proof-Concept

A key aspect of calibration based defect detection is the suitable parameterization of the defect of interest. For instance, rock pocket defects result in localized reduction in stiffness while the change in mass, relatively speaking, is negligible. However, honeycomb defects result in reduction in both stiffness and mass. If the defect is strictly known to be a rock pocket through local non-destructive methods, then the defect parameterization may exclude the mass-related parameters from calibration. However, for honeycomb defects to be detected, the mass of the structure must also be parameterized.

A first approach to the defect detection problem is through calibration of individual stiffness and mass properties of all seventeen units. Such an approach can be considered as a total reconstruction of the elemental matrices of the beam. Assuming a lump mass and stiffness model, a total reconstruction approach yields a total of 34 calibration parameters (i=1:17; k(i) and m(i)). A total of 34 calibration parameter would result in non-unique calibration solutions unless the available experimental information is in sufficiently high quantity and quality. From a practical stand-point, this kind of high demand on experimentation is commonly not feasible.

An alternative formulation is to define the defect in terms of its ordinate across the length of the beam *(x)* and two reduction percentages: one for stiffness *(α)* and one for mass *(β)* [Figure 3]. As mentioned earlier, the consequence of rock pocket defects is primarily some localized reduction in the stiffness of the member. Therefore, β parameter will be kept constant and equal to 1. In the presence of one defect, this approach reduces the model calibration to three parameters (x, α and β) only. However, when a higher number of defects is present in the member, the number of calibration parameters increases in proportion to the number of defects. This approach accepts the stiffness and mass properties of the beam to be identical throughout the member except for at certain locations where the damage is present.

The consequence of internal defects can be represented in different precision levels varying from a very refined finite element or a substructure consisting of several finite elements. The desired precision will be reflected in the parameterization of the defect, and naturally a larger amount of experimental information will be necessary for improved precisions. In this study, because only lower order vibration modes were available, the defects are represented with a relatively crude model. However, a more sophisticated and precise approach than the one adapted herein is unlikely to lead to practical tools for rapid assessment of existing concrete structures.

The defect is assumed to be contained in one of the 17 segments of the FE model, and the structural effect of the defect is modeled with a reduction in stiffness of that bin. This approach reduces the number of calibration parameters significantly. This approach however, has the drawback that a good estimate of the number of defects should be available *a priori*. This drawback is more pronounced especially when experimental evidence in support of calibration is limited to the first few natural frequencies of the member. This statement is illustrated in the Scenario #1 given below.

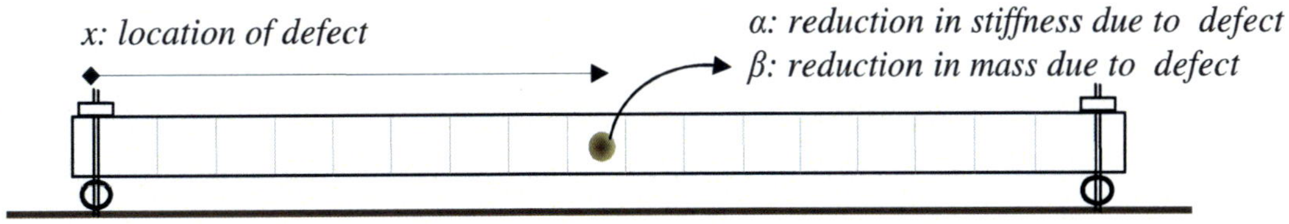

Figure 3: Scenario #1: Defect is defined with a location parameter and a reduction parameter.

A FE model of a simply supported beam is developed in ANSYS. The beam has a cross-section of 15 cm x 22.5 cm and a length of 1.95 m. The beam is assumed to be simply supported at both ends. A uniform 30 GPa Young's Modulus and 2400 kg/m^3 density are assumed. The beam model has a total of 17 sections, each of which are assigned an individual mass (density) and stiffness (Young's modulus) property [Figure 3]. The mesh of the beam FE model is given in Figure 4. A more refined mesh was deemed unnecessary because an element with a quadratic shape function was used. For defect indicators, the first bending modes in the Y direction are considered [Figure 5].

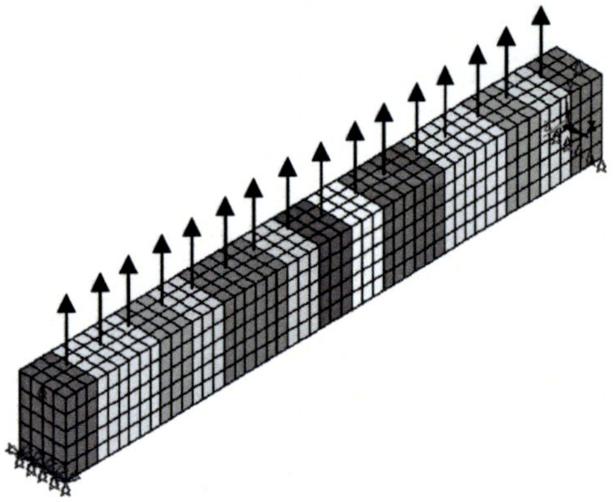

Figure 4: Finite element model of a simply supported beam.

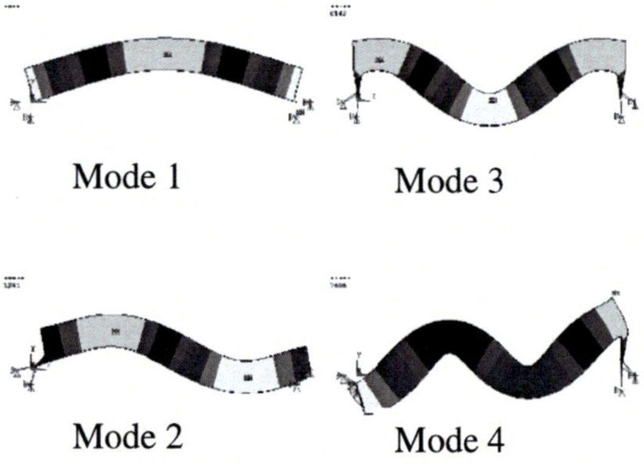

Figure 5: The first four bending modes of the beam are used for defect detection.

- *Scenario #1: Detecting single rock-pocket defect using only natural frequencies.*

Let's assume a single defect close to the mid-length of the beam [Figure 6]. We will assume to know the exact number and type of the defect: one rock-pocket defect. Therefore, the calibration parameters will be reduced to two parameters only: x and α. This scenario represents the ideal, in other words the easiest, condition for defect detection approach proposed herein.

Virtual experimental data of the natural frequencies of the first four bending modes are simulated by the FE model with the prescribed defect (x = 7, α = 0.75 and β = *1.0*). Virtual experiments are deterministic in nature, however to account for potential variability during physical experiments, a 2.5 percent experimental uncertainty is assumed. The goal is to retrieve the 'correct' x and α value used to generate the virtual experiments.

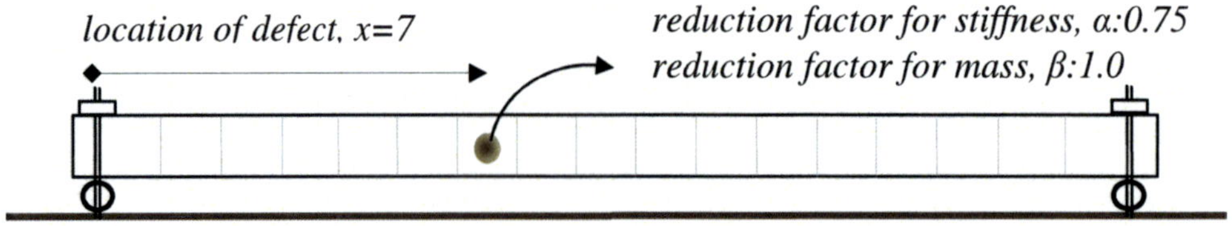

Figure 6: Characteristics of the virtual defect simulated using the finite element model.

In this framework, the maximum allowable changes for the calibration parameters are assigned through the domain within which the MCMC algorithm performs random walks. The defect can be located in any of the 17 bins of the beam FE model. Therefore, the x parameter is allowed to be (x = 1:17). The defect is expected to reduce the Young's Modulus of concrete, therefore a range from 21 GPa to 30 GPa is considered for the stiffness reduction parameters (α = 0.7:1). Five different levels of stiffness reduction are considered between these lower and upper bounds.

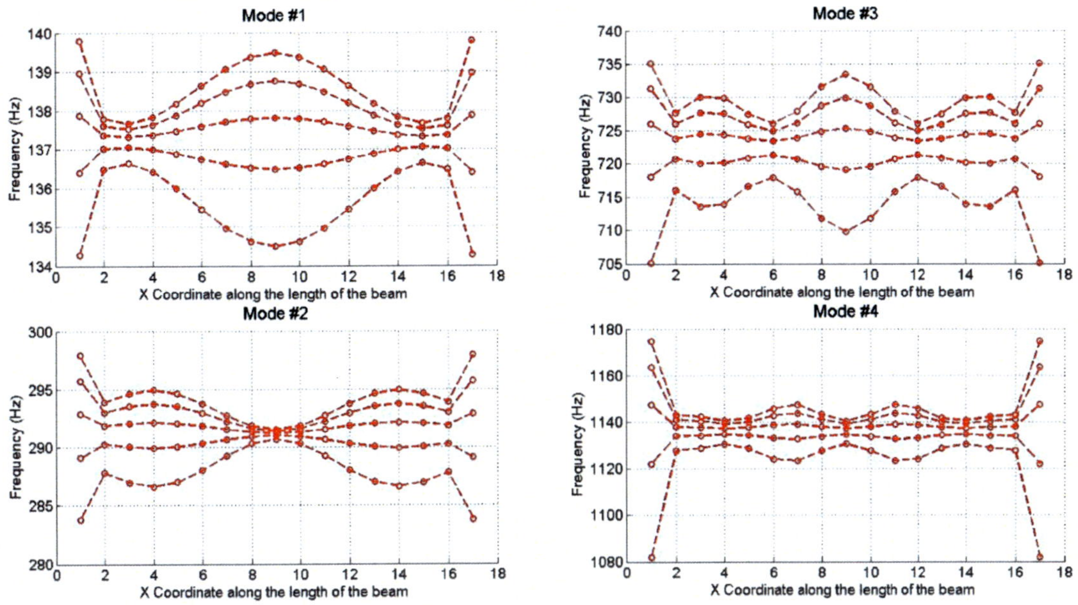

Figure 7: The influence of the defect location on the first four natural frequencies.

A total of 85 computer runs are obtained each time varying the location of defect and the reduction factor for stiffness. The natural frequencies of the first four bending modes are observed to vary depending on the defect location along the beam [Figure 7].

A GPM is trained using the first four natural frequencies of the beam. Because only the natural frequencies are adapted herein, we cannot expect to locate the defect in a symmetric structure without incorporating spatial information, such as mode shapes. Clearly, two possible symmetric locations of defect will be equally plausible; and MCMC algorithm would converge to one of these plausible solutions. It also is possible that the posterior distribution may yield two poles with equal probability for the location parameter (x).

Then, according to the procedure described in Section 3, the two parameters, which describe the location and extent of damage, are characterized. The posterior distributions of these two parameters after calibration are given in Figure 8.

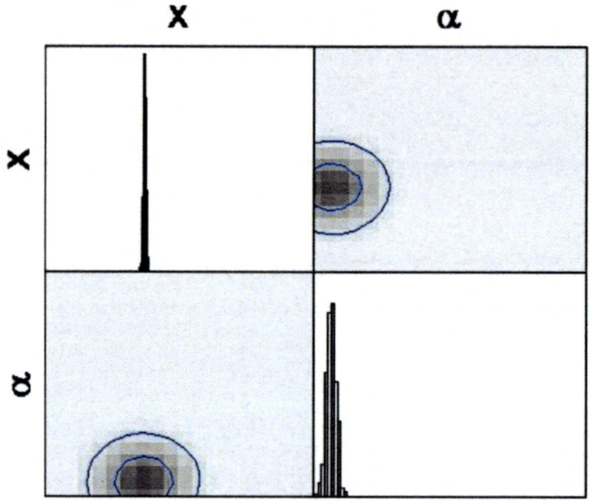

Figure 8: The posterior distributions of the two calibration parameters: location and severity of rock pocket defect.

Table 1: The calibrated values of the parameters that define the location and severity of rock pocket defect.

Calibration Parameter	True Value	Calibrated Value (mean)
Location of Defect (x)	7.0 units	6.92 units
Extent of Defect (α)	0.750	0.753

As seen in Table 1, the calibration approach successfully identified the location and extent of the rock-pocket defect – once the type of defect and the number of defects are known. It must be emphasized that these results present the best case scenario and, in practical applications, we cannot expect this knowledge to be available. Because the true number of defects in a member will not be known, the defect detection method should be able to indicate distinct poles in the posterior distribution of the location parameter (x) for each defect. The same approach is applied to the case where there were two distinct rock pocket defects, illustrated in Figure 9, using only the first four natural frequencies. The calibration is observed to yield an incorrect defect location which was somewhere between the two actual defects. The method predicts one single defect with a higher severity- in other words; it incorporates the effects of two defects into a single, but larger one. This is an example of a very general problem where the influence of a defect at one location on the measured quantities is identical to the influence of a defect of different severity or even type located at some other location in the system. These results, not shown here in the interest of brevity, raise the problem of uniqueness and emphasize the importance of the quality and quantity of experimental information.

- *Scenario #2: Detecting two Rock-pocket defects using natural frequencies and mode shapes.*

Perhaps the most important problem during calibration based damage or defect detection is the uniqueness: several different combinations of calibration parameters can make the FE model match the experiments. Such a problem is observed in Scenario #2 while attempting to detect more than one defect using only the natural frequencies. One obvious remedy to the problem is increasing the quantity of experimental information. In this scenario, the FE model will be calibrated using both natural frequency and mode shape information to detect two separate rock pocket defects as illustrated in Figure 9.

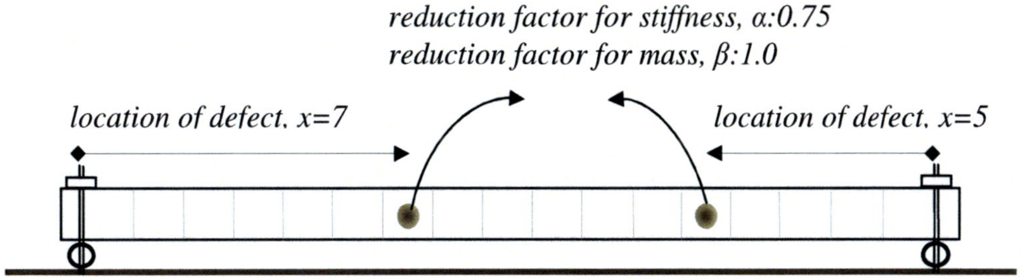

Figure 9: Scenario #2: Characteristics of the two virtual defects simulated using the finite element model.

The beam FE model with two virtual rock pocket defects, as illustrated in Figure 9, is used to generate simulated experimental data. The defects are located at the seventh and thirteenth compartments. Consistent with the previous scenario, the first four bending modes are used as defect indicators. The mode shapes were defined for the vertical deflections of the 18 equidistant points at the center of the beam width. Mode shape vectors are high dimensional and make the statistical inference computationally expensive. Therefore, singular value decomposition is applied to reduce the dimensionality of mode shape vectors. A total of four singular values are used to represent the mode-shape vectors.

The FE model is calibrated to retrieve the 'correct' values for x and α parameters used to generate the virtual experiments. The posterior distributions, given in Figure 10, illustrate two poles for the location of defect- proving the possibility of detecting more than one defect even though the true number of defects is initially unknown. Also, the posterior distribution that defines the severity of defects displays a correct trend.

The difference in the posterior distributions of the two scenarios, with one defect versus two defects, emphasizes the challenge in detecting multiple defects in the absence of knowledge about the true number of defects. However, in the second scenario, although the levels of uncertainty in the posterior distributions are significantly increased – meaning that the influence of defects are spread over a larger area- the proposed method still delivers viable information regarding the location and severity of defects. In cases where more than one defect is present, the outcome of most concern is the ability to determine the presence and the number of defects. The results presented in Figure 10 can be considered satisfactory for such purposes. Perhaps one can repeat the calibration procedure defining two distinct parameters (x1 and x2) as well as two distinct reduction factors (α1 and α2) to better identify the location as well as the severity of the defects.

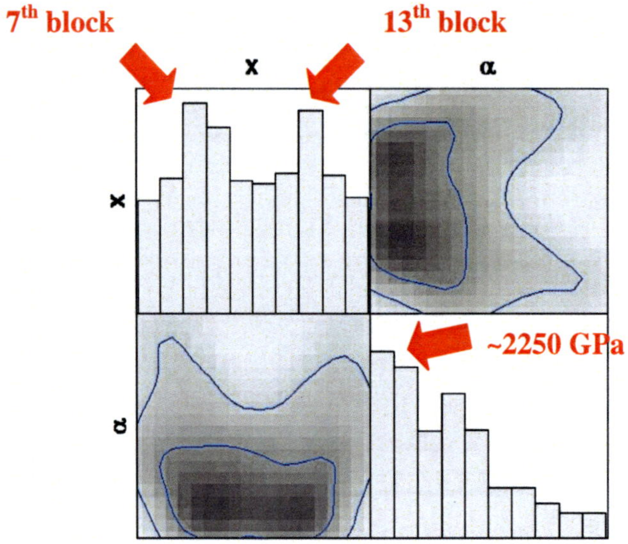

Figure 10: The posterior distributions of the two calibration parameters: location and severity of defect.

As a result, using simulated data, we can successfully detect, locate and quantify defects in concrete members. Clearly, it is easier to identify defects using simulated experimental information. The effectiveness of inverse methods must, however, be tested using real physics measurements, because methods which perform well with simulated data may struggle when dealing with measurement errors and inadequate simulation models.

In the next section, the practical application of this approach is illustrated on scaled beam specimens built and tested under laboratory conditions. When dealing with real measurements the following problems are anticipated to raise: (a) the defect present in the test beams is unavoidably different than the defect model represented in the FE model; (b) the boundary conditions of the beam during testing is, again unavoidably, different then the idealized boundary conditions in the FE model.

5. Test Beams:

Test beam consists of a lightly reinforced concrete beam, 15 cm by 22.5 cm in cross section and 1.95 m in length. The beam is supported with roller suppers at both ends. To prevent rattling, the beam is clamped by two 5/8" steel rods running between the roller supports provided both at the bottom and top surface of the beam. The rock pocket defects are created with aggregates placed in a net before the concrete is placed, see Figure 11. Approximate locations of these rock pockets are provided in Figure 12.

Figure 11: Rock pocket defects are intentionally located in the concrete beam.

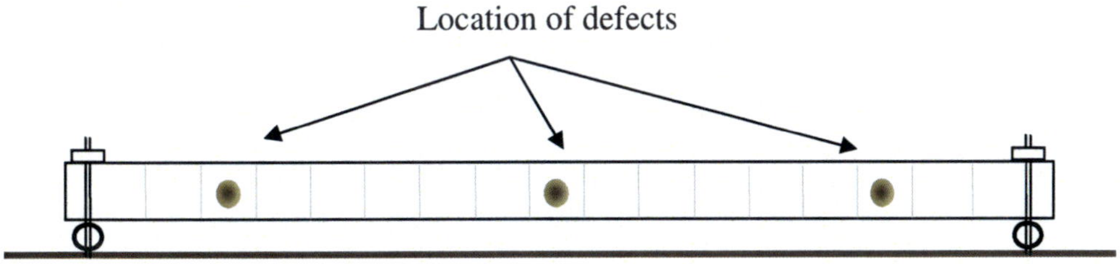

Figure 12: The location of rock pocket defects.

6. Experiments:

The experiments are conducted to obtain the time history response of the beams to a hammer impact excitation. Accelerations at 16 points located equidistant are measured using accelerometers. Vibration measurements are strictly limited to the vertical vibration of the beam to be relevant to practical application where only the top of the beam would be accessible for testing. Two locations, one approximately at quarter-length (excitation #1 in Figure 13) and one at approximately at mid-length (excitation #2 in Figure 13) of the beam, are excited with an impact hammer (see Figure 14).

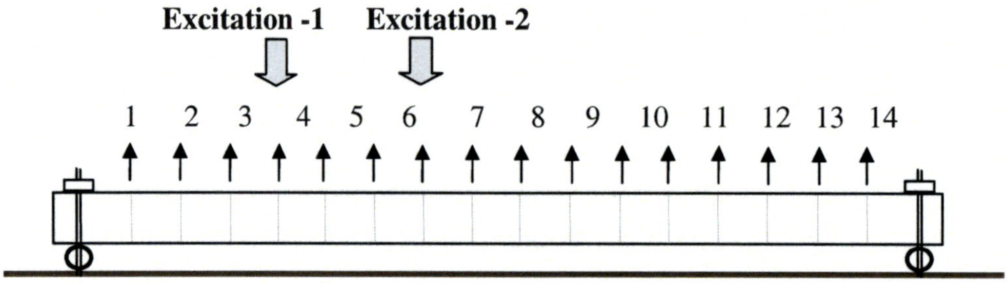

Figure 13: Experimental set-up includes 16 measurement location and two excitation locations.

Figure 14: The boundary conditions of the beams are designed to provide simple support and at the same time to prevent rattling.

7. Feature Extraction:

The Frequency Response Functions (FRF), obtained exciting point #7, are illustrated in Figure 15. A number of modes are observed between 40 Hz and 450 Hz. Because of the complexity of the boundary conditions, some of these modes are noted to be a result of the coupling between the beam flexural modes and the supporting system. These coupled modes are not included during defect detection.

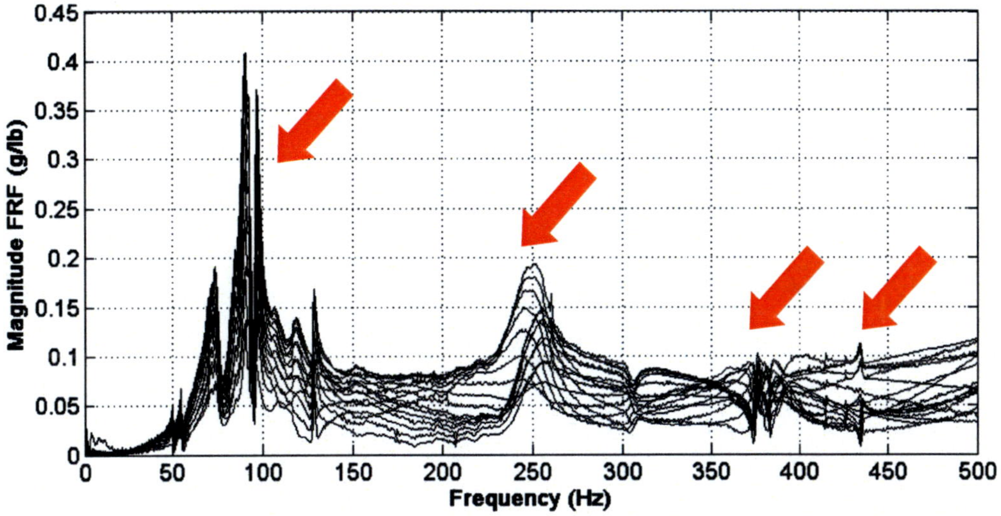

Figure 15: The FRF of beam with rock pockets identified frequencies are illustrated with read dashed lines: (top) excitation 1, (bottom) excitation 2.

Feature extraction is completed using rational fraction polynomials method (Richardson and Formenti, 1982). A total of four bending modes are identified within the tested frequency range. The frequencies of these modes are given in Table 2, while the mode shapes are presented in Figure 16.

Table 2: The natural frequencies obtained from the beam with rock pockets.

Mode	Frequency
1	89.2 Hz
2	242 Hz
3	380 Hz
4	435 Hz

The variability of physical experiments can only be studied by repeated experiments. However, the experiments are typically costly and time-consuming relative to the FE models. It is a common application to repeat only a portion of the experiment. For instance, in a test setup with k measurement points, repeating the experiments at a select few measurement points (<<k) can reduce the required resources yet can still yield reasonable information about the inherent variability in the experiment. The driving point measurements at point #7 are repeated 16 times to quantify the experimental variability, which is incorporated during calibration.

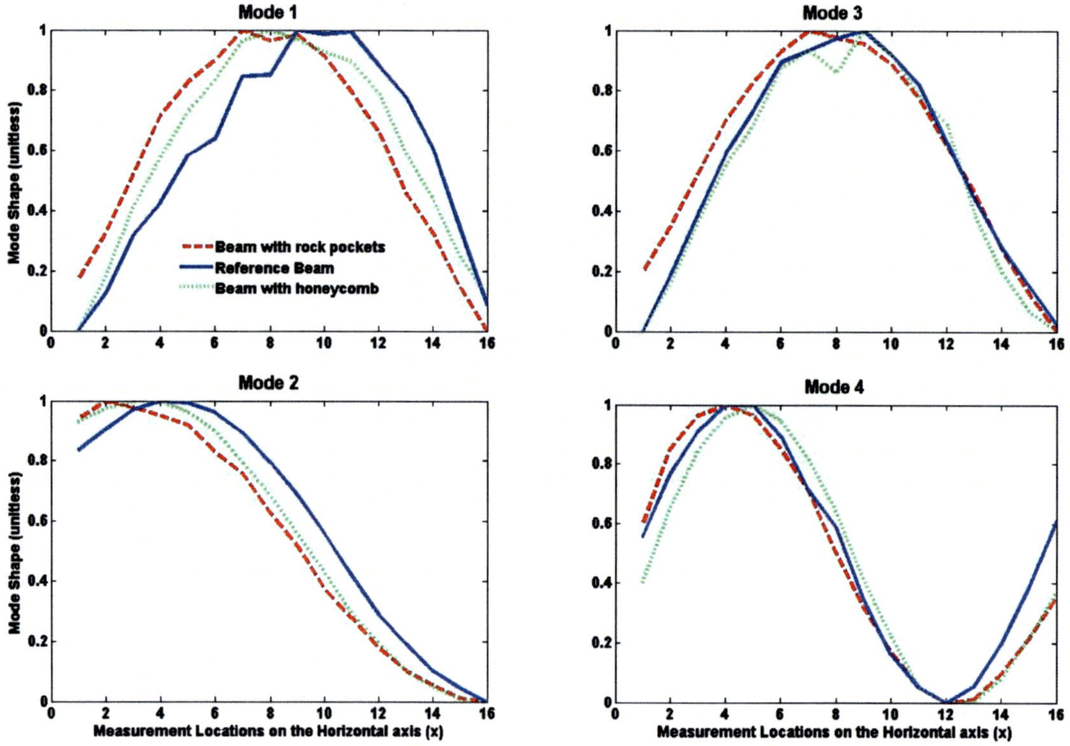

Figure 16: The first four mode shapes of the beams with rock pocket defects are given in red dashed lines (the blue and green lines represent two nominally identical beams. The blue solid line is the mode shapes of the reference beam without defects and the green dotted line is the beam with honeycomb defects. The results obtained from these beams are not included herein.)

8. Numerical Models:

Defect detection of a simply supported beam is a simple problem that can be tackled using closed form solutions developed for continuous systems; however a discrete model approach adds versatility in the application of the procedure to systems of various geometry and complexity. Discretization of the model does not influence the nature or formulation of the defect detection problem. Discrete numerical models are built as explained earlier in Section 4. The beams are built based on linearly elastic behavior assumption. This assumption is reasonable as the vibration levels induced during the tests are low. Moreover, the defects are not anticipated to induce nonlinearity.

The boundary conditions are represented with two springs exerting vertical restraints both at the top and bottom surface of the beam [Figure 17]. The stiffness constants are estimated based on the axial stiffness of the steel rods to be approximately 1.04×10^7 kN/m. The spring constants are assumed to be symmetric for both ends– which is with all likelihood not the case, because variation in assembly is a routinely experienced practical problem. However, it is very difficult to quantify this variability.

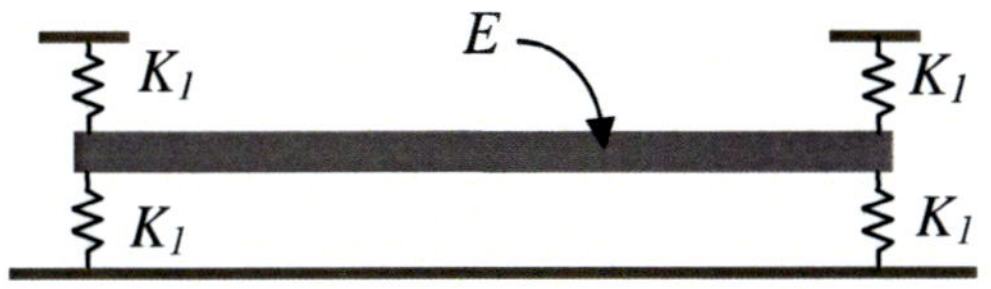

K1: Linear spring
E: Young's Modulus

Figure 17: The configuration of the finite element model boundary conditions.

The FE model is build to have 17 sections each of which can have independent young's modulus and density. The corresponding mode shapes obtained by the beam FE model are illustrated in Figure 18.

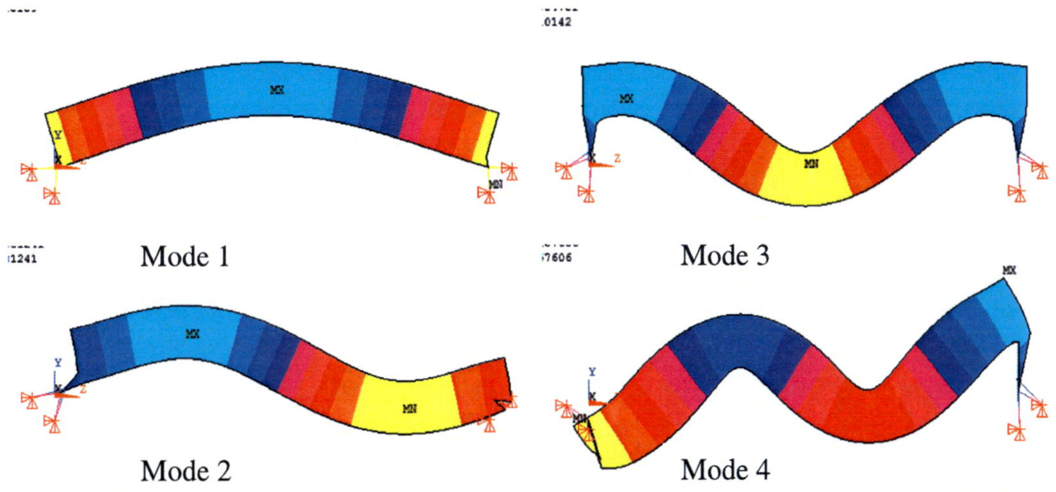

Figure 18: The first four mode shapes predicted by the finite element model.

9. Determination of Defects in Test Beams:

The particular difficulty while detecting defects in the test beams is due to complexity of the in situ boundary conditions of the test beams [Figure 14]. Typically, beam vibrations are measured under free-free boundary conditions to alleviate the problems due to difficulty of reproducing the theoretical fixed or pinned boundary conditions in laboratory conditions. However, to have any practical value, a defect detection approach must be applicable under conditions commonly found in, for instance, existing concrete structures as well as the prestressed precast reinforced concrete beam manufacturing industry. The significant differences between the initial FE model predictions and the experimental measurements are shown in Table 3.

The x and α parameters are calculated using various levels of experimental information: (1) only the first three natural frequencies (see scenario 1 in section 4), (2) the first three natural frequencies along with the singular values of the first three mode shapes (see scenario 2 in section 4) and (3) the first three natural frequencies along with the three mode shape vectors. In any of these three cases, the correct location of the defects is not obtained.

Table 3: The significant differences between the initial FE model predictions and experimental measurements.

Mode	Experimental Frequency Beam with rock pockets	FE Model (mean of 85 runs) before Calibration Without Defects
1	89.2 Hz	95 Hz
2	242 Hz	255 Hz
3	380 Hz	390 Hz

The posterior distributions obtained using the first three natural frequencies and mode shape vectors; that is, the largest available dataset used in calibration, are illustrated in Figure 19. Although the severity of the defect seems to be identified relatively accurately, the effects of three defects are lumped into one single location. The reasonable success obtained when two defects were simulated in Section 4 is not present in this particular case. The explanation may be because (1) defects in numbers larger than two are significantly more difficult to distinguish, (2) the imprecision of physical experiments degrade the success of the inference process, (3) the inaccuracies in the numerical model through the boundary conditions is interfering with the inference process.

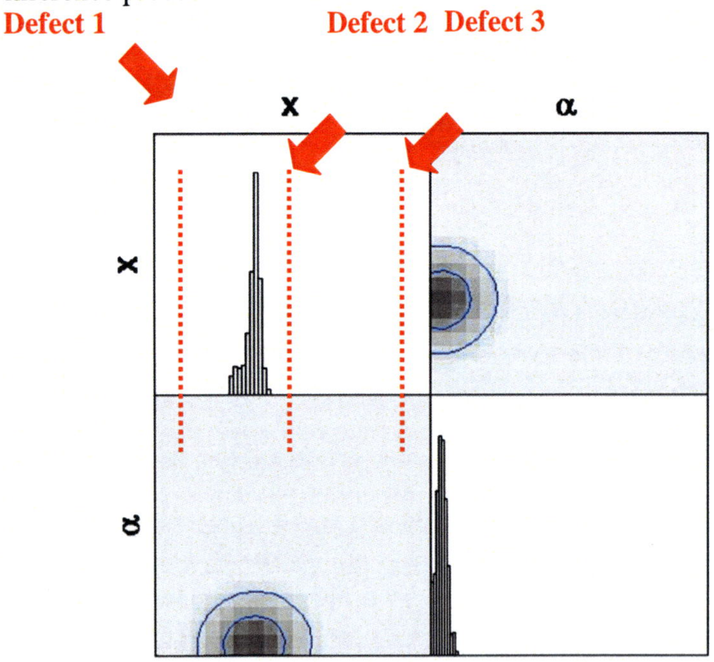

Figure 19: The posterior distributions obtained for the test beam defect locations.

As mentioned earlier, the spring constants representing the boundary conditions are approximated by calculating the axial stiffness of the steel rods used to clamp the beam at both ends. Although, this calculated value is expected to be imprecise, it is not defined as a calibration parameter. Perhaps by calibration, the boundary condition parameters simultaneously with the defect detection may yield better results.

10. Discussion and Conclusion:

In civil engineering practice, a method, which can detect the location, severity and type of damage in concrete construction, is of great value. In this paper, a general procedure for defect detection is applied to concrete beams with both simulated experiments and physical experiments conducted under laboratory conditions. .

The adapted approach does not depend upon the type of the defect, the structure of interest or the measured quantities as it relies on the simple fact that a structure with and without defects would yield different response characteristics. By coupling these response characteristics with a simulation model, one can, theoretically, determine the presence of defects. Therefore, it can be concluded that FE modeling coupled with nondestructive testing can provide a viable solution to detection of internal defects in concrete members. This statement is supported by the findings obtained using simulated experiments. Although the actual implementation of the proposed methodology will be application specific, its generality allows it the use of various forms of experimental information. Moreover, the use of discrete method add further versatility such that the method can be deployed to structural forms of varying geometry and complexity.

Although the same success is not observed when the method is exercised on a scaled laboratory specimen, this does not necessarily demonstrate the limits of the adapted method. Due to the limited test cases, it is difficult to know the exact cause of the problem. We tabulate three possible explanations: (1) defects in numbers larger than two are significantly difficult to distinguish, (2) the imprecision of physical experiments degrade the success of the inference process, (3) the inaccuracies in the numerical model through the boundary conditions is interfering with the inference process. The physical experiments must be repeated for a variety of scenarios to evaluate the potential of the method to be a rapid, practical assessment tool that can be implemented in industry.

Moreover, if a defect is smaller than the defect size which would result in detectable and measurable change in the dynamic response, its influence will be undistinguishable within other environmental effects, such as ambient vibration, measurement noise, etc. Therefore, it is important to note that a defect detection approach based on vibration measurements, as in every nondestructive technique, has a lower defect size limit for which it can be deployed. Also, for two closely spaced cracks, unless a large amount of accurate experimental information is available, the calibration approach will, with all likelihood, smear the effects of these two defects into a larger defect located between the two. For defects which are spaced farther apart, the problem of non-unique solutions poses lesser of a problem.

Also, existing structures and their operational conditions are subject to statistical variations. Therefore in practical applications of this method to existing civil engineering systems, it is anticipated that the posterior distributions would yield higher uncertainty compared to those presented herein.

11. Acknowledgements:

Elizabetta Pistone's help during vibration experiments is appreciated. The author thanks Paul Kremer and the work-study students of Pennsylvania State University for their assistance during the casting of the concrete beam specimens. The author also thanks Geoffrey Kurgan for his editorial help.

12. References:

Acciani, G., Fornarelli, G., Giaquinto, A., Maiullari, D., Brunetti, G. (2008), "Non-Destructive Technique for Defect Localization in Concrete Structures Based on Ultrasonic Wave Propagation," ICCSA 2008, Part II, LNCS 5073, pp. 541–554.

Aoki, T., Sabia, D., Rivella, D., Muto, H., (2005), "Dynamic Identification and Model Updating of the Howa Brick Chimney, Tokoname, Japan," Structural Studies, Repairs and Maintenance of Heritage Architecture, IX, Vol. 83, pp. 265- 275.

Doebling, S., Farrar, C. R., Prime, M. B. (1998), 'A summary review of vibration-based damage identification methods," Shock and Vibration Digest, Vol. 30, No. 91, pp. 91- 105.

Friswell. M. I., (2008), "Damage Identification using Inverse Method," Dynamic Methods for Damage Detection in Structures, Springer-Verlag, Austria.

Higdon, D., Gattiker, J., Williams, B., Rightley, M., (2008), "Computer Model Calibration Using High-dimensional Output," Journal of the American Statistical Association, Vol. 103, No. 482, pp. 570-583.

Hogg, R. V., and Craig, A. T., (1978), "Introduction to Mathematical Statistics," Macmillan.

Kennedy, M., O'Hagan, A., (2000), "Predicting the Output from a Complex Computer Code When Fast Approximations are Available," *Biometrika*, Vol. 87, pp. 1-13.

Mori, K., Spagnoli, A., Murakami, Y., Kondo, G., Torigoe, I., (2002), "A new non-contacting non-destructive testing method for defect detection in concrete," NDT&E International, Vol. 35, pp. 399- 406.

Richardson M. H., Formenti, D. L., (1982), "Parameter Estimation From Frequency Response Measurements Using Rational Fraction Polynomials," Proceedings of the 1st International Modal Analysis Conference, Orlando, FL.

Sansalone, M., (1997), "Impact-Echo: The Complete Story", ACI Structural Journal, Vol. 94, No. 6, pp. 777-786.

Sohn, H., Farrar, C. R., Hemez, M. F., Shunk, D. D., Stinemates, D. W., Nadler, B. R., Czarnecki, J. J., (2004), "A review of Structural Health Monitoring Literature: 1996- 2001" Los Alamos National Laboratory Report, LA-13976-MS.

Vestroni, F., (2008), " Structural Identification and Damage Detection," Dynamic Methods for Damage Detection in Structures, Springer-Verlag, Austria.

Worden, K., Manson G., Allman D. J., (2003), "Experimental validation of a structural health monitoring methodology. Part I: novelty detection on a laboratory structure," Journal of Sound and Vibration Vol. 259, pp. 323–343.

Proceedings of the IMAC-XXVIII
February 1–4, 2010, Jacksonville, Florida USA

A Comparison of Direct Velocity, Direct and Compensated Acceleration Feed-back Control Systems in Mitigation of Low-frequency Floor Vibrations

Erfan Shahabpoor, Paul Reynolds, Donald Nyawako
Department of Civil and Structural Engineering, University of Sheffield, Sir Frederick Mappin Building, Mappin Street, Sheffield, S1 3JD, United Kingdom
Corresponding author: Tel: +919916842525; E-mail: e.shahabpoor@gmail.com

NOMENCLATURE

x	Displacement vector (n×1)	K	Stiffness matrix (n×n) (N/m)
$\dot{x}$	Velocity vector (n×1)	k	Modal stiffness (N/m)
$\ddot{x}$	Acceleration vector (n×1)	ζ_n	Modal damping ratio
F	Force vector (n×1)	ω_n	Natural modal frequency (rad/s)
M	Mass matrix (n×n)	g	Feed-back gain
M	Modal mass(kg)	$G_A(j\omega)$	Actuator's transfer function
C	Damping matrix (n×n)	$G_s(j\omega)$	Sensor's transfer function
C_m	Modal damping	$G(j\omega)$	Plant's transfer function

ABSTRACT

Numerous control systems, passive, active and semi-active, have been developed in past research to mitigate undesirable vibrations in civil engineering structures. Active vibration control (AVC) is emerging as a realistic option for mitigation of human-induced vibrations in office floors. Some AVC control laws commonly associated with vibration mitigation of human-induced office floor vibrations that have been investigated within this research include direct velocity feedback (DVF), direct acceleration feedback (DAF) and compensated acceleration feedback (CAF). The research presented in this work evaluates the bandwidths of effectiveness of the vibration mitigation performances of each of the aforementioned AVC control laws for idealised single-degree-of-freedom (SDOF) models of a floor structure with frequencies ranging from 1 Hz to 20 Hz. Modal masses and damping ratios of these SDOF floor representations have been kept constant. It was found that DVF performs best in low frequency floors while DAF performs well at much higher frequency floors. CAF performs quite well in both low and high frequency floors used in these studies.

Keywords: Active control, compensator, human induced vibrations.

1. Introduction

Recent advancements in construction materials and design technologies have led to lighter and more slender structural elements that are more appealing from an architectural point of view. Slender floor structures, for example, being a consequence of such advancements, often do not meet the desired vibration serviceability requirements. Problems have been reported in numerous structures like office floors, shopping malls, airports concourses and restaurants to name but a few (1).

There has been a vast amount of research in the past to mitigate human induced vibrations in floors. Passive technologies were the typical first generation approach. They employed some mechanical elements and devices to dissipate part of the energy in floors and thereby reduce the vibration levels. These technologies do not need any external energy sources and are quite reliable. However, they have limited effectiveness in some instances and have to be tuned to specific dominant modes of vibration (2).

To improve the efficiency of passive techniques, semi-active control was developed. With these techniques, some features of passive vibration mitigation systems can be adjusted in real-time to increase their vibration mitigation effectiveness for a broad spectrum of disturbances. Examples of such properties that can be adjusted include the fluid viscosity of viscous dampers as seen in magneto-rheological (MR) and electro-rheological (ER) dampers. However, there are still a number of issues with regards to their effectiveness in various scenarios, particularly for mitigation of human induced vibrations in floor structures.

Active vibration control (AVC) has been developed to make vibration mitigation, for example of floors under human induced vibrations, to be more efficient and adaptive, although it is not the ultimate solution and has some potential drawbacks in itself. An active control system in simplest form comprises of a sensor to monitor the structural response, a control algorithm to compute the required control signal and an actuator to introduce the control force (3).

Within AVC studies several control laws exist, typical examples of which make use of feedback loops. Some of the AVC control laws that have been implemented by researchers for mitigation of human induced vibrations in floors include DVF, DAF and CAF. Each possesses some advantages and disadvantages, which make a proper selection of which one to implement in a particular application complicated.

This paper investigates the vibration mitigation performance of DVF, DAF and CAF control laws on some idealised SDOF floor slabs with variable dynamic properties i.e. frequencies ranging from 1 Hz to 20 Hz. The modal masses and modal damping ratios have been kept constant in all the models. This would give an indication of what type of control system can be implemented in a floor structure of given modal properties. Actuator dynamics and sensor dynamics have been considered in all the cases, where the sensor dynamics refers to the dynamics of the integration circuit. Root locus studies are used to evaluate optimal control gains that are implemented in the analytical studies. Analytical studies comprise of heel-drop tests making use of an idealised triangular pulse and walking excitation measured from treadmill tests. For heel-drop excitation, the time for the response to decay to 5% of its peak is used as the performance indicator. For walking excitation, the peak value of the 1s running RMS (MTVV) has been as the performance indicator. Some root locus plots are presented to elaborate on the reasons for the vibration mitigation performances attained.

2. Dynamics of Control System Components

2.1. Floor Models

In this research, SDOF models as shown in Fig. 1 and with linear, constant coefficient differential equation taking the form of Eq. 1, have been selected to represent the dominant modes of some 20 idealised simply supported floors of different modal properties. The dominant modes of these floors have been assumed to have natural frequencies ranging from 1 Hz to 20 Hz as highlighted in Table 1. The modal masses and damping ratios of each of these models have been assumed to be constant. This aids in evaluating the possible bandwidths of efficiency of each of the AVC control laws noted in the abstract for vibration mitigation of floors of different modal properties. For a collocated sensor/actuator pair in each idealised floor representation, the transfer functions of each of these models can be represented by Eq. 2,

$$m\ddot{x}(t) + c\dot{x}(t) + kx(t) = f(t) \quad (1)$$

$$G(s) = \frac{\chi_i s^2}{s^2 + 2\zeta_i \omega_i s + \omega_i^2} \quad (2)$$

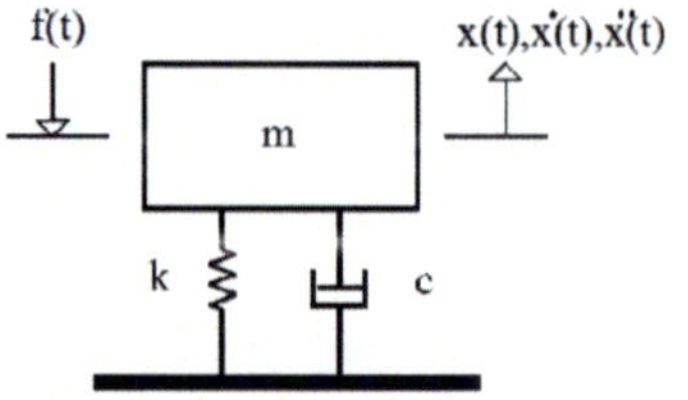

Figure 1 – SDOF Model

where m is the modal mass, c is modal damping and k is the modal stiffness, s = jω, ω is the frequency, χ_i , ζ_i and ω_i are the inverse of the modal mass, the damping ratio and natural frequency associated with the ith mode, respectively.

SDOF Model Properties												
Model ID	m (Mass)	ζ (Damping ratio)	f (Natural Frequency)	ω (Natural Frequency)	K (Stiffness)		Model ID	m (Mass)	ζ (Damping ratio)	f (Natural Frequency)	ω (Natural Frequency)	K (Stiffness)
	Kg	-	Hz	rad/sec	N/m			Kg	-	Hz	rad/sec	N/m
M01	20000	0.03	1	6.283185307	789568.352		M11	20000	0.03	11	69.1150384	95537771
M02			2	12.56637061	3158273.41		M12			12	75.3982237	113697843
M03			3	18.84955592	7106115.17		M13			13	81.681409	133437052
M04			4	25.13274123	12633093.6		M14			14	87.9645943	154755397
M05			5	31.41592654	19739208.8		M15			15	94.2477796	177652879
M06			6	37.69911184	28424460.7		M16			16	100.530965	202129498
M07			7	43.98229715	38688849.3		M17			17	106.81415	228185254
M08			8	50.26548246	50532374.5		M18			18	113.097336	255820146
M09			9	56.54866776	63955036.5		M19			19	119.380521	285034175
M10			10	62.83185307	78956835.2		M20			20	125.663706	315827341

Table 1 – Properties of System SDOF Models

2.2. Actuator Dynamics

The proof-mass actuator considered in this research is an APS dynamics Model 400 mechanical shaker. It has a 30.4 kg reaction mass and its estimated natural frequency is 1.8 Hz. The dynamics can closely be modelled by the linear third-order model in Eq. 3. $K_A>0$, and ξ_A and ω_A are, respectively the damping ratio and natural frequency of the actuator and the pole at $-\varepsilon$ provides the low-pass characteristics of this actuator (2).

$$G_A(s) = \left(\frac{K_A s^2}{s^2+2\xi_A\omega_A s+\omega_A^2}\right)\left(\frac{1}{s+\varepsilon}\right) \tag{3a}$$

$$G_A(s) = \left(\frac{12600s^2}{s^3+62.16s^2+728.6s+6573}\right) \tag{3b}$$

2.3. Sensor Dynamics

The sensor dynamics i.e. the dynamics of the integrator circuit used in these studies can be modelled by the filter in Eq. 4.

$$G_S(s) = \left(\frac{s}{s^2+8s+100}\right) \tag{4}$$

3. Control Laws

Three different control laws used for mitigation of human-induced floor vibrations in previous research include DVF, DAF and CAF (2). The vibration mitigation performance bandwidths of these control laws for human induced vibrations are investigated in this AVC research on some idealised floors whose modal properties have been presented in section 2.1. The human excitation forces that have been used in the analytical studies, briefly described in section 4, comprise of an idealised heel-drop excitation force (1) as well as real walking excitation forces obtained from treadmill tests at The University of Sheffield.

A voltage saturation level of V_L= 0.5V has been introduced in the analytical studies. In principle, this is the saturation voltage that would be implemented in the experimental implementation of the AVC system. The three control laws noted previously will now be briefly introduced.

3.1. Direct Velocity Feedback (DVF)

With DVF control law, for a collocated sensor and actuator control system, sensor outputs (velocity) are multiplied by pre-selected control gains and these fed back to the collocated actuators as expressed in Eq. 5.

$$F_c = -g.\dot{y}_s \quad (5)$$

Where $\dot{y}_s$ is the collocated velocity output signal, g is the control gain and F_c is the control voltage that is fed to the actuator amplifier to drive the actuators. The control set-up for DVF control law is presented in Fig. 2. This figure incorporates a voltage saturation stage.

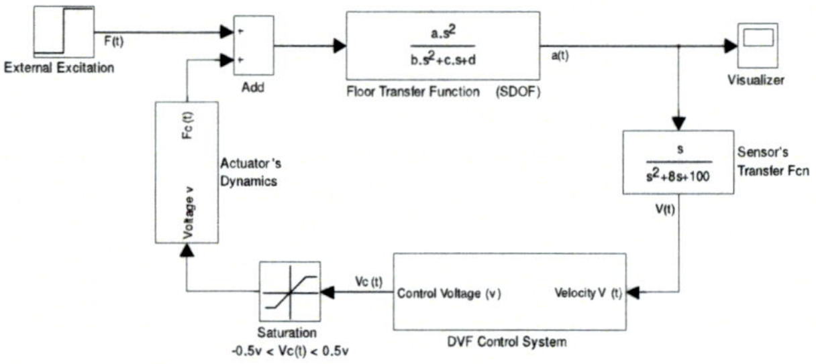

Figure 2 - DVF Scheme (with saturation Voltage-limiter)

3.2. Direct Acceleration Feedback (DAF)

For DAF control law, also considering a collocated sensor and actuator set-up, sensor outputs (acceleration measurements in this case) are multiplied by pre-selected control gains to generate the control signal that is then fed to the actuator amplifier to drive the actuators. This is presented in Eq. 6 and with the control set-up being presented in Fig. 3.

$$F_c = -g.\ddot{y}_s \quad (6)$$

Where $\ddot{y}_s$ is the collocated acceleration output, g is the control gain and F_c is the control voltage signal to amplifier.

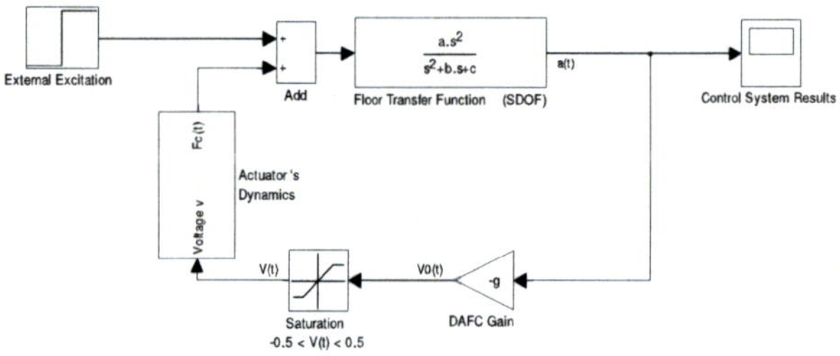

Figure 3 - DAF scheme

3.3. Compensated Acceleration Feedback (CAF)

To increase the stability of DAF system and make it more amendable to the introduction of significant damping, a phase-lag compensator can be designed. The CAF set-up as designed in a previous research (2) is represented in Fig. 4.

The compensator takes the form of $\frac{S+a}{S}$. By increasing the constant 'a', the compensator acts as an integrator $\frac{a}{S}$, and similarly, by selecting small values for the constant, 'a', the compensator behaves more like DAF. In this paper, constant 'a' = 55. This is applicable for the frequency bandwidth of interest in these studies presented.

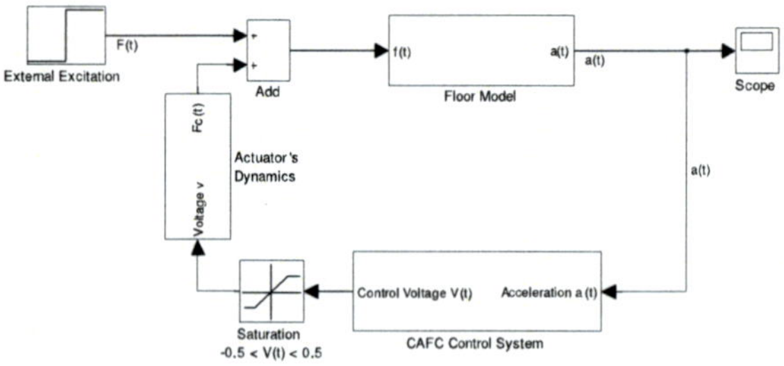

Figure 4 – CAFC system using phase-lag compensator

4. Human Excitations Forces

The excitation forces used in the analytical studies presented in this paper include an impulsive excitation force and a real walking force time history obtained from treadmill walking tests.

4.1. Heel-drop Force

The heel-drop excitation force can be idealised as a triangular pulse with a peak force of 2670 N, reducing to zero in a time of 0.05 s, which corresponds to an impulse of 67 Ns (1). This is shown in Fig. 5.

4.2. Walking Excitation

Walking force time histories obtained from treadmill walking tests at The University of Sheffield were used in the analytical studies. The walking force specifications are: Body weight: 886 N, Walking Speed = 1.2085 m/s equivalent to 1.7 Hz pacing rate (and further assuming a 70 cm step length). This is shown in Fig. 6.

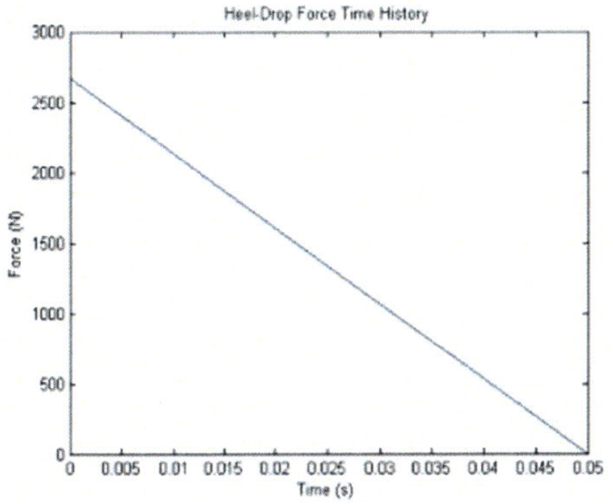

Figure 5 - Heel-drop force time history

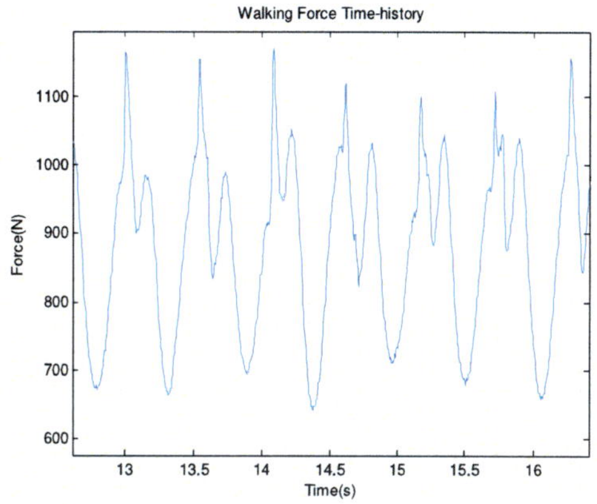

Figure 6 – Part of walking force time history

5. Simulation Results

The three techniques chosen to evaluate the efficiencies of DVF, DAF and CAF control schemes for vibration mitigation of the idealised floor systems noted earlier include:

(a) A study of the damping introduced by the control systems (DVF, DAF and CAF) into the primary structure by a study of the root locus studies of the closed-loop systems.

(b) A study of the MTVV, described as the peak of the 1s running RMS of the uncontrolled and controlled (DVF DAF and CAF) responses for walking excitation only.

(c) A study of the time it takes for the uncontrolled and controlled responses to decay to 5% of the initial peaks for impulsive excitation only.

5.1. Damping Ratio

The closed-loop systems for each of the setups in Figs. 2, 3 and 4 can be described by the transfer function representation in Eq. 7. This incorporates the primary structure, actuator and sensor (integrator circuit) dynamics. $G(j\omega)$ is the transfer function of the floor, $G_A(j\omega)$ is the transfer function of the actuator, $G_s(j\omega)$ is the transfer function of the sensor and $G_c(j\omega)$ is the transfer function of a compensator, if any.

$$G_T(j\omega) = G(j\omega)G_A(j\omega)G_s(j\omega)G_c(j\omega) \tag{7}$$

By making use of root locus studies for the transfer function representation in Eq. 7 for each of the idealised structural models, optimum gains were evaluated as presented in section 5.4 as well as maximum damping introduced into each of the structural models.

Fig. 7 shows a plot of the damping ratio introduced into the structural models for DVF, DAF and CAF control laws for optimum gains. DVF exhibits the best performance at low frequencies while CAF exhibits better performance at higher frequencies. The performance here is with reference to the amount of damping introduced into the structural models for each of the control laws.

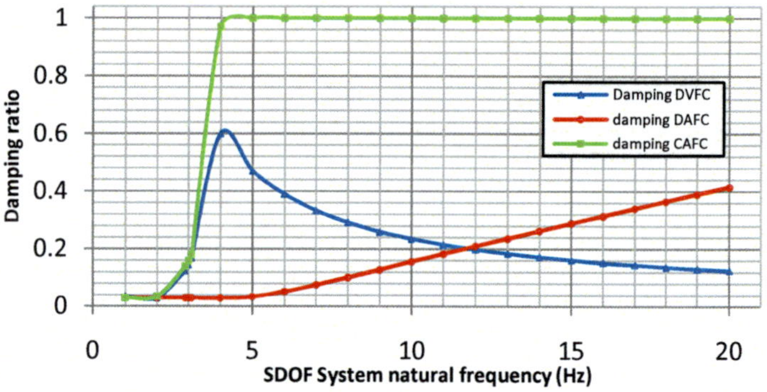

Figure 7 - Structural damping ratios for DVF, DAF & CAF control for each of the idealised structural models

5.2. MTVV Studies (For Walking Excitation Only)

The optimum gains evaluated in the root locus studies in the previous section and presented in section 5.4 are implemented in these analytical studies for DVF, DAF and CAF control laws. The MTVV is defined as the peak of the 1s running RMS acceleration response of both the uncontrolled and controlled structures. A typical plot of responses of an open and closed loop system and the corresponding 1s running RMS acceleration are shown in Fig. 8.

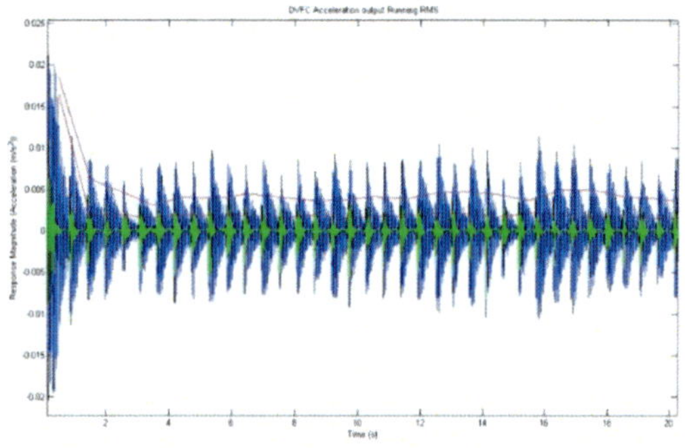

Figure 8 - A typical open-loop and closed-loop system acceleration response and the corresponding 1 sec running RMS accelerations

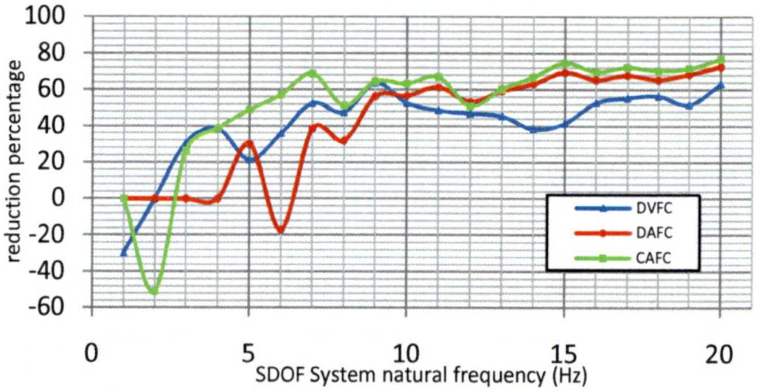

Figure 9 – Acceleration MTVV Reduction of DVFC, DAFC and CAFC systems

The trends of the reduction in MTVV of the closed loop (for DVF, DAF and CAF) responses in comparison with the open-loop system responses are plotted in Fig. 9. Again, it can be see that DVF exhibits better vibration mitigation performances at low frequencies and CAF exhibits better vibration mitigation performances at high frequencies.

5.3. Reduction in Decaying Time (For Heel-drop Excitation Only)

The optimum gains determined from the root locus studies in section 5.1 presented in section 5.4 are used in the analytical studies in this section for each of the idealised structural models. The time for the responses to decay to within 5% of the initial peak is used to evaluate the vibration mitigation performance of DVF, DAF and CAF for each of the structural models subjected to an impulsive excitation force. Fig. 10 shows a typical open-loop and closed-loop response. Fig. 11 shows the vibration mitigation performances for each of the control laws implemented on each of the idealised structural models as percentage improvements. Again, it can be seen that DVF performs best at low frequencies while CAF performs better at higher frequencies.

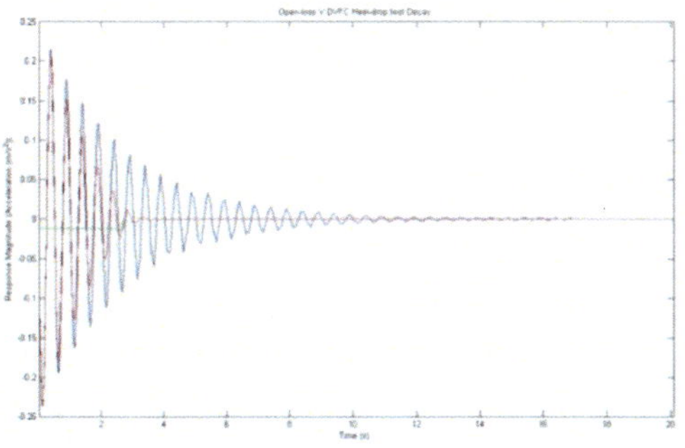

Figure 10 - Typical open-loop and closed-loop acceleration responses to an impulsive excitation

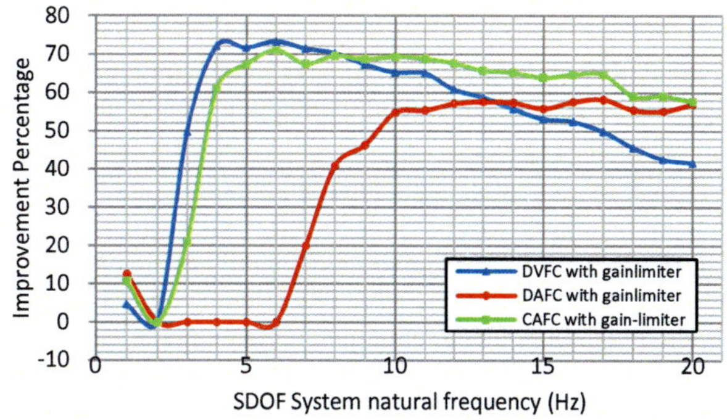

Figure 11 - Decrease of Reduction Time of Peak Responses to 5% of its Value from Open-loop to Closed-loop system (Heel-drop Test)

5.4. Optimum Gains

The optimum gains used in the above analytical studies that have been determined from the root locus studies are presented in Fig. 12 for DVF, DAF and CAF control laws.

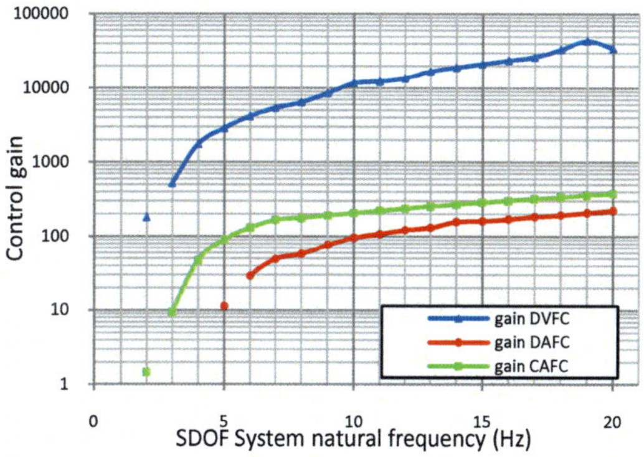

Figure 12 - Optimal gains for DVF, DAF & CAF control laws

6. Results from Analytical Studies

6.1. DVF Control Law

From Figs. 7, 9, and 11, it can be seen that DVF is more effective in predominantly low frequency floors i.e. between 3-10 Hz. Fig. 7 exhibits a sharp peak between 3-4 Hz and there is a gradual decrease in higher frequencies. DVF does not have any significant effect on the very low-frequency floors i.e. with natural frequencies ranged 1-3 Hz as a result of the actuator dynamics lying within this range.

Some root locus plots of some of the idealised floor models in Figs. 13, 14, 15, 16 and 17 aid in justifying the observed vibration mitigation performances with DVF control law.

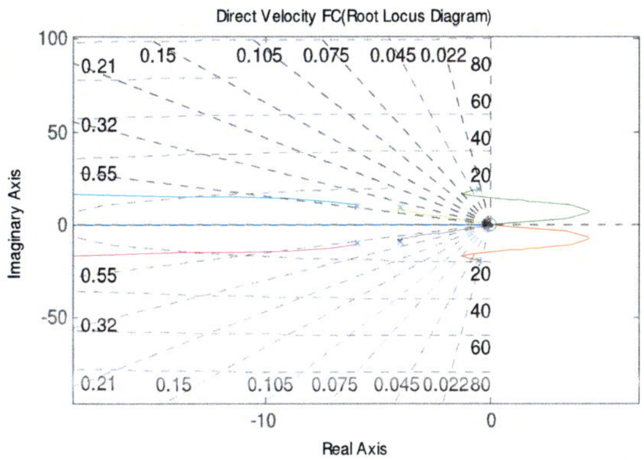

Figure 13 - Root locus plot for structural model with f = 3 Hz for DVF control law

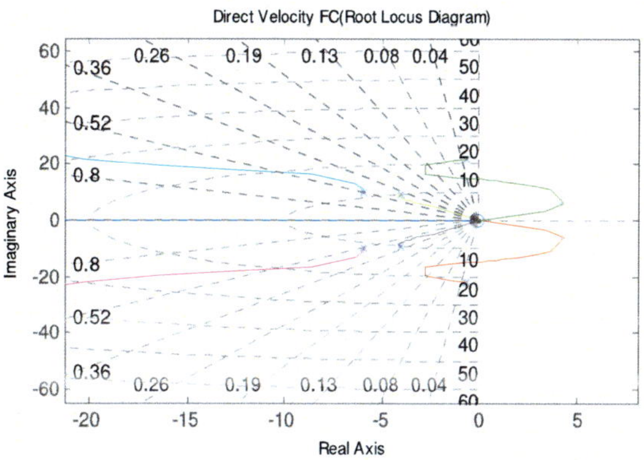

Figure 14 - Root locus plot for structural model with f = 3.5 Hz for DVF control law

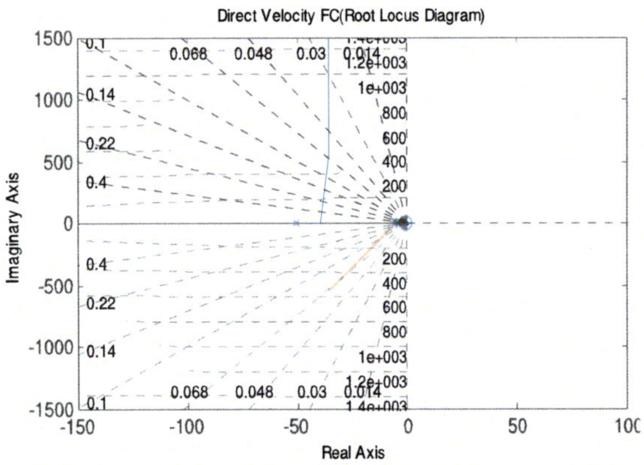

Figure 15 - Root locus plot for structural model with f = 3.8 Hz for DVF control law

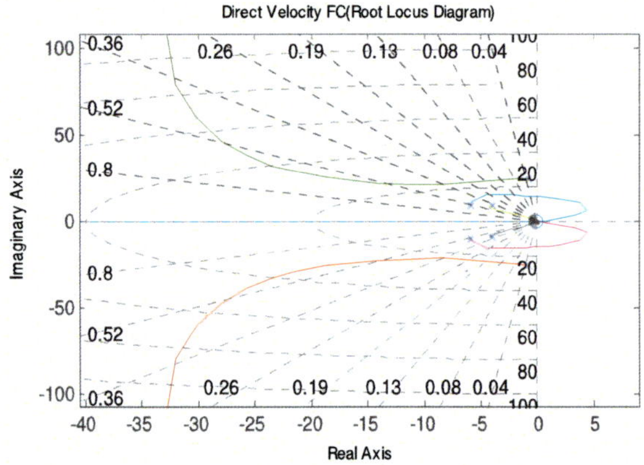

Figure 16 - Root locus plot for structural model with f = 4 Hz for DVF control law

The structural models comprise of a pair of very low-damped poles and a zero in the origin. A pair of high-damped poles represents the actuator dynamics and a pole in ε = -50.26 represents the low-pass element of the actuator. An obvious pole flipping between floor and actuator transfer function poles can be seen about f = 4 Hz, and the loci suddenly gain a departure angle very close to the negative real axis after 4 Hz which implies higher structural damping can be attained after this frequency. This accounts for the sharp peak between 3.5-4 Hz in Fig. 7.

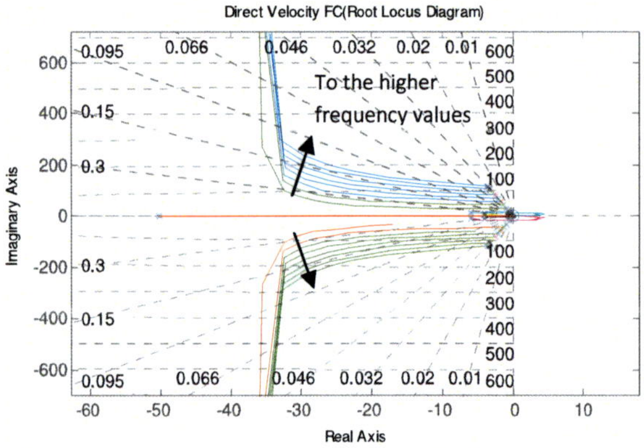

Figure 17 - DVFC system Rlocus diagram evolution for f = 5 Hz to f = 20 Hz

6.2. DAF Control Law

From Fig. 7, it can be seen that DAF control law does not offer any significant influence on the dynamic properties of the idealised models with natural frequencies below 5 Hz. After 5 Hz, however, the damping introduced into the system gradually increases almost linearly. This can be explained considering the results from the root locus studies presented in Figs. 18, 19, 20 and 21.

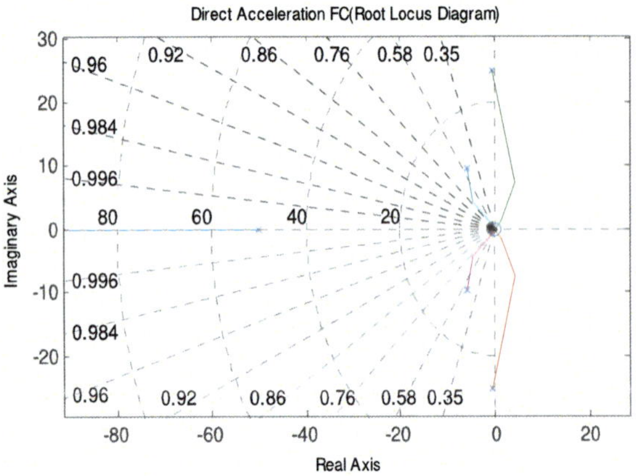

Figure 18 – f = 4 Hz DAFC system Rlocus diagram

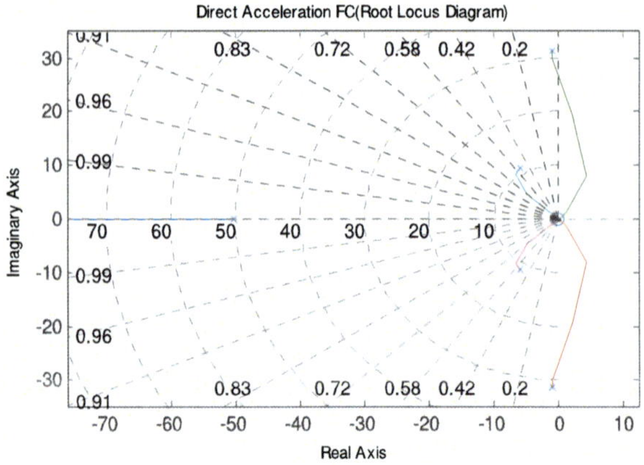

Figure 19 – f = 5 Hz DAFC system Rlocus diagram

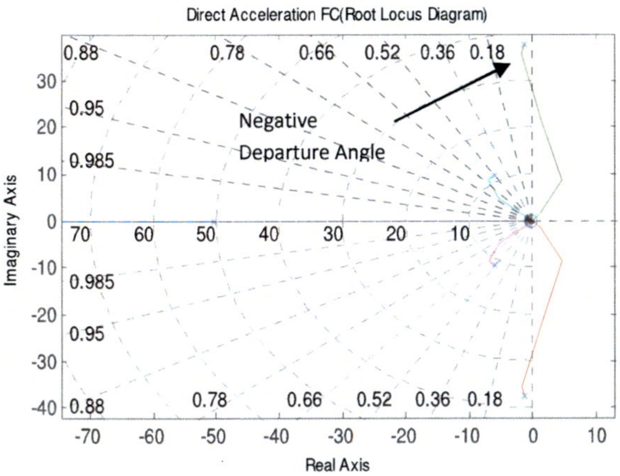

Figure 20 – f = 6 Hz DAFC system Rlocus diagram

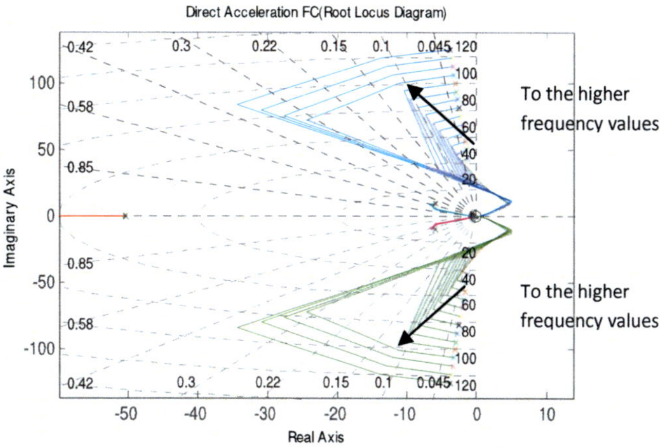

Figure 21 – Evolution of f = 7 to 20 Hz DAFC system Rlocus diagram

No pole flipping occurs in these closed loop systems. For idealised structural models with modal properties between 1 Hz to 5Hz, the departure angles of the loci of models are located in the right half plane towards the positive real axis. This indicates a near-unstable scenario with a very low damping value. This positive departure angle introduces much less damping than the open loop damping of the structural models. For idealised structures with natural frequencies above 5 Hz, the departure angles of the loci gradually change clockwise from positive to negative values. A gradual evolution of the root locus plot for structural models above 7 Hz can be seen in Fig. 21.

6.3. CAF Control Law

CAF control law is an integrated version of DVF and DAF control schemes. It aims at utilising the advantages of DVF in lower frequencies and DAF in higher frequencies. This can be seen in Fig. 7. The zero of the compensator at '–a' improves systems stability by inducing higher damping as well as attracting the system poles after pole flipping. Figs. 22, 23, 24, 25, 26 and 27 highlight the evolution of root locus plots of the closed-loop systems for the idealised structural models noted earlier.

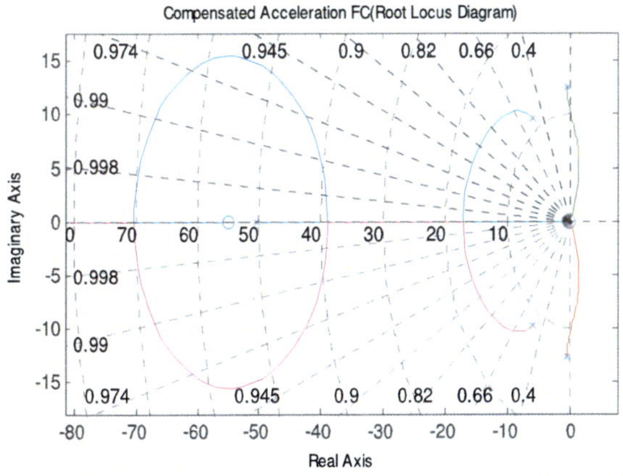

Figure 22 - Root locus plot of system model with f = 2 Hz

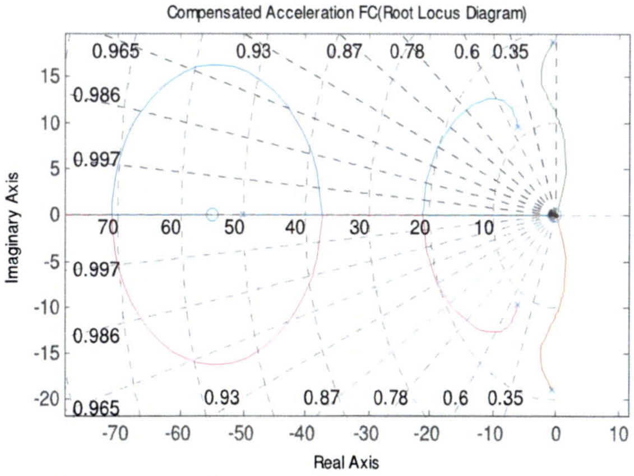

Figure 23 - Root locus plot of system model with f = 3 Hz

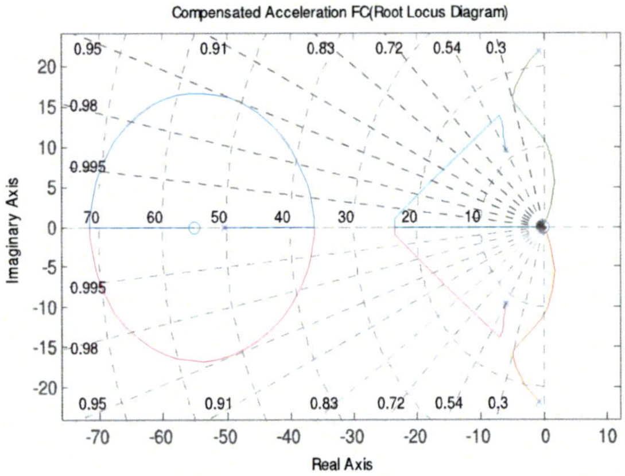

Figure 24 - Root locus plot of system model with f = 3.5 Hz

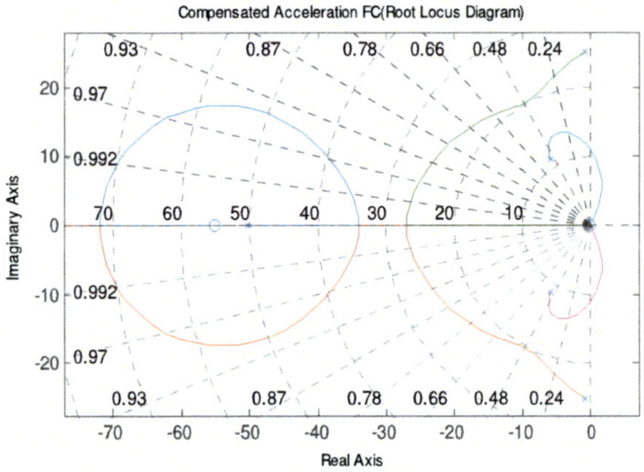

Figure 25 - Root locus plot of system model with f = 4 Hz

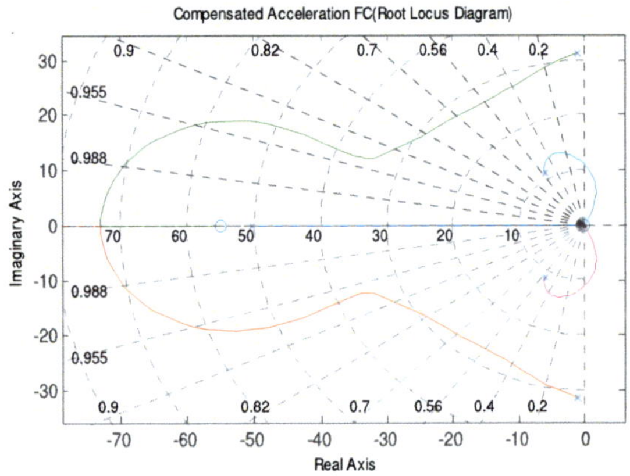

Figure 26 - Root locus plot of system model with f = 5 Hz

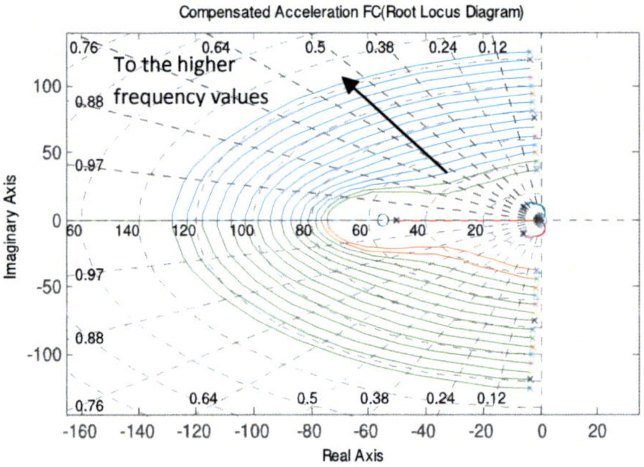

Figure 27 - Root locus plots of system models with f = 6 - 20 Hz

Two pole flipping phenomena can be seen in Figs. 25 and 26. The first one happens between f = 3.5 Hz to f = 4 Hz, where the new floor loci show higher stability and damping. The second one happens between f = 4 Hz and f = 5 Hz where phase-lag compensator zero at '-50.26' attracts one of the system poles that makes the system highly stable and damped.

7. Conclusion

DVF performs best for low frequency structural models while DAF has a tendency to perform better at higher frequencies. CAF acts quite well in both in mid-range and high frequency floors; there is still a problem with implementations in low-frequency floors. Finding proper compensators for mitigation of vibrations in very low frequency structural models is an exciting challenge and an ongoing area of research.

Acknowledgement

The first author is very grateful to Mrs. Legha Shahabpoor for her kind care and support.

Bibliography

1. **Linda M. Hanagan, Thomas M. Murray.** *Active Control Approach for Reducing Floor Vibrations.* s.l. : Journal of structural engineering, November 1997, Vol. 123, pp. 1497-1505.

2. **Ivan M.Diaz, Paul Reynolds.** *Acceleration Feedback control of human-induced floor vibrations.* 2009.

3. **A. Preumont.** *Vibration Control of Active Structures: An introduction.* Dordrecht, The Netherlands : Kulwer Academic, 1997.

4. **Reynolds P, Diaz IM, Nyawako DS.** *Vibration testing and active control of an office floor.* Orlando, Florida, USA. : In: proceedings of the 27th International Modal Analysis Conference, 2009.

5. **Diaz IM, Nyawako DS, Reynolds P.** *On-Off nonlinear velocity feedback control for cancelling floor vibrations.* St.petersburg, Russia : proceeding of the 4th Europen Conference on structural Control, 2008. Vol. 1, pp. 175-82.

Proceedings of the IMAC-XXVIII
February 1–4, 2010, Jacksonville, Florida USA

Practical Implications of Optimizing an Active Floor Vibration Controller

Linda M. Hanagan
Associate Professor of Architectural Engineering
The Pennsylvania State University
104 Engineering Unit A
University Park PA, 16802

ABSTRACT

Lightweight steel floor systems, utilizing typical bay sizes while supporting few non-structural elements, are prone to disturbing levels of floor vibration due to walking excitation. Even though design criteria exist to assess and avoid this problem, decisions based on cost are often made to construct a floor that is prone to excessive vibration. These floors are very difficult to fix by conventional means without great disruption to the occupants.

Active control, using a velocity feedback loop and a proof mass actuator to deliver the force, has proven to be a very effective repair with little disruption to the occupants. Perhaps the greatest barrier to wider implementation of this repair technique is cost. To keep costs down while obtaining the best performance possible, it is important to understand the dynamic behavior and interaction of all the elements. This paper explains the important parameters in designing an active control system. Then specific hardware is described and utilized in implementing an active control system in an occupied building.

INTRODUCTION

Excessive floor vibrations are usually caused by either equipment or by normal occupant activities. Occupant activities that have a repetitive motion, such as walking, dancing, and jumping, can cause objectionable floor motion due to resonance. Resonance exists when the frequency of the floor disturbances are the same as a natural frequency of the floor system. When resonance occurs, the amplitude of the vibration is largely affected by weight and damping characteristics of the structure. Damping is an effect that causes a dissipation of energy so that resonant vibrations cannot build to large levels. At resonance, decreases in weight and damping result in higher vibration amplitudes given a certain level of disturbance. Such decreases in weight and damping are occurring more often due to recent trends in building construction and use resulting in an increase in the instances of problem floor vibrations.

Modern construction practices are producing buildings that have natural frequencies more prone to resonance, support less weight, and have less damping than buildings of the past. Additionally, many modern offices have fewer partitions, fewer paper files, etc. and therefore fewer non-structural mechanisms for providing damping and weight. In some instances, owners of office buildings in successful service for 30 years are reporting problems that seem to be the result of a change from fully enclosed offices to cubicles. Also, structural materials have become stronger and more lightweight, thus allowing longer spans with less weight. Longer spans tend to vibrate at lower frequencies, thus resulting in natural frequencies more prone to resonance. All of these trends lead to an increased likelihood of problem floor vibrations. Several case studies describing problem floors are presented in Ref. [1].

Active control of floor vibration was originally implemented with off-the-shelf components and a computer controlled feedback circuit. In April 2005, implementation improvements were described in US Patent No. 6,874,748. The claims of this patent are primarily optimization recommendations and were primarily documented to protect a future device developer.

DOI 10.1007/978-1-4419-9831-6_20, © The Society for Experimental Mechanics, Inc. 2011

OVERVIEW OF THE CONCEPT

Control of resonant vibrations is fundamentally accomplished by adding damping to the system. By using direct velocity feedback, the control force is proportional to velocity, thus adding the effect of damping to the system as shown by the single degree of freedom system represented in Eq. 1:

$$M\ddot{y}(t) + C\dot{y}(t) + Ky(t) = F(t) - G\dot{y} \quad [1]$$

where M, C, and K are the system mass, damping, and stiffness, and $y, \dot{y},$ and $\ddot{y}$ are the displacement, velocity, and acceleration of the mass. $G\dot{y}(t)$ is the control force and when moved to the left hand side of the equation, it increases the damping term to $C + G$.

This simple concept becomes a bit more complicated when practically implemented. This paper discusses the practical implementation issues that should be considered when building an efficient active floor vibration control system.

PRIMARY COMPONENTS

The practical implementation issues and recommendations are discussed in terms of the three primary components of the control system.

a) a proof-mass actuator (comprising a linear electric motor, a moving mass, a linear spring, and a damping mechanism) to generate the control force from an input signal,
b) a sensor capable of producing an output signal proportional to vibration velocity, and
c) a feedback controller to generate an appropriate control signal, that is a function of the velocity signal, to drive the proof mass actuator.

Proof-Mass Actuator

A proof-mass actuator, also known as a shaker, creates a force through the inertial effect of a moving mass. This mass can be moved up and down, by an electric motor, to create the required force. An electro-magnetic motor has proven most suitable to create the driving force because it is quiet and responds quickly at low frequencies (low phase distortion). When selecting the electric motor, the force and stroke limits are the most important consideration and are strongly related to cost. When the motor is configured into a proof-mass, the stiffness and damping should be compatible with those limits. Otherwise, needless expense has gone to underutilized force or stroke. Because motor force is expensive, a spring system attached to a frame should be utilized to support the static weight of the moving mass at the center of the stroke. A sketch and schematic of this system is shown in Fig. 1.

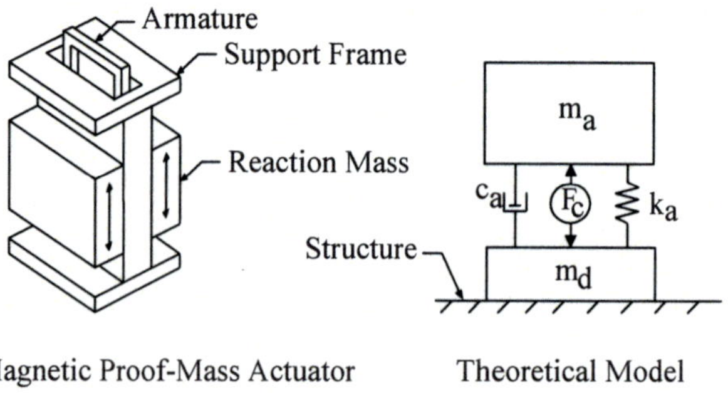

Fig. 1. Illustration and theoretical model of shaker

The required motor force level limit necessary to provide significant control of resonant vibration can be determined from vibration excitation levels predicted by AISC Design Guide 11 [2]. For walking vibration, the peak resonant force (lbs) at the fundamental natural frequency of the floor bay is predicted by Eq. [2].

$$F_{walking} = 157(0.83)\exp(-0.35f_n) \quad [2]$$

where f_n is the fundamental frequency of the floor system. Since many problem floors have fundamental frequencies in the range of 5 – 8 Hz, the peak control force for walking induced vibration should be in the range of 10 – 25 lbs. Similar predictions can be made for rhythmic excitations due to activities like dancing and aerobics. The reader is cautioned that significant vibration levels of vibration can occur at off-resonant frequencies when groups engage in rhythmic activities. A damping solution, such as that proposed here, is ineffective in these instances.

There is interplay between the stroke and force limits and the passive parameters (m_a, c_a and k_a) of the proof-mass actuator that should be understood to optimize the system. First, there is a relationship between the motor force in and the control force out. The force varies with damping and the ratio of the forcing frequency to the natural frequency as shown in Fig. 2. The red lines represent a damping value that provides adequate stability of the actuator mode without too much phase or amplitude distortion between to input signal (assumed proportional to the motor force, Fin) and the resulting control force (Fout). The stability of the actuator mode is described more thoroughly in Ref. [3]. A damping ratio (β) of 0.3 and 0.4 of critical ($c_{cr} = 2\sqrt{k_a m_a}$) is recommended. The vertical line in each plot is located at a frequency ratio of 2. This line represents the recommended cutoff of floor frequencies that can be controlled without significant phase distortion between the input signal and the control force. That is, the fundamental floor frequency should be twice that of the actuator to maintain adequate stability and control effectiveness. The force limit of the actuator can be maintained by a clipping circuit which limits the input signal.

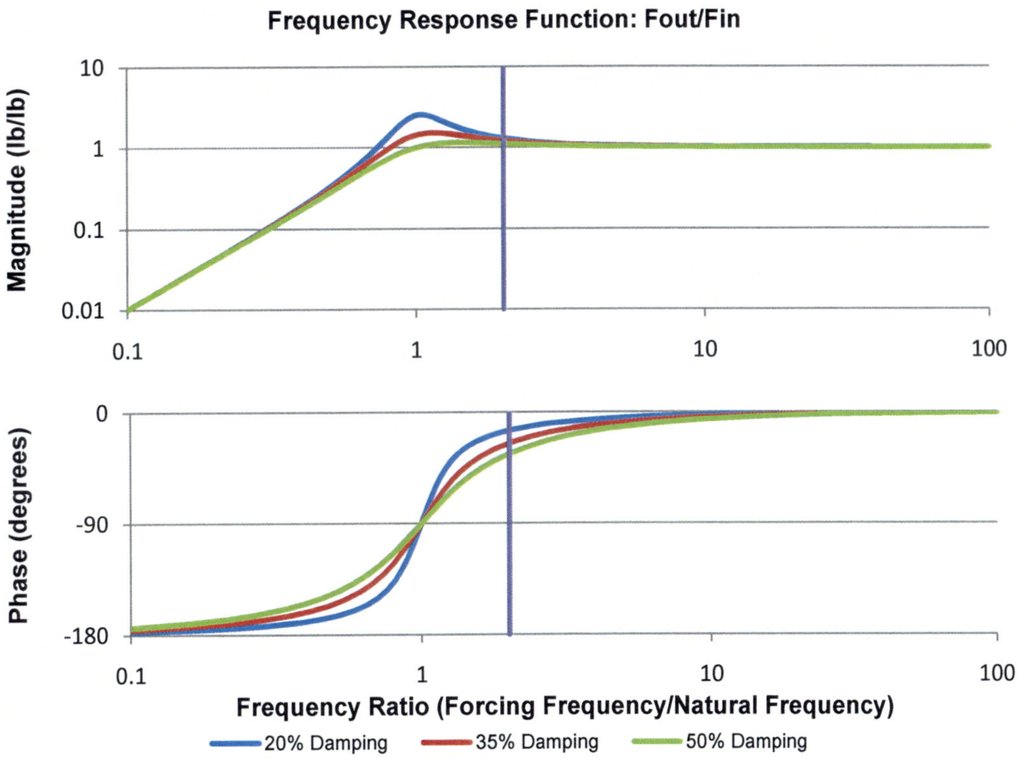

Fig. 2. Frequency Response Function of Proof-Mass Actuator: Fout/Fin

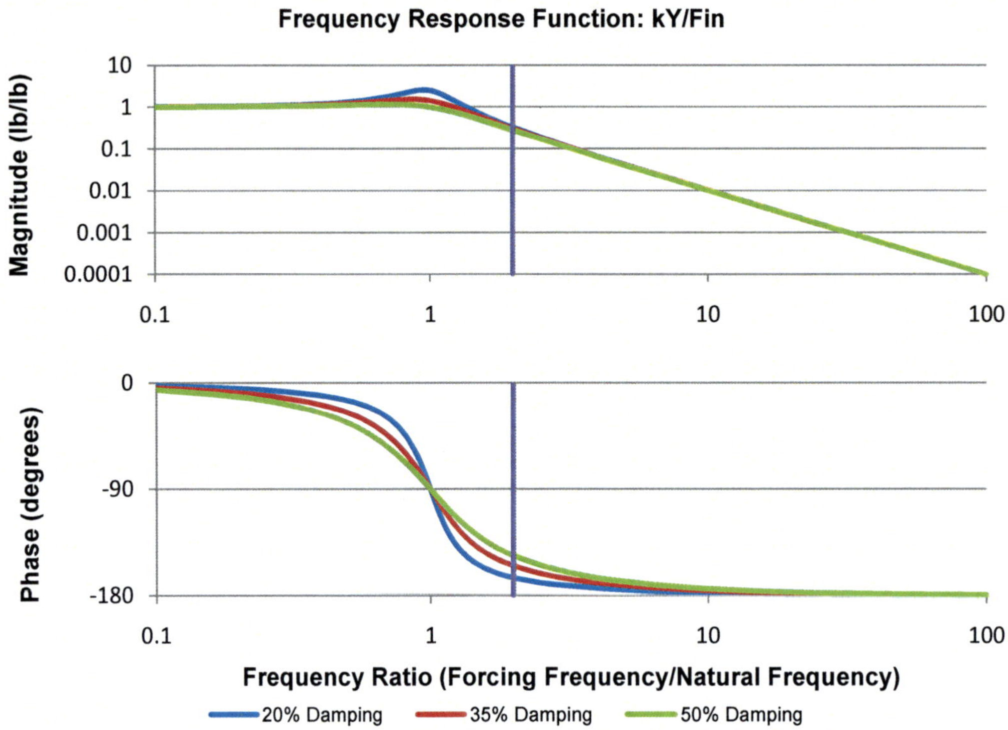

Fig. 3. Frequency Response Function of Proof-Mass Actuator: kY/Fin

For a simple control circuit, where the control force is proportional to the floor velocity and the stroke and force limits are not exceeded, the stroke limit should be maintained by the signal clipping circuit also. Achieving stroke control within a force limit is best understood from a frequency response function of the displacement times the stiffness verses the input force as shown in Fig. 3. From the magnitude graph, it can be noted that the displacement amplitude is frequency dependant and peaks near the damped natural frequency of the actuator frequency for a steady-state input. Eq. 3 below shows the inequality that must be satisfied for a simple clipping circuit to achieve both force and stroke control in the event there is a large velocity response.

$$\frac{1}{\sqrt{4\beta^2(1-\beta^2)}} \leq \frac{k_a y_{max}}{F_{inmax}} \qquad [3]$$

where β = damping ratio for the actuator = c_a/c_{cr}; k_a = linear spring stiffness; y_{max} = maximum stroke of the actuator (half of the total stroke of the actuator); F_{inmax} = force limit set by the clipping circuit in the controller. A well configured actuator will have properties such that the inequality is close to equal with F_{inmax} set near the force limit of the actuator. Otherwise expense associated with either excess force capacity or excess stroke capacity is wasted.

Velocity Sensor

A velocity sensor can be used to provide a velocity feedback signal to the control system. Velocity sensors provide a relatively accurate measure of velocity magnitude for frequencies within the published bandwidth of the sensor. The lower limit on the bandwidth is the result of sensor dynamics. Typically piezoelectric sensors accomplish a velocity output measurement by applying a filter to an accelerometer that acts as an integrator circuit. Fig. 4 shows the estimated sensor dynamics of a PCB sensor that was used successfully in a control circuit. The vertical line at 2.5 Hz indicates the lower bandwidth of the sensor. The straight line in each graph represents the relationship of velocity to acceleration if an infinite sensor bandwidth was possible. It should be noted that the magnitude is reasonably predicted at frequencies above the lower bandwidth line. This might imply that a good measure of velocity can be achieved for frequencies above the lower bandwidth. When used in a control configuration, the phase is more important than the amplitude. If this sensor were to be used to measure resonant frequencies near the lower limit of the bandwidth, the phase would be about 66 degrees instead of 90 degrees. At 66 degrees, the projection on to the imaginary axis (sin(33)=0.55) would be only 55% of the actual velocity, whereas, at 2x the lower bandwidth, the projection on to the imaginary axis would be about (sin(66) = 0.91) 91% of the actual velocity. This has proven to be an acceptable distortion of the actual velocity. It is therefore recommended that piezoelectric velocity sensors only be used to control floors with fundamental natural frequencies at least 2x the lower limit of the sensor bandwidth.

Feedback Controller

The feedback controller takes the signal from the velocity sensor and adjusts it to provide an input signal to the proof-mass actuator. Minimally, the controller must apply an appropriate gain to the velocity feedback signal and this signal must be clipped if it would overdrive the actuator. It is recommended that the control gain be set to approximately one half of the control gain that would cause unstable behavior. The effectiveness of the control system with varying feedback gains is described in Ref. [3]. To find the unstable control gain, the gain should be adjusted until an increasing oscillation is observed near the natural frequency of the actuator. Although this is unstable, possibly harmful behavior in a purely linear feedback circuit, the clipping circuit will keep the actuator from being overdriven. The feedback controller may also need an offset correction circuit. Some sensors have a slight DC offset that should be corrected to avoid any DC input to the control actuator. Finally, piezoelectric sensors require a charge amplifier to produce a measureable signal. This circuit can also be included in a feedback controller box. All of the circuit components can be purchased on the internet and configured by someone with some knowledge about analog circuit components.

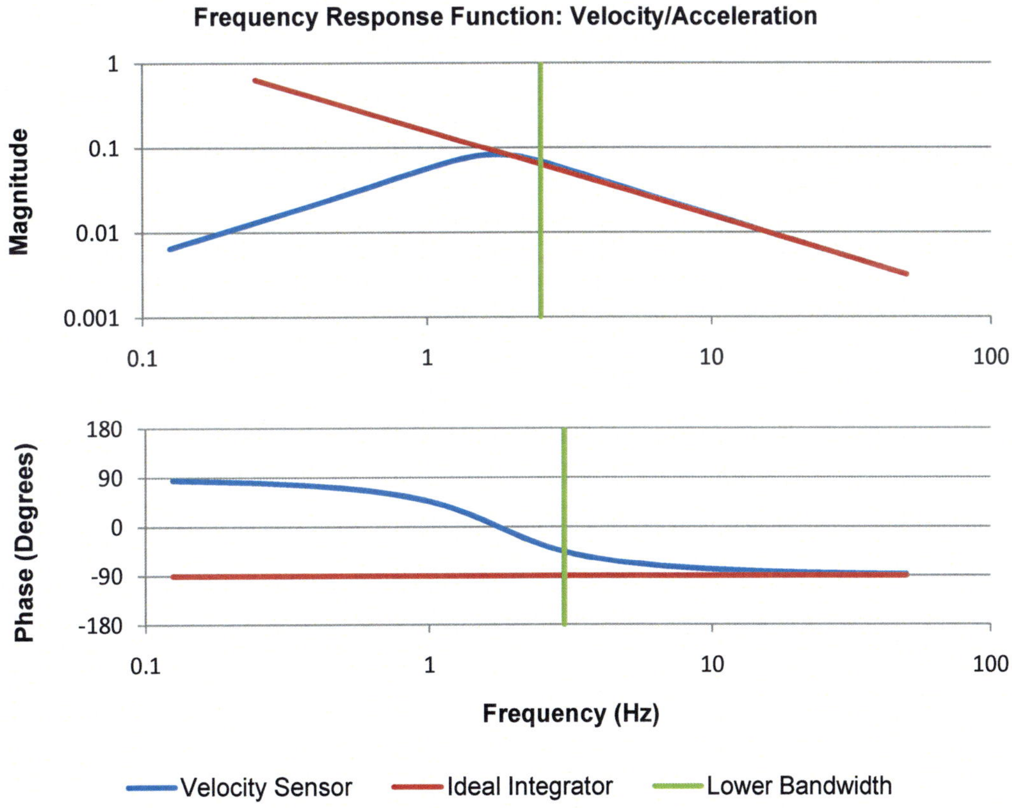

Fig. 4. Frequency Response Function of Piezoelectric Velocity Sensor

CONFIGURING A SYSTEM

The system described above was built and utilized in several temporary and one permanent installation. These installations are described in Refs. [4] and [5].

APS Dynamics sells two "shaker" models with a proof-mass option. One system is capable of delivering a 30 lb peak control force. The larger system is capable of delivering a 100 lb peak force. Both systems have a total stroke of 6" and an estimated system damping of approximately 0.30. The larger shaker was used in all of the installations. This shaker does not satisfy the requirement of Eq. [3]; therefore, the control force must be clipped below 100 lbs. Using Eq. [3] and the shaker properties listed in Table 1 below, the control force should be limited to approximately 26 lbs. Currently, the 100 lb shaker with matching amplifier is just under $18,000. The smaller shaker could be configured with added mass and stiffness to more efficiently utilize the motor capability. Alternately, a cheaper linear motor could be configured into a proof-mass system. It was the hope that the patent would inspire device development.

Table 1. APS Shaker Properties for Model 100

m_a	67 lb/386.4 in/sec^2
k_a	15.4 lb/in
β	0.30

PCB Industrial Division sells several piezoelectric velocity sensors with a bandwidth starting at 3 Hz. The sensor used in previous studies does not seem to be available. A system made with these components could deliver a 26 lb control force to a floor with a fundamental frequency at 6 Hz or above.

CONCLUSION

Active control has only proven a reasonable solution when cost is not an issue. Since this is rarely the case, the development of a system that is cost effective is necessary. Following the claims of US Patent No. 6,874,748 and the recommendations contained herein, a more efficient device than currently available could be developed. The key to this development is a cheaper proof-mass actuator that could be mass produced.

REFERENCES

1. Hanagan, L.M. (2005), "Walking-Induced Floor Vibration Case Studies," *Journal of Architectural Engineering*, ASCE, 11(1),14-18.
2. Murray, T.M., Allen, D.E. and Ungar E.E. (1997). "Floor Vibrations Due to Human Activity," *AISC Steel Design Guide #11*, AISC, Chicago, IL.
3. Hanagan, L. M., and Murray, T. M. (1997). "Active Control Approach for Reducing Floor Vibrations," *Journal of Structural Engineering, ASCE,* 123(11), 1497 - 1505.
4. Hanagan, L.M. and Murray, T.M. (1998). "Experimental Implementation of Active Control to Reduce Annoying Floor Vibrations." *Engineering Journal, AISC,* 35(4), 123-127.
5. Lichtenstein, H. A. (2004). "All Shook Up," *Modern Steel Construction*, AISC, Chicago, IL.

Proceedings of the IMAC-XXVIII
February 1–4, 2010, Jacksonville, Florida USA

Dynamic Testing and Analysis of a Football Stadium

F. Necati Catbas [1], Mustafa Gul [2] and H. Ozerk Sazak [3]

Department of Civil, Environmental and Construction Engineering

University of Central Florida, Orlando

[1]catbas@mail.ucf.edu

[2]mgul@mail.ucf.edu

[3]sazakozerk@gmail.com

ABSTRACT

The enthusiasm of people during popular events may cause excessive vibration in stadium structures especially if this movement is coordinated, e.g. if a song is played in the stadium. For certain cases, this vibration may even exceed the acceptable limits defined in different standards. These dynamic vibration levels are generally very difficult to simulate and determine at the design stage. Therefore, the stadia are sometimes monitored to determine the performance of the existing structure during real events such as games and concerts. In this study, dynamic response monitoring of a steel football stadium is presented. The objective of this study is to determine the vibration levels in the context of human-comfort levels in design codes in a comparative way for different events. Especially, the dynamic response caused by human-stadium interaction is evaluated for specific events such as scoring and playing certain songs. The characteristics of these events are also explored in terms of frequency content in relation to the dynamic characteristics of the stadium.

INTRODUCTION

Crowd excitation during the stadium events may result excessive vibration, which might create serviceability problems at the stadiums. Modern stadium structures are usually more slender because of the architectural aspects and improved technologies in construction. Although this type of a design brings many advantages, the level of vibration during certain events might get high, especially if the crowd is involved in a sort of coordinated motion. High level of vibrations may even exceed the limits, which are defined by international codes. Therefore stadiums should be carefully investigated in terms of vibration characteristics.

Different researchers from different countries conducted various tests related to stadium monitoring. These stadiums are analyzed based on human comfort, serviceability where the vibration levels are compared to the standards, e.g. ISO2631 [1]. Laboratory investigations on the evolution of human vibration-perception and comfort states with the change of vibration levels are conducted previously [2]. Acceleration values are used to maintain the consistency with international standards. Another approach is modal parameter identification from forced or ambient tests. For example, Reynolds and Pavic [3] presented analysis results of modal testing and in-service monitoring of a large contemporary cantilever grandstand. They showed that crowd occupation can affect the modal parameters of a stadium. They also indicated that previously proposed methods for assessment of vibration serviceability can lead to inconsistent results due to their sensitivity to the data acquisition and analysis techniques used. Reynolds et al. [4] conducted forced and ambient excitation tests to analyze the dynamic

T. Proulx (ed.), *Dynamics of Civil Structures, Volume 4*, Conference Proceedings of the Society for Experimental Mechanics Series 13,
DOI 10.1007/978-1-4419-9831-6_21, © The Society for Experimental Mechanics, Inc. 2011

characteristics of a stadium. They mentioned that calculated frequencies are lower than the measured frequencies and this might be the result of the extra stiffness caused by the non-structural elements that were not included in the design. Caprioli et al. [5] tested the G. Meazza stadium during soccer matches and live concerts because of some complaints about the high vibration levels in some buildings close to the stadium during big live concerts. They showed that vibration inside the stadium is nearly the same with vibration outside of the stadium. They also mentioned that vibration characteristic is determined by synchronized loadings. Salyards and Hanagan [6] analyzed the effect of coordinated motion on a large football stadium. They observed that coordinated motion vibration level during game is more substantial that reacting to big events during the game.

Objective and Scope

The authors had previously presented the preliminary results of a monitoring study where the dynamic characteristics and response of a stadium during different games were investigated [7]. Examples of two different data sets recorded in two games, and preliminary results of these tests were presented. The objective of this paper is to analyze the effects of the vibration to human comfort and to investigate the dynamic characteristics of the same stadium structure during different events. Monitoring is conducted in two different games before, during and after each game. In both games, nearly the number of spectators watching the game was almost the same .

Figure 1 summarizes the overall objective of this on-going study where different experimental and analytical components are combined for the analysis. In the current study, the preliminary experimental results are presented in the context of human comfort and modal analysis. For the future studies, the FE model will be improved and calibrated by using the experimental analysis results (e.g. natural frequencies and mode shapes). Finally, different simulations will be conducted with the calibrated model under different type of loadings for performance evaluation.

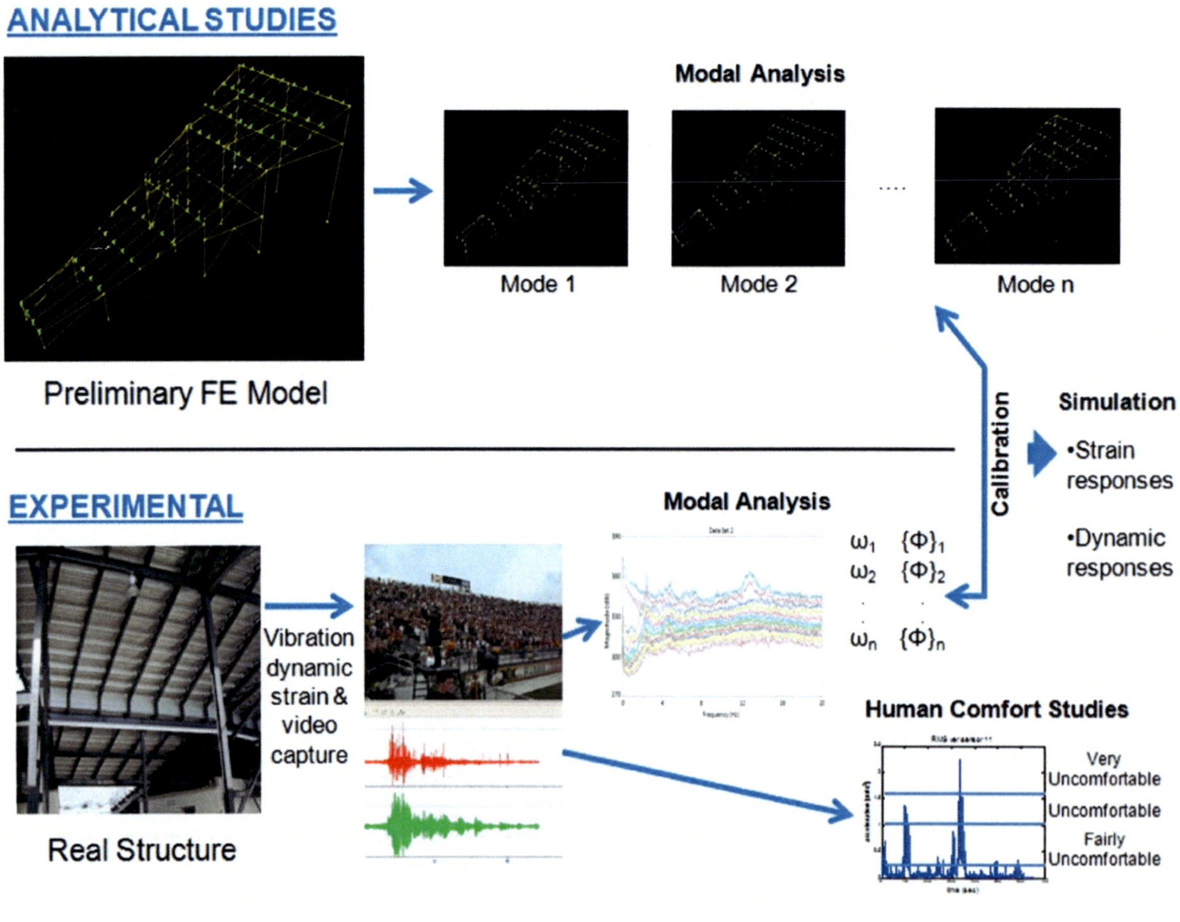

Figure 1: The general objective of the on-going study

INSTRUMENTATION AND DATA ACQUISITION

The stadium is a steel frame type structure sitting over 25 acres and the seating capacity is approximately 45,000. The construction of the stadium was completed in 2007 and it was opened to service in the same year. Stadium has 38 lower level and 36 upper level sections. The sensors are installed at the student sections where vibration level is usually higher than other sections of the stadium. The investigated section is one of the rotating corners of the stadium as marked in Figure 2 and has a narrowing down frame section.

Figure 2: The section under investigation

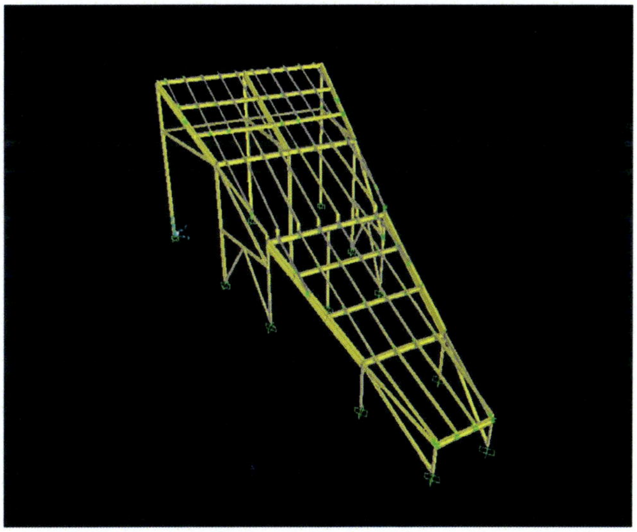

Figure 3: General view of the preliminary FE model

Rear end of the frame section has 60 feet opening and about 55 ft height. On the other hand, the front end has about 15 ft opening and about 7.5 feet height. Figure above shows the general view of the preliminary FE model (Figure 3). Stadium is mainly based on two sections in terms of seating numbering which are upper section and lower section. Accelerometers are placed to both upper section and lower section for the monitoring of stadium. Twelve accelerometers in total are used in that section (Figure 4). Eight of those accelerometers were in the upper section (Figure 5) and remaining four accelerometers were in the lower section (Figure 6). Accelerometers are placed in all three directions to measure the vibrations levels in all directions. For the accelerometers middle points of the frames or stingers are chosen for the sake of maximum vibration.

Figure 4: Instrumentation of Upper Section Sensors

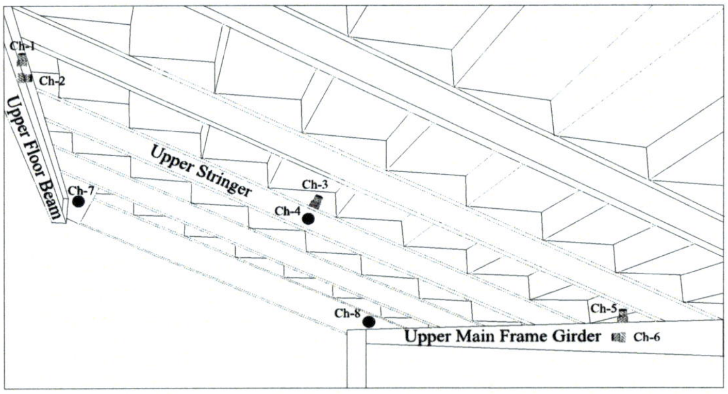

Figure 5: Locations of Sensors (Upper Section)

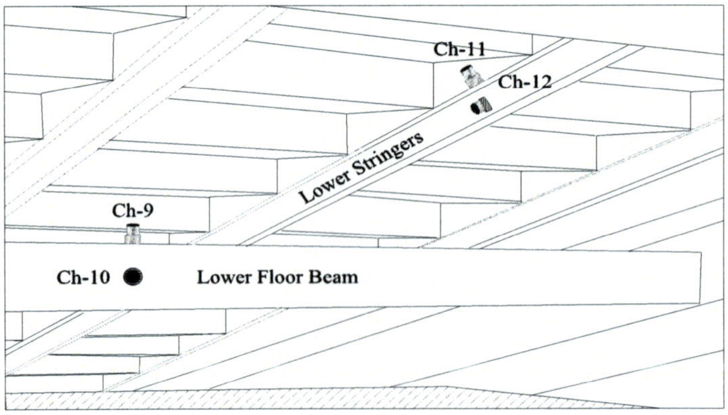

Figure 6: Location of Sensors (Lower Section)

VXI and Agilent Technologies acquisition system is used to record the vibration data (Figure 7). The sampling frequency 100 Hz and data is collected for 10 minute intervals (approximately 25 data sets per game). In addition, video recording was carried out intermittently in order to synchronize some of the typical events such as interceptions and touchdowns.

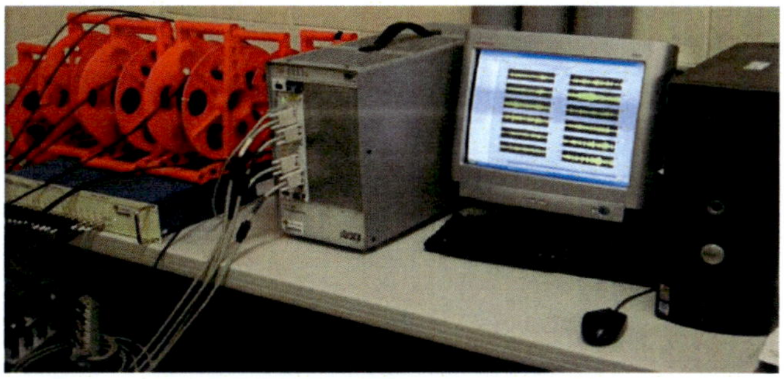

Figure 7: Data Acquisition System

HUMAN COMFORT ANALYSIS

For the human comfort analysis, the general procedure in ISO 2631-1:1997 (ISO 1997) is followed. First, the root mean square (RMS) values are obtained using the weighted acceleration. The time domain acceleration data is converted to frequency domain to apply the filters and to obtain weighted acceleration at the end. In frequency domain, filters such as high pass filter, low pass filter, acceleration-velocity transition filter, upward step filter are applied. After applying these filters simultaneously (the final filters are in the following figure), the data is converted to time domain and the weighted acceleration values are obtained (Figure 8).

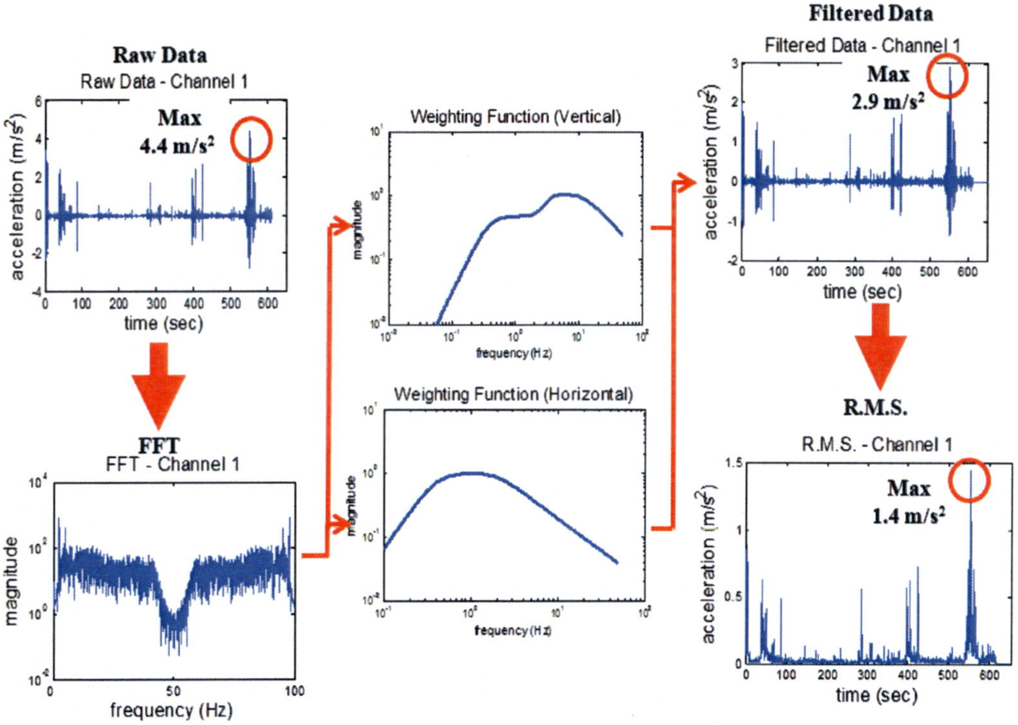

Figure 8: Path of Obtaining Weighted Acceleration

When the acceleration is converted to weighted acceleration (a_w), RMS is calculated as shown in Eqn. 1 to quantify the severity of vibration where a_w is the weighted acceleration and τ is the integration time which is selected as 1 second in this study as recommended in ISO 2631-1:1997 (ISO 1997). Acceptable limit values for human comfort depend on many factors with each different situation however, the RMS values usually give a very good idea about the vibration levels and corresponding human comfort levels and they are presented in Table 1.

$$a_w(t_0) = \left\{ \frac{1}{\tau} \int_{t_0-\tau}^{t_0} [a_w(t)]^2 dt \right\}^{\frac{1}{2}} \qquad (Eq\ 1)$$

Table 1: RMS Comfort Levels

RMS Values	Corresponding Situation
<0.315	Not Uncomfortable
0.315-0.63	A Little Uncomfortable
0.5-1	Fairly Uncomfortable
0.8-1.6	Uncomfortable
1.25-2.5	Very Uncomfortable
>2	Extremely Uncomfortable

EVENT BASED ANALYSIS

The vibration measurements are taken in different games to be able to compare the performance of the structure . As can be seen, same type of events selected from different dates are analyzed. Both games were played at same stadium and nearly the same number of people attended. The vibration levels measured during different events e.g. popular song, touchdown, and interception are shown in Figure 9.

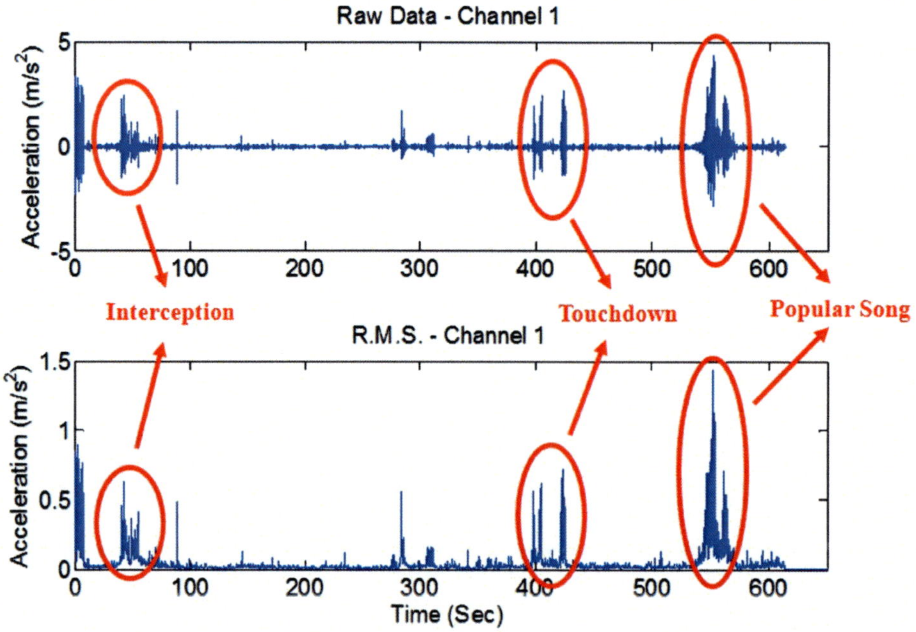

Figure 9: Raw Data vs. RMS values

Table 2: Raw Acceleration (m/s²)

		Upper Floor Beam (Vertical)	Upper Main Frame Girder (Longitudinal)	Upper Floor Beam (Lateral)	Lower Floor Beam (Vertical)	Lower Stingers (Vertical)
Game 1	No Event	0.49	0.32	0.05	0.98	1.07
	Interception	2.44	1.17	0.35	1.97	2.23
	Touchdown	2.65	0.69	0.17	1.03	1.37
	Popular Song	4.39	1.03	0.32	2.19	3.54
Game 2	No Event	0.06	0.14	0.02	0.51	0.53
	Interception	0.51	2.05	0.37	1.52	1.58
	Touchdown	1.31	0.62	0.32	0.96	1.15
	Popular Song	2.51	1.79	0.59	3.84	5.15

1. 10.20.2007 Game 1 (45,510 attendance, 44-23 Win)
2. 10.04.2008 Game 2 (43,147 attendance, 31-17 Win)

The data was recorded before and during the games as well as after all the spectators left the stadium. During these periods, notes were taken by the researchers so that the data sets could be analyzed accordingly. The vibration levels for no event, interception, touchdown and popular song cases are shown in Table 2.

Investigating five different locations (upper floor beam, upper main frame girder, upper floor beam, lower floor beam, lower stringers) and in three different directions, vertical sensors are seen to be the critical sensors. It can easily been seen that popular song event which is a coordinated motion is the critical event for the stadium.

Table 3: RMS Weighted Acceleration (m/s²)

		Upper Floor Beam (Vertical)	Upper Main Frame Girder (Longitudinal)	Upper Floor Beam (Lateral)	Lower Floor Beam (Vertical)	Lower Stingers (Vertical)
Game 1	No Event	0.10	0.04	0.01	0.23	0.31
	Interception	0.63	0.11	0.02	0.61	0.87
	Touchdown	0.62	0.06	0.01	0.29	0.61
	Popular Song	1.44	0.12	0.05	1.07	1.67
Game 2	No Event	0.03	0.02	0.01	0.21	0.24
	Interception	0.27	0.18	0.03	0.51	0.63
	Touchdown	0.53	0.07	0.02	0.55	0.53
	Popular Song	1.54	0.22	0.05	1.57	2.25

When RMS results are analyzed, again vertical sensors show the big difference from the longitudinal and lateral direction placed sensors (Table 3). Vibration levels give the idea of how comfortable the section after comparing the results with the RMS comfort levels given in Table 1. During the popular song vibration levels reach to the level of 1.57 m/s^2. This level of vibration refers to uncomfortable situation level. Even depending on the duration of the vibration, it can be considered as very uncomfortable situation level.

FREQUENCY DOMAIN MODAL ANALYSIS

As mentioned above, the data is collected with a 100 Hz sampling rate. For the modal analysis, the data is analyzed in 0-20 Hz since most of the important vibration modes of the structure lie in this range. Three different data sets are compared and the first six modes identified using Enhanced Frequency Domain Decomposition algorithm [8] are presented in Table 4. In each data set nearly the same frequencies are obtained and an example Singular Value Decomposition plot is shown in Figure 10.

Table 4: Modal Frequencies

	Set 1 (Hz)	Set 2 (Hz)	Set 3 (Hz)	Direction
Mode 1	2.34	2.44	2.32	Coupled (Vert., Long., Lat.)
Mode 2	2.44	2.54	2.49	Long.
Mode 3	2.81	3.71	2.88	Lat.
Mode 4	4.66	4.91	4.66	Coupled (Vert., Long., Lat.)
Mode 5	7.01	7.01	7.00	Coupled (Vert., Long., Lat.)
Mode 6	12.94	12.72	12.67	Vert., Long.

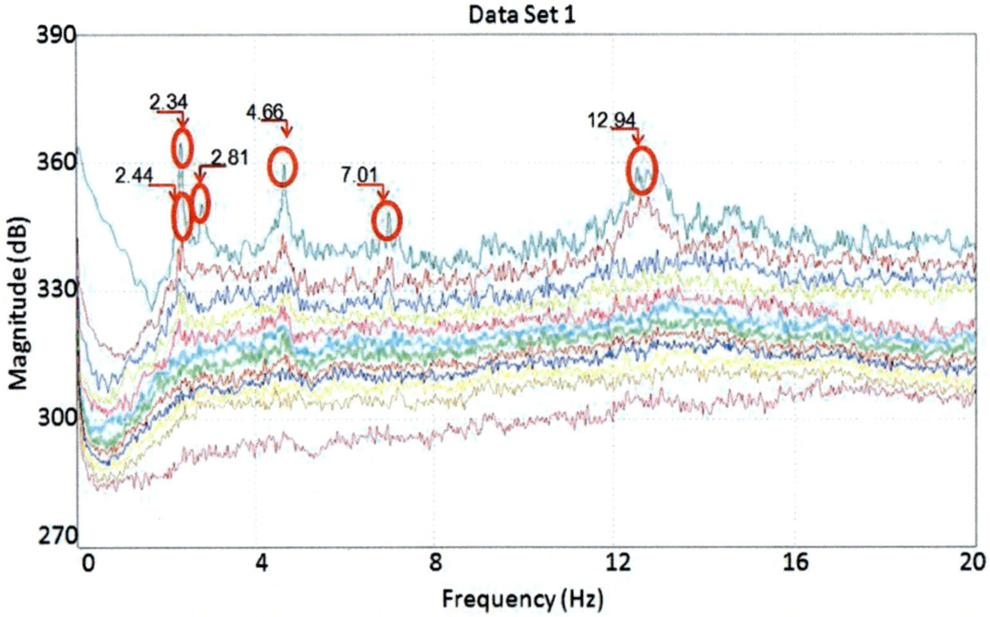

Figure 10: A sample Singular Value Decomposition plot of the FRFs

CONCLUSION

In this paper, preliminary results of an on-going monitoring study of a stadium are presented. First human comfort analysis is conducted based on vibration data collected during different games. Different types of events are analyzed. For event-based analysis, RMS values are obtained with the procedure defined in the ISO 2631 code. While analyzing different events, it is observed that popular song event that corresponds to coordinated motion is the most critical event for human comfort. In the raw data set, the maximum acceleration was recorded as 4.39 m/s^2 in the first game and 5.15 m/s^2 in the second game. When the RMS values calculated by using the weighted accelerations they are found as 1.44 m/s^2 for the first game and 2.25 m/s^2 for the second game for the same events. This level of vibration is considered as very uncomfortable in the standards. The overall vibration levels at the stadium seem to be considerable higher than the other examples reported in the literature suggesting that a more detailed analysis may be warranted. It is also important to know the dynamic loading characteristics and the modal parameters of the structure. For the modal analysis of the structure, the data is analyzed using a frequency domain ambient vibration analysis method. The first six modes are obtained for three different data sets and it is shown that these results are in good agreement, indicating that the extracted modal parameters are acceptable. For future work, the FE model will be calibrated by using the analysis results of the experimental data. By using the calibrated FE model, different types of analyses will be conducted to investigate the performance of the stadium under different dynamic loading conditions.

REFERENCES

1. Standardization, International Organization for, ISO 2631-1. Mechanical Vibration and Shock-Evaluation of Human Exposure to Whole-Body Vibration. Part 1: General Requirement. 1997: Geneva.

2. Nhleko S. P., Williams M.S., Blakeborough A., Vibration Perception and Comfort Levels for an Audience Occupying a Grandstand with Perceivable Motion. in 27th International Modal Analysis Conference (IMAC XXVII). 2009. Orlando, FL.

3. Reynolds, P. and Pavic, A., Vibration Performance of a Large Cantilever Grandstand during an International Football Match. Journal of Performance of Constructed Facilities, 2006. 20(3): p. pp. 202-212.

4. Reynolds, P., Pavic, A., and Carr, J., Experimental Dynamic Analysis of the Kingston Communications Stadium. The Structural Engineer, 2007. 85(8): p. pp. 33-39.

5. Caprioli, A., Castellani, A., Cigada, A., and Vanali, M. Vibration Monitoring of the G. Meazza Stadium in Milano during Concerts and Football Matches. in 23rd International Modal Analysis Conference (IMAC XXIII). 2005. Orlando, FL.

6. Salyards, K. A. and Hanagan, L. M. Analysis of Coordinated Crowd Vibration Levels In a Stadium Structure. in 25th International Modal Analysis Conference (IMAC XXV). 2007. Orlando, FL.

7. Catbas, F.N., Gul, M., Dynamic Response Monitoring and Correlation to Crowd Movement at a Football Stadium. in 27th International Modal Analysis Conference (IMAC XXVII). 2009. Orlando, FL.

8. ARTeMIS Structural Vibration Solutions A/S

Proceedings of the IMAC-XXVIII
February 1–4, 2010, Jacksonville, Florida USA

Verification of Crowd Dynamic Excitation Estimated from Image Processing Techniques

C.A. Jones*, P. Reynolds*, E. Zappa**, S. Manzoni**, A. Cigada**
* University of Sheffield
Department of Civil and Structural Engineering, Sheffield, United Kingdom, S1 3JD
** Politecnico di Milano
Dipartimento di Meccanica, Via La Massa, 34 20156, Milano, Italy
c.a.jones@sheffield.ac.uk

ABSTRACT

Dynamic structural responses of a major UK stadium have been measured during a series of high-profile events. In order to verify and validate recent approaches to the determination of vibration responses, a series of models have been developed and comparisons between measured and predicted results made.

Using advanced image recognition and data processing techniques, video records of the events were used to determine approximations to the real forces induced by the crowds. The image processing is based on a digital correlation method and on post-processing aimed at decreasing image noise disturbance (due for example to camera flashes in the video). The output is the velocity field of the grabbed area. These data are further exploited to estimate the load on the structure.

The forces estimated from the image processing have then been used, in conjunction with a highly detailed and accurate FE model, to determine improved estimates of the responses. Comparisons are made between design, measured and back-analysis results using both codified and real crowd forces. The accuracy of the crowd dynamic forces determined from image processing is assessed.

1 PROBLEM OVERVIEW

Stadia are increasingly lightweight and flexible structures due to increased use of slender cantilevers and optimised design procedures. These structures are further subject to occupancy by high density crowds which can generate large loads, and may result in strongly perceptible vibrations. Much current research is aimed at improving methods for the prediction of such vibrations. In particular, the load models used in analysis are very simplistic and do not accurately represent the forces induced by real motions of actual crowds. This paper investigates levels of response predicted using more realistic dynamic loads estimated from image processing of real crowd motions.

In 2000, the City of Manchester Stadium (Figs. 1(a) & (b)) was designed for, amongst other criteria, the avoidance of uncomfortable or panic- inducing vibrations. The stadium itself is of concrete construction with a steel roof considered to act independently of the main bowl of the structure. With a capacity of 47,726 the stadium is the fourth largest in the English Premier league and has a reputation as a major international venue. Several high profile events, including the 2008 UEFA cup finals, the 2002 commonwealth games as well as multiple major concerts and international sports matches, have cemented this image.

Between 2004 and 2005 researchers at the University of Sheffield, UK, performed dynamic testing and long term vibration monitoring using a remote system [1] on the west stand. As well as identifying the characteristic dynamic

T. Proulx (ed.), *Dynamics of Civil Structures, Volume 4*, Conference Proceedings of the Society for Experimental Mechanics Series 13,
DOI 10.1007/978-1-4419-9831-6_22, © The Society for Experimental Mechanics, Inc. 2011

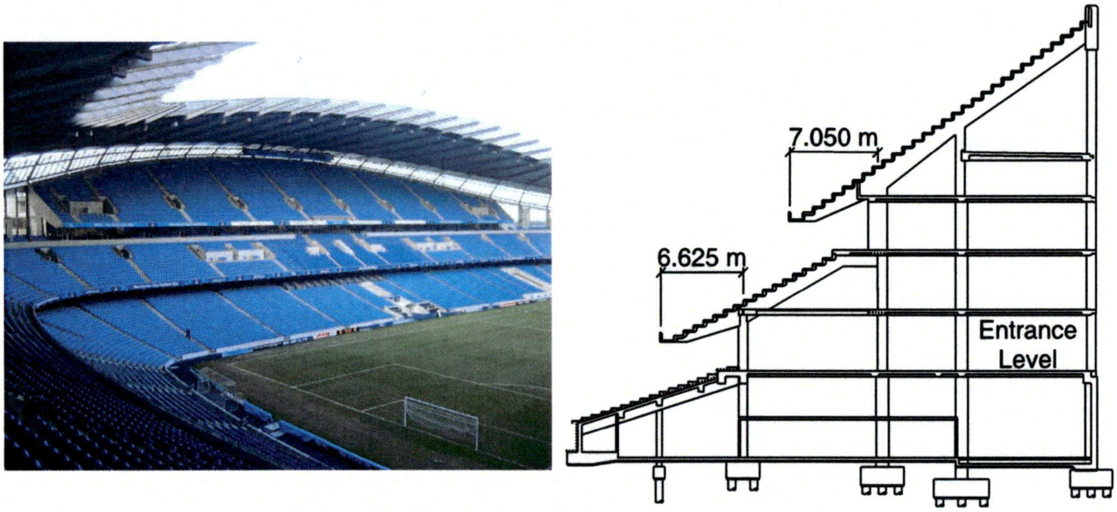

Figure 1: (a) City of Manchester Stadium and (b) a typical cross section of the west stand

properties of the structure, the actual performance of the stand was recorded in the form of structural accelerations and videos of crowd activity. This investigation focuses on recreating the responses of the structure to crowd actions observed during a concert headlined by U2 which took place on 14 June 2005.

A technique has been developed [2] which allows for the estimation of crowd accelerations from video records. With some processing, these data can be converted into time varying forces which can be applied as part of several different response analysis techniques. This paper describes the image/video processing techniques used, the derivation of the associated forces (put into context against other published values), the development of a system model for the structure and the estimation of responses of said system to the aforementioned loading. Subsequently, discussion of the predicted responses in comparison with measured values and the implications of using 'real' force histories leads into the conclusions of the investigation. The paper concentrates on the middle tier of the structure as that is where the initial video processing efforts were focused.

2 VIDEO PROCESSING

The aim of image acquisition and processing is to estimate the acceleration time histories of people in the crowd. This information will be used to estimate the exciting force produced by jumping people.

The authors have already developed a technique to estimate the crowd velocity, which is then used to derive acceleration [2–4]. The method is based on Digital Image Correlation (DIC) but it relies upon good resolution images (22 mm/px or better) to get reliable results. The images considered in this work are instead of relatively poor resolution, as can be seen in Fig. 2. This is due to the need to carry out imaging and monitoring a large area of the stadium. Because of the low resolution images available, the acceleration time histories estimated with the DIC technique are corrupted by a non-negligible level of noise. Therefore only the Root Mean Square (RMS) value of the crowd acceleration are estimated, as described later in this section. Moreover, as can be seen in Fig. 2 due to the perspective, the resolution of the images strongly changes as the region of the image changes, therefore image processing has not been carried out on the full image, but just on a single portion (highlighted in Fig. 2). The considered portion is that where the image resolution is the best possible.

Before discussing the whole adopted procedure to estimate the crowd motion, brief remarks are given about the DIC technique. The DIC technique consists of splitting a movie image into different rectangular Regions of Interest (ROI) with a fixed size. Then, for each ROI, the most similar region of the subsequent image of the movie is searched for (i.e. the region which minimizes a figure of merit; the definition of the figure of merit depends upon the adopted algorithm [5]). Once the displacement of each ROI has been estimated, it is possible to find the velocity field because the time lag Δt, elapsed between the acquisition of the two images, is known. The same procedure

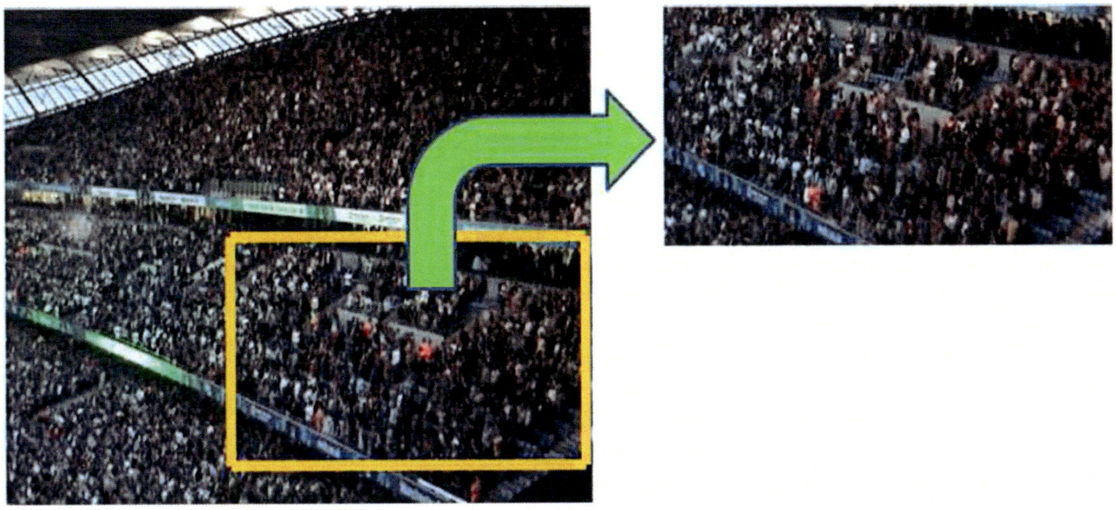

Figure 2: The analysed image area

can be applied to each pair of subsequent images (j and $j+1$ with $j = 1, 2, 3 \ldots$) to estimate the velocity fields at time $t_j = j\Delta t$. More details can be found in [2]. In [4] the authors analysed the effect of the size of ROIs on the results uncertainty; it was demonstrated that ROIs size should be at least 8x8 pixels. In this paper the size of the ROIs was therefore fixed at 8x8 pixels.

2.1 DIC SIGNALS POST PROCESSING

Once the velocity time history has been estimated for all the ROIs, some signal processing is required. In particular, one of the main difficulties in the post-processing of the obtained velocity time histories was the treatment of noise related to flashes of photo-cameras visible in the images. It has been estimated that each flash was present in one or in two successive images. These flashes often cause large (and meaningless) peaks in the estimated vertical velocity results. A method therefore was required to define an automatic method able to detect the spurious peaks of acceleration and to exclude them from the analysis.

The technique selected to exclude camera flashes was as follows. The numerical difference D was calculated for the velocity time-history of each ROI:

$$D(n \times \Delta t) = [\text{velocity}((n+1) \times \Delta t) - \text{velocity}(n \times \Delta t)] \quad \text{for} \quad n = 1, 2, 3 \ldots$$

where $\Delta t = \frac{1}{f_s}$ and f_s is the sampling frequency. The ROIs were analysed individually and from each of them a D record was obtained. The Kurtosis coefficient was then calculated for each D record (one D record for each ROI). The Kurtosis coefficient of a time signal x is calculated as:

$$\text{Kurtosis} = \frac{1}{m} \times \frac{\sum_{i=1}^{m}(x_i - \bar{x})^4}{s^4} - 3$$

where m is the number of samples, $\bar{x} = \frac{\sum_{k=1}^{m} x_k}{m}$ and $s = \sqrt{\frac{\sum_{k=1}^{m}(x_k - \bar{x})^2}{m-1}}$. The Kurtosis coefficient increases in the presence of many spikes within the considered signal x. When this parameter was higher than a threshold T, the larger peak (in module) of the D signal was marked as a number not to be considered in the following calculations. The same operation was applied to the sample of the velocity time history causing the peak in the D record. At this point the Kurtosis parameter was recalculated and the comparison with the threshold T was performed once again, and so on until the Kurtosis parameter became lower than T. Different threshold values T were considered through trial and error. Finally T was fixed to 1 as this value guarantees to eliminate most flashes and outliers while preserving the dynamics of the signals.

A further processing was required to take into account the effect of perspective in the collected images. The DIC technique gives as output a velocity time history in px/s for each ROI. A conversion factor between millimetres and pixels was then computed as a function of the considered area of the image; in this way it was possible to calculate the velocity time histories in mm/s.

Once the above mentioned task was completed for all ROIs, a RMS of the processed velocity time histories was calculated for all of them. Each point of the running RMS was calculated on 25 velocity time samples and the adopted overlap was 50%. This means that the time resolution for the RMS history is 0.5 s (the image sampling frequency is 25 Hz). Finally a total RMS (named RMS_{tot}) was estimated on an interesting area containing different ROIs:

$$RMS_{tot}(t_k) = \sqrt{\frac{\sum_{i=1}^{N}(RMS_i(t_k))^2}{N}}$$

where N is the number of ROIs within the considered area and $RMS_i(t_k)$ is the RMS value at time t_k for the ith ROI in the considered area.

2.2 REPRESENTATIVES OF THE ESTIMATED MOTION

Due to both the large number of people included in the imaged stand portion and the poor image resolution, it is not possible to estimate the jumping amplitude and frequency of each person; an average computation is therefore required. Since in the end the interesting data is the overall dynamic loading due to the crowd, the averaging procedure does not constitute a limit in the application of this measuring technique, providing that the averaging area respects two conditions: 1) it is large enough to produce an estimated acceleration with acceptably low uncertainty and 2) it is small enough to describe the distribution of loading on the structure.

Since the stadium stands are almost uniformly crowded, the second condition is actually not relevant. On the contrary the uncertainty linked to the estimated acceleration data has to be analysed. Because there is no practical way to estimate the motion of a large number of people with another measuring technique to be used as a reference, the only consideration that can be done is regarding DIC-estimated data repeatability. This can be done by comparing the results obtained in different areas of the stand on the condition of nominally uniform crowd motion, for example during the refrain of a very popular song, when most of the spectators jump following the beat frequency. Also, later in this paper, the acceleration estimated with the DIC will be used to predict the dynamic loading due to the spectators and, using a FE model of the hosting structure, the vibration response will be calculated. If the estimated vibration is comparable with that measured, another indirect validation of the predicted acceleration uncertainty will be given.

To perform the repeatability analysis the area of the image highlighted in Fig. 2 was divided into 16 sub-regions (see Fig. 3) and the mean acceleration time history was computed for each sub-region. Finally for each of them the RMS_{tot} was calculated. It was verified that these 16 areas have similar and stable values of the mentioned RMS_{tot} parameter. Moreover other tests were done defining a small number of larger sub-regions and the RMS_{tot} parameter does not change significantly even if the size of the region is increased.

It is worth pointing out that the RMS_{tot} parameter is calculated in mm/s. The inertial force exciting the structure due to the crowd motion is instead linked to acceleration. Therefore the RMS_{tot} of the 16 areas have been transformed to acceleration by multiplying by a coefficient $2\pi f$ where $f = 2.1$ Hz. In fact it has been estimated, by auto-correlation operations of the estimated crowd motion, that in the considered condition the people movement is at about 2.1 Hz when the motion is significant. Of course the frequency of motion is linked to the rhythm of the music and it changes as the song changes, however it is quite straightforward to estimate it as soon as the crowd velocity is estimated with the DIC. Once the motion frequency is known the acceleration time history of each of the 16 areas can be

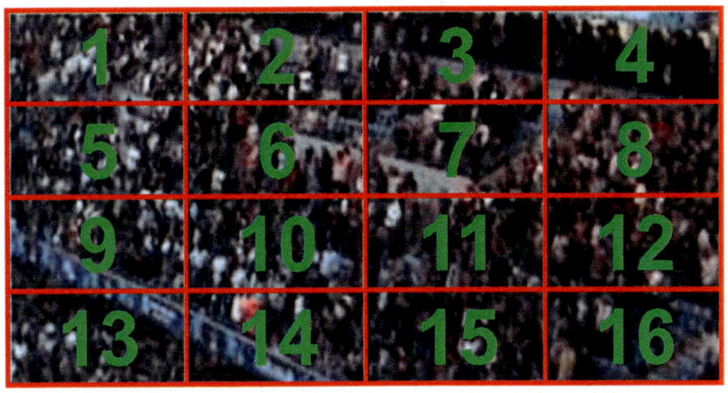

Figure 3: The 16 considered areas

calculated by multiplying the 16 RMS_{tot} by the coefficient $\sqrt{2}\cos(2\pi f t)$. Thus it was possible to estimate a mean crowd acceleration for the 16 areas of Fig. 3.

3 FORCE ESTIMATION

As the crowd accelerations were derived primarily from the middle tier, the rest of the analysis focuses on that tier, with less emphasis being placed upon other regions of the structure.

Force estimates were obtained by multiplying the accelerations obtained in section 2 by the apparent mass of the crowd in each considered area, assuming that force is equal to mass times acceleration. To obtain forces per unit area, these values were then divided by the estimated plan area beneath the crowd for each area. Tables 1(a) & (b) reproduce the estimated values for mass and area, whilst Table 1(c) lists the absolute maximum amplitude of the estimated forces per unit area.

Table 1: Estimated properties of video capture regions (a) plan area of area [m^2], (b) mass of crowd in area [kg], (c) Inferred maximum pressures for each area [N/m^2]

35.00	38.25	24.00	16.50
32.00	31.50	25.00	22.75
30.00	26.25	24.50	21.00
40.50	18.75	20.00	17.50

(a)

1680	1440	1440	1120
2480	1120	1200	1760
1120	1760	1920	960
1840	880	1200	560

(b)

492	350	424	466
705	246	354	522
290	446	540	349
420	272	394	209

(c)

Areas 7 and 9 were found to best reflect the mean of all areas and were thus considered for use as representative data sets. Area 7 was selected for further analysis over area 9 as it was entirely enclosed within a single tier. During the considered sub-event it was typical for 40% of the crowd to be obviously active, or about 0.9 people per square meter over the entirety of the structure. The result forces thus compare well with those generated by models for crowd bouncing activities. Fig. 4(a) shows a comparison between the pressures obtained from utilising 0.9 people per square meter with a model proposed by Parkhouse & Ewins [6] for crowds of 50 people bouncing at 2 Hz. It can be seen that the overall magnitude of the force histories is of a similar magnitude.

Although Fig. 4 illustrates that the derived and design loads are of a similar magnitude, the actual time histories, and their frequency domain representations, are quite different. Fig. 4(c) in particular indicates that the fundamental frequency of crowd motion is around 2.1 Hz as there is a strong peak in the data. Fig. 5(a) shows the overall time histories as derived from the available crowd acceleration data, whilst Fig. 5(b) presents the 240 second period of excitation investigated in detail from area 7, after normalisation to a unit weight of active crowd members.

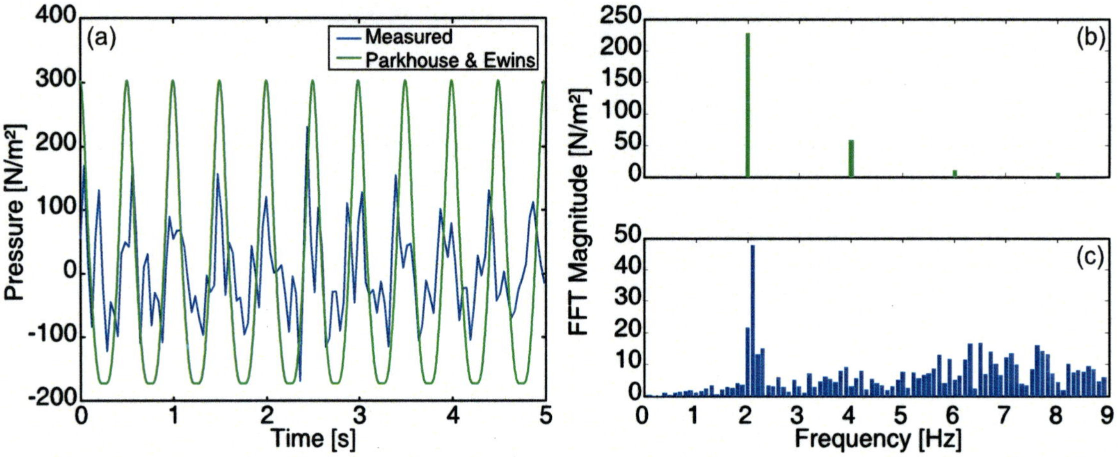

Figure 4: Comparison of measured and design load histories in (a) the time domain (b)&(c) the frequency domain

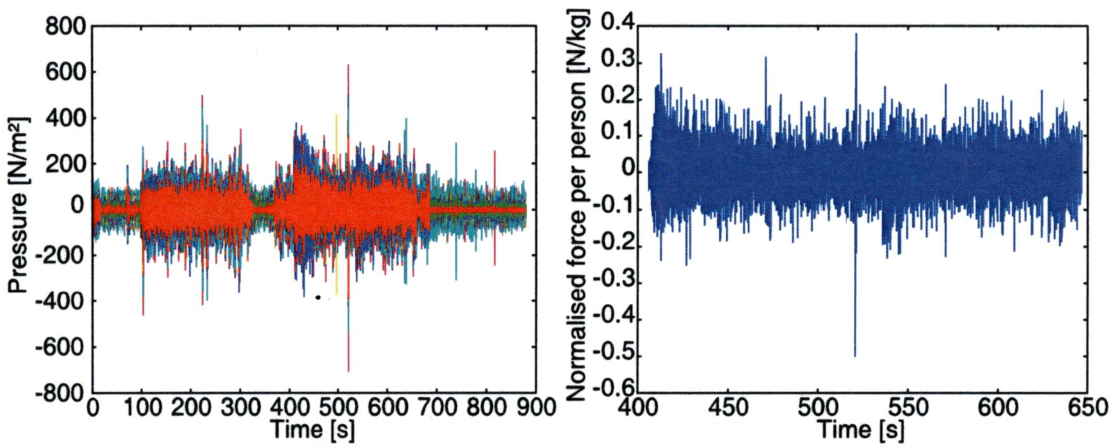

Figure 5: (a) Full excitation time histories derived from video data (b) 240 second of greatest excitation from area 7 used for further analysis, normalised to a unit weight of active crowd members

4 SYSTEM IDENTIFICATION

The dynamic properties of the empty structure were determined through a series of experimental measurements. The obtained properties were further expanded upon through the development of supplementary FE models which offered more detailed spatial information than could be obtained in the limited time available for measurements.

4.1 EXPERIMENTAL MODAL ANALYSIS

Experimental modal analysis of the structure was a two-phase process comprising of controlled excitation testing with shakers and a series of ambient vibration surveys. The shaker testing was used to identify the local cantilever modes under levels of excitation which are feasible in-service. The ambient vibration surveys acted both as a quality assurance check for the shaker testing and as a method for identifying global modes. In the case of the City of Manchester stadium, the vertical component of the global modes was small and hence can be safely ignored for vertical load analysis. On other structures, especially stadia with long cantilevers and short backspans, combined front-to-back and vertical 'nodding' modes can prove critical.

The test scheme identified the modes shapes and modal properties outlined in Fig. 6.

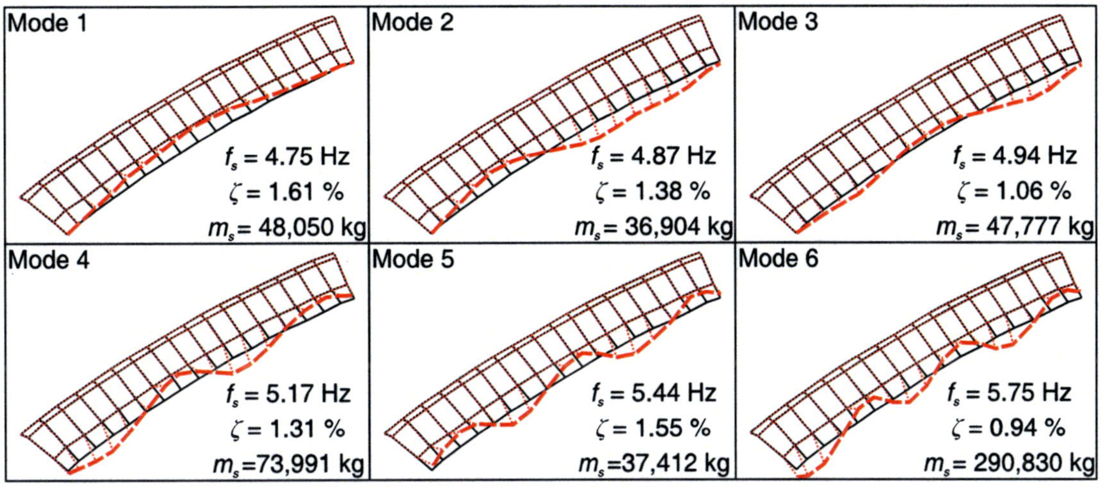

Figure 6: Middle tier modes

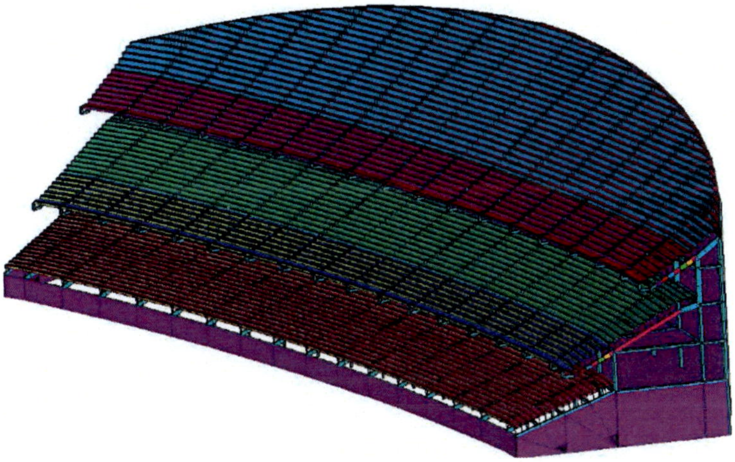

Figure 7: Final FE model

4.2 FE MODELLING

In order to supplement the experimental modal analysis a series of FE models were developed in ANSYS [7]. The final 3D model (Fig. 7) encompassed the entire west stand to a high level of detail and gave some consideration to non-structural elements. A finite element updating scheme, utilising a mixture of manual and automated processes on both the macro (geometry tweaking) and micro (property tuning) scales, helped to identify sources of error and improve the representativeness of the model.

Post-updating MAC values reached above 65% average, which for such a large and complex structure is reasonable, whilst frequency errors were smaller than 5% (Table 2). The largest single source of error was the stiffness of the concrete, especially in the rakers, which had been reduced significantly by cracking.

The final FE model served to supplement the experimental modal analyses, primarily through a greatly increased resolution of mode shapes, which would prove useful for determining responses of different locations to a dispersed crowd.

Table 2: Middle tier EMA-FE Correlation

Natural Frequencies			Mode Shapes
EMA [Hz]	FE [Hz]	Error [%]	MAC [%]
4.75	4.73	-0.43	86
4.86	4.86	-0.01	77
4.94	5.03	1.81	69
5.19	5.14	-0.97	87
5.44	5.35	-1.67	93
5.72	5.60	-2.11	86

5 CALCULATION OF RESPONSES

Simulated responses were calculated using a methodology similar to the one presented by Pavic & Reynolds [8]. That is, a modal space approximation to the problem was made whereby occupying humans are considered as additional modal-space degrees of freedom. Structural DOF properties were taken from Fig. 6, supplemented by mode shapes calculated in section 4.2. The crowd was, based on observations of the whole structure, assumed to be uniformly distributed with approximate density 2.25 people per square meter, of whom 40% were considered to be active.

The modified modal space model approach (Fig. 8) has been investigated by multiple authors and shown to capable of improving estimates of structural responses [9–11].

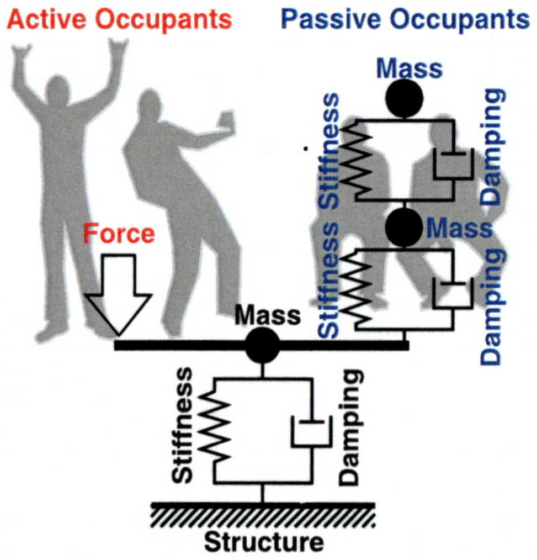

Figure 8: Example occupied modal space model utilising human body model 2a from [12]

A total of 16 different simulations were run, considering 6 different load models and 7 slightly different methods of considering the effects of human occupancy upon the final system. Table 3 summarises the maximum transient vibration values (MTVV), as obtained from the maximum of running one-second RMS average acceleration responses at mid-span on the tip of the middle tier cantilever. Table 4 describes the load and crowd models utilised in more detail.

The loads proposed by Parkhouse & Ewins [6] result in responses with magnitudes close to measurements, especially once the effects of human occupancy are considered. In comparison the loads proposed in ISO 10137:2007 [14], which are admittedly intended as extreme design limits, result in several times the measured response magnitude even when accounting for human-structure interaction. The loads estimated using the video processing techniques outline herein lie somewhere in the middle ground - prior to consideration of human-structure interaction the responses are higher than obtained utilising the design loads from Parkhouse & Ewins and hence the UK design guidelines. Inclusion of human-dofs greatly reduces the responses and brings them more into line with measure-

Table 3: Measured and Simulated responses from mid-span, middle tier

Scenario	Load Model	Energy Absorption	MTVV1 [m/s^2]	Error [%]
Measured	-	-	0.38	-
1	L1	A1	0.23	-39
2	L1	A2	0.22	-42
3	L1	A3	0.42	11
4	L1	A4	0.37	-3
5	L1	A5	0.30	-21
6	L1	A6	0.21	-45
7	L2	A7	0.53	39
8	L3	A7	0.35	-8
9	L4	A3	1.84	384
10	L5	A3	0.80	111
11	L6	A1	1.26	232
12	L6	A2	0.35	-8
13	L6	A3	0.33	-13
14	L6	A4	0.21	-45
15	L6	A5	0.18	-53
16	L6	A6	0.56	47

ments. Despite this, the variability is between each scenario is high, requiring that parametric and sensitivity studies be performed in the future.

Fig. 9 illustrates the measured responses as well as the time histories obtained when investigating scenarios 1 and 11, which do not consider the occupant dynamic properties. Fig. 10 focuses more closely on the average response levels, considering the measured responses and the values obtained from simulation scenario 12, which best matched the measured MTVV levels.

Table 4: Identifers used in Table 3

Load Model	Description
L1	Parkhouse & Ewins [6], coefficients for 50 people bouncing at 2 Hz
L2	UK Guidance [13], Scenario 4 (highly active events where stand motion expected), based on L1
L3	UK Guidance [13], Back substitution of observed events, based on L1
L4	ISO 10137 [14], Jumping actions of crowds, poor coordination.
L5	ISO 10137 [14], Vertical actions of seated crowds.
L6	As inferred from video processing
Damping/HSI Model	**Description**
A1	Measured modal damping only
A2	Additional (seated) 2dof passive crowd, properties from Wei & Griffin's [12] model 2(a)
A3	Additional (standing) 2dof passive crowd, properties from Matsumoto & Griffin's [15] model 2(a)
A4	Additional passive crowd sdof, properties from UK guidance [13], based on A2/A3
A5	A4 plus additional active crowd sdof, properties from UK guidance [13], based upon curve fitting by Dougill et al. [16]
A6	Modal damping set to 6% for all modes
A7	A5, but the active crowd DOF is actuated (equal and opposite force on active crowd DOF and structure instead of just structure)

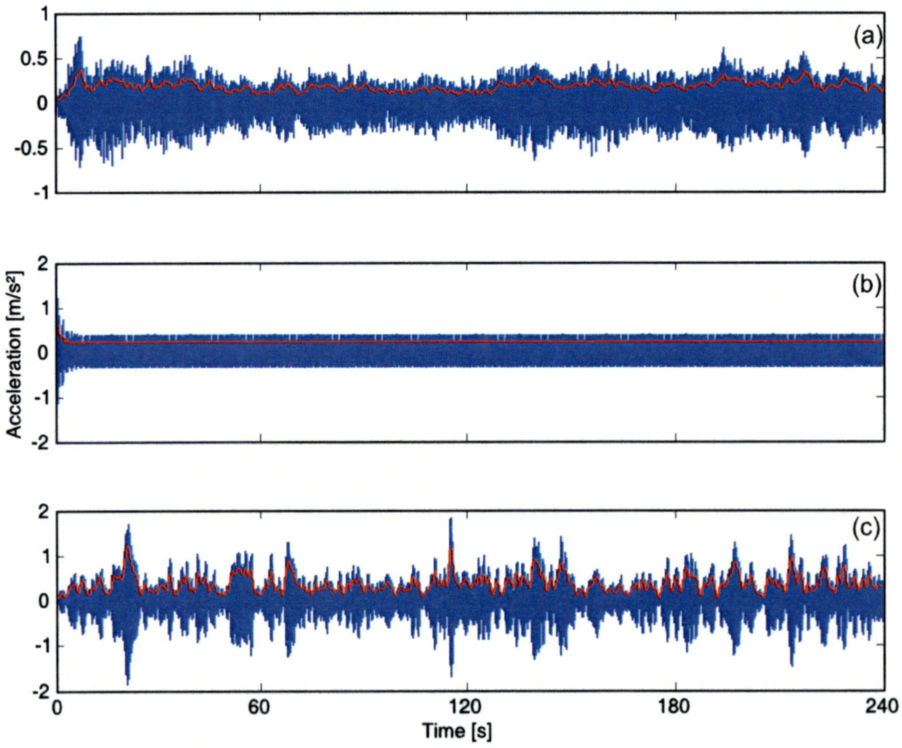

Figure 9: Mid-span, middle tier responses (a) measured and (b) simulated by scenario 1 (c) simulated by scenario 11

6 CONCLUSIONS

This paper has presented a series of forces derived from video records of crowd activities upon a major UK stadium. These forces have subsequently been used in simulations involving modal space approximations of a large scale complex FE model. The responses obtained are closer to measurements than traditionally predicted by design

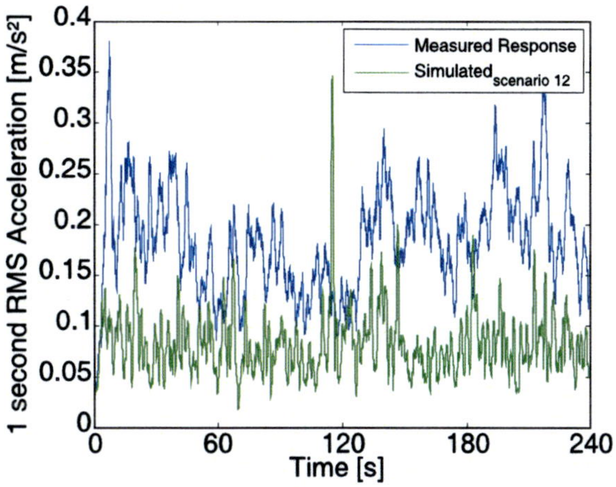

Figure 10: RMS responses at mid-span, middle tier, as measured and best simulated (scenario 12)

codes. Further tweaking of the simulation approach in line with current research into human-structure interaction yielded, although the actual time histories still differed, simulated responses which well match the measured ones.

Although there are obvious differences between data used in simulations attempting to recreate measurements and data for use in safely conservative design codes, better understanding of the overall phenomenon will eventually yield optimised design methodologies. The results obtained in this paper indicate that one potential avenue for further research is a more in-depth study of the load generated by real, large, crowds of people which otherwise remain largely unknown.

In particular, further work with the techniques described in this paper would allow for a better understanding of transient excitation and responses in the context of active crowds on stadia grandstands to be developed.

ACKNOWLEDGEMENTS

The authors would like to acknowledge the UK Engineering and Physical Sciences Research Council (EPSRC) and Arup for CASE voucher 06000921, which allowed this study to take place. In addition, EPSRC grants GR/S73761/01 & GR/R21998/01, as well as the stadium management, facilitated the experimental work. The further general support of members of the Vibration Engineering Section at the University of Sheffield and the Arup Northwest structures team is gratefully acknowledged.

REFERENCES

[1] **Reynolds, P., Pavic, A.** and **Ibrahim, Z.**, *A remote monitoring system for stadia dynamics*, Proceedings of ICE: Structures and Building, Vol. 157, No. 6, pp. 385–393, 2004.

[2] **Caprioli, A., Manzoni, S.** and **Zappa, E.**, *People-induced vibrations of civil structures: image-based measurement of crowd motion*, Experimental Techniques, 2009.

[3] **Caprioli, A., Manzoni, S.** and **Zappa, E.**, *Crowd movement measurement based on image processing*, *26th International Modal Analysis Conference (IMAC XXVI)*, 2008.

[4] **Caprioli, A., Manzoni, S.** and **Zappa, E.**, *Crowd motion estimation by means of a PIV algorithm*, *Structural Dynamics EURODYN 2008*, July 2008.

[5] **Gui, L.** and **Merzkirch, W.**, *A comparative study of the MQD method and several correlation-based PIV evaluation algorithms*, Experiments in Fluids, Vol. 28, 2000.

[6] **Parkhouse, J. G.** and **Ewins, D. J.**, *Crowd-induced rhythmic loading*, Structures and Buildings, Vol. 159, No. SB5, pp. 247–259, 2006.

[7] **ANSYS Inc.**, *ANSYS Academic Research 11.0*, Computer Program, 2007.

[8] **Pavic, A.** and **Reynolds, P.**, *Experimental verification of novel 3DOF model for grandstand crowd-structure dynamic interaction, 26th International Modal Analysis Conference (IMAC XXVI)*, 2008.

[9] **Sachse, R., Pavic, A.** and **Reynolds, P.**, *Parametric Study of Modal Properties of Damped Two-Degree-of-Freedom Crowd-Structure Dynamic Systems*, Journal of Sound and Vibration, Vol. 274, No. 3-5, pp. 461–480, 2004.

[10] **Sim, J. H. H., Blakeborough, A.** and **Williams, M.**, *Modelling of joint crowd-structure system using equivalent reduced-DOF system*, Shock and vibration, Vol. 14, No. 4, pp. 261–270, 2007.

[11] **Pedersen, L.**, *Experimental Investigation of Human-Structure Interaction, 24th International Modal Analysis Conference (IMAC XXIV)*, 2006.

[12] **Wei, L.** and **Griffin, M. J.**, *Mathematical models for the apparent mass of the seated human body exposed to vertical vibration*, Journal of Sound and Vibration, Vol. 212, No. 5, pp. 855–874, 1998.

[13] **IStructE/DCLG/DCMS Joint Working Group**, Dynamic performance requirements for permanent grandstands: recommendations for management design and assessment, Institution of Structural Engineers, London, 2008.

[14] **International Organization for Standardization**, *ISO 10137:2007, Bases for design of structures - Serviceability of buildings and walkways against vibration*, 2007.

[15] **Matsumoto, Y.** and **Griffin, M. J.**, *Mathematical Models for The Apparent Masses of Standing Subjects Exposed to Vertical Whole-Body Vibration*, Journal of Sound and Vibration, Vol. 260, No. 3, pp. 431–451, 2003.

[16] **Dougill, J. W., Wright, J. R., Parkhouse, J. G.** and **Harrison, R. E.**, *Human structure interaction during rhythmic bobbing*, The Structural Engineer, Vol. November 2006, No. 22, pp. 32–39, 2006.

Proceedings of the IMAC-XXVIII
February 1–4, 2010, Jacksonville, Florida USA

Mathematical modelling of near-periodic jumping force signals

Vitomir Racic[1], Aleksandar Pavic[2], James M. W. Brownjohn[3]

[1]Research Student, [2]Professor, [3]Professor - Vibration Engineering Section
Department of Civil and Structural Engineering, The University of Sheffield
Sir Frederick Mappin Building, Mappin Street, Sheffield, S1 3JD, UK

ABSTRACT

A mathematical modelling procedure has been developed to generate synthetic vertical force signals induced by a single person jumping. The ability to replicate much of the temporal and spectral features of real jumping loads gives this model a definite advantage over the conventional half-sine models coupled with Fourier series analysis. This includes modelling of the omnipresent lack of symmetry of individual jumping pulses and beat-by-beat variations in amplitudes and timing. The model therefore belongs to a new generation of synthetic narrow-band jumping loads which simulate reality better. The proposed mathematical concept for characterisation of irregular jumping pulses may be utilised in vibration serviceability assessment of civil engineering assembly structures, such as grandstands, spectator galleries and gym floors, to estimate realistic dynamic structural response due to people jumping.

NOMENCLATURE

m	body mass	μ	mean value
g	gravity	σ	standard variation
$z(t)$	fitting function	$A_{\overline{P}}(f)$	Fourier amplitudes of $\overline{P}_i$
α_i	dynamic impact factors	$A'_{\overline{P}}(f)$	Fourier amplitudes of $\overline{P}'_i$
W_i	Gaussian peak amplitudes	$S_{\overline{P}}(f)$	ASD of $\overline{P}_i$
ω	angular frequency	$S'_{\overline{P}}(f)$	ASD of $\overline{P}'_i$
ω'_k	synthetic angular frequency	$\sigma^2_{\overline{P}}$	variance of $\overline{P}_i$
P_i	measured peak amplitudes	x, y	Cartesian coordinates
P'_i	synthetic peak amplitudes	ρ, θ	polar coordinates
$\overline{P}_i$	measured variations of peak amplitudes	t_i, θ_i, c_j	Gaussian centres
$\overline{P}'_i$	synthetic variations of peak amplitudes	b_i, β_i, δ_j	Gaussian widths
$\varphi(x)$	Normal pdf		

DOI 10.1007/978-1-4419-9831-6_23, © The Society for Experimental Mechanics, Inc. 2011

1 INTRODUCTION

Civil engineering structures, such as grandstands, spectator galleries and floors accommodating sports facilities and concert venues, are expected to be fit for purpose when occupied and dynamically exited by multiple persons and ultimately large number of crowds at special events. There have been numerous high-profile examples of excessive vibration being generated by people dancing, bouncing or jumping in unison [1-3]. These failures have clearly indicated shortcomings in current provisions and the level of uncertainty with which civil structural engineers are faced nowadays when designing any of the above mentioned types of structure which require vibration performance assessment. Of all steps involved in vibration serviceability design, such as establishing acceptance criteria and creating a structural model, determining design loads has the greatest uncertainty and to this end there have been numerous attempts to provide their reliable and practical descriptions [4-6]. Jumping action is generally considered as the most severe loading scenario. Therefore, the aim of this paper is to propose an improved synthetic jumping force model that can be used to simulate more reliably dynamic response, thereby assessing vibration serviceability of civil engineering structures (for which such dynamic excitation is relevant) convincingly at the design stage.

Dynamic forces generated by people moving on a structure are commonly determined by direct measurement of the interface forces between the feet and the structure itself, hence they are known as ground reaction forces (GRFs). The typical vertical force time history induced by a single person jumping takes the form illustrated in Figure 1 and is characterised by a series of single-peak pulses separated by zero-force periods. These periods correspond to the aerial phase of jumping, where both feet leave the ground.

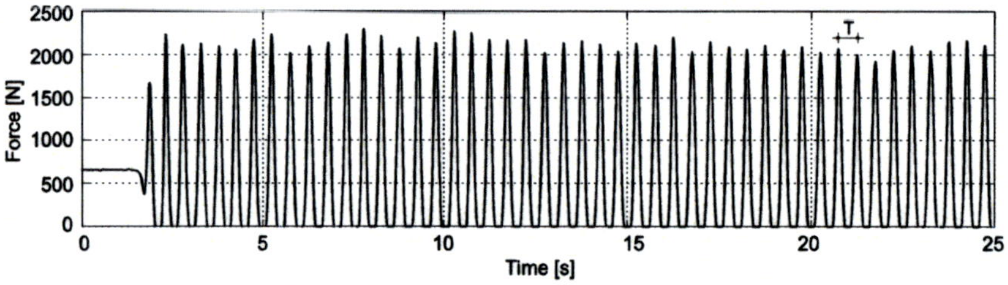

Figure 1: Example of a vertical jumping force record due to a single person jumping at 2 Hz. T=0.5 s is the average period of jumping cycle.

The common engineering practice is to idealise a continuously measured jumping force signal as periodic with the period being the average time between two consecutive jumps i.e. the average duration of a single jumping cycle [1]. This means that actual forces due to continuous jumping can be recreated by adding a sequence of identical pulses on a cycle-by-cycle basis. For dynamic analysis, a set of periodic jumping pulses can be represented more efficiently if expressed in terms of Fourier series with the fundamental harmonic having frequency identical to the jumping rate [4, 5]. Such models, utilising up to three harmonics, can be found in the current British Standard BS 6399-1 [7] and Commentary D of the National Building Code of Canada [8]. In these jumping loads are defined for use in the design of structures likely to be subjected to significant vertical occupant motion, such as footbridges, grandstands, and concert or gym floors, where human induced serviceability issues may govern design.

The ability of Fourier series to represent measured continuous pedestrian forces due to walking has been examined relatively recently by Brownjohn *et al.* [9]. It was shown that there were significant differences between the resonant responses due to real walking forces (which are narrowband and therefore 'imperfectly' periodic) and the equivalent periodic simulation. The effect is more pronounced for higher harmonics where the simulated vibration response is regularly overestimated. This issue has not been researched in great detail for jumping excitation, so the available literature is limited. Nevertheless, a similar analysis can demonstrate that 'synthetic' periodic jumping forces generate a considerably higher resonant response compared with its realistic narrowband counterpart [10].

Based on a method for replicating measured electrocardiogram (ECG) signals [11], this paper introduces a novel modelling strategy, in the civil structural engineering context, to generate a vertical jumping force time history with realistic varying pulse-to-pulse periods, peak amplitudes and morphology. The ability to match near-periodic nature of real jumping force leading to good compatibility in the harmonic components is the key advantage of the proposed approach. All modelling parameters determining the pattern of the synthetic forcing function can be derived easily from measured jumping force signals induced by different individuals chosen at random and asked to jump under a wide range of conditions. As this has not been done before, the paper presents a feasibility study for establishing a probability-based stochastic model of jumping loads which can account for the statistically diverse human population. This will enable analytical treatment of jumping forces similar to other stochastic loads dynamically exciting civil engineering structures, such as wind, waves and earthquakes.

2 DATA COLLECTION

Jumping GRFs were recorded by an AMTI BP-400600 force plate [12] rigidly fixed to a laboratory floor (Figure 2). Apart from the safety reasons, the platform built around the force plate gave the impression of a bigger jumping space to avoid test subjects deliberately targeting a relatively small plate surface area, which might influence natural variability of the GRF. A single male test subject (age 25, mass 65 kg) was prompted by a metronome to jump at a constant beat of 120 beats per minute i.e. 2 Hz (Figure 2). The test lasted 25 seconds and the force time history was sampled at 200 Hz. Figure 1 shows a typical force record filtered using the eighth order lowpass digital Butterworth filter with cutoff frequency 80 Hz.

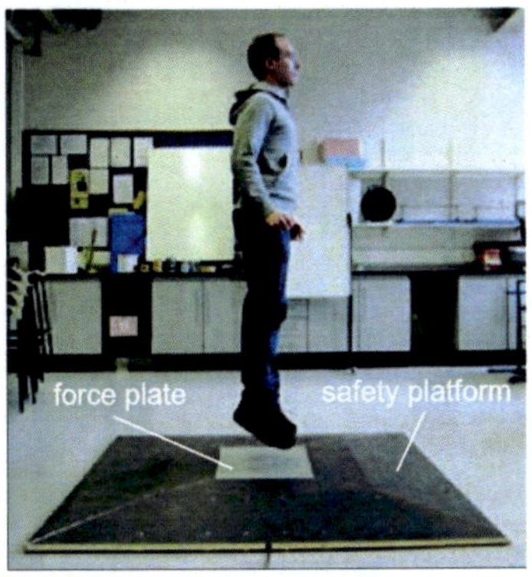

Figure 2: A trained test subject jumping on the force plate.

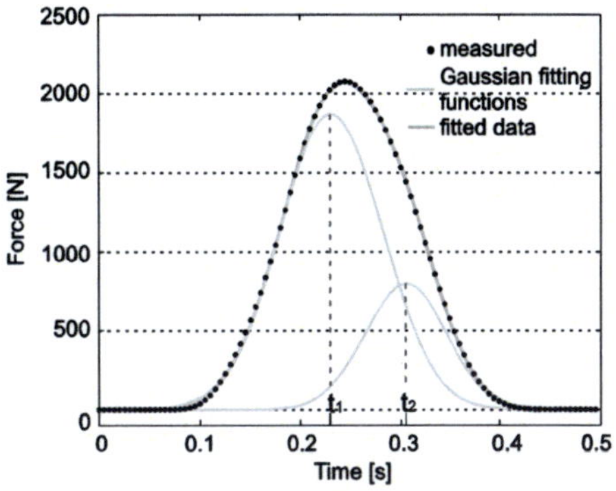

Figure 3: Fitting the average measured jumping beat by a series of two Gaussian functions.

Table 1: Fitting parameters of the average jumping pulse:

parameter	fit	95% confidence interval
α_1 [-]	2.795	(2.628, 2.962)
b_1 [s]	0.070	(0.068, 0.073)
t_1 [s]	0.232	(0.228, 0.236)
θ_1 [rad]	2.915	(2.865, 2.966)
β_1 [rad]	0.880	(0.855, 0.917)
α_2 [-]	1.4588	(1.221, 1.697)
b_2 [s]	0.060	(0.058, 0.063)
t_2 [s]	0.306	(0.301, 0.311)
θ_2 [rad]	3.845	(3.782, 3.908)
β_2 [rad]	0.754	(0.729, 0.792)

3 JUMPING FORCE MORPHOLOGY

Assuming that the average jumping pulse can represent the individual's long-term performance, a set of 44 successive jumping cycles have been cut into half-second segments and aligned by their peaks to calculate the mean jumping force beat (Figure 3). This underlying pattern can be mathematically modelled as a sum of two Gaussian functions:

$$z(t)=\sum_{i=1}^{2}(\alpha_i mg)e^{-\frac{(t-t_i)^2}{2b_i^2}},\quad t\in[0,0.5] \tag{1}$$

Here, m is the mass of the test subject, g is gravity (g=9.81 m/s^2), α_i are dynamic impact factors (yielding heights of the i-th Gaussian peaks), t_i is the position of the centre of the peak, and b_i controls the width (i.e. time duration) of the corresponding bell-shaped curve (Figure 3). The sum of two Gaussian exponentials having different parameters α_i, b_i and t_i is chosen to model better the apparent lack of vertical symmetry of the jumping pulse about its peak, hence they are placed asymmetrically relative to the measured pulse time history. Fitting parameters given by Equation (1) are presented in Table 1.

4 VARIABILITY OF HUMAN JUMPING FORCE

To analyse variations in real jumping force on a beat-by-beat basis (Figure 1), the set of 44 successive jumping beats is used further to identify peak-to-peak intervals T_i (Figure 4a) and peak amplitudes P_i (Figure 5a). Corresponding histograms are shown in Figure 4b and Figure 5b.

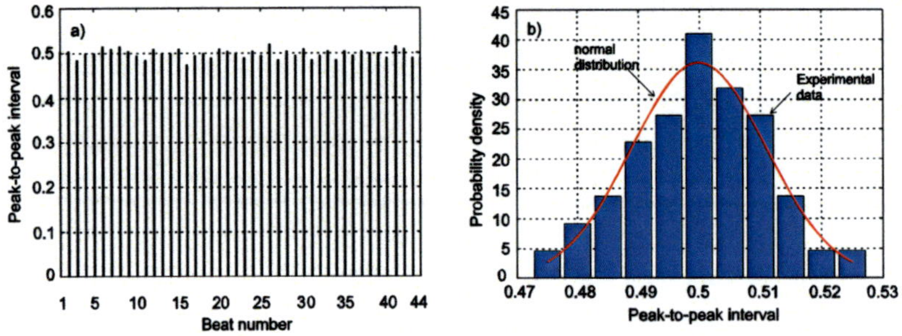

Figure 4: a) Peak-to-peak intervals T_i and b) corresponding histogram.

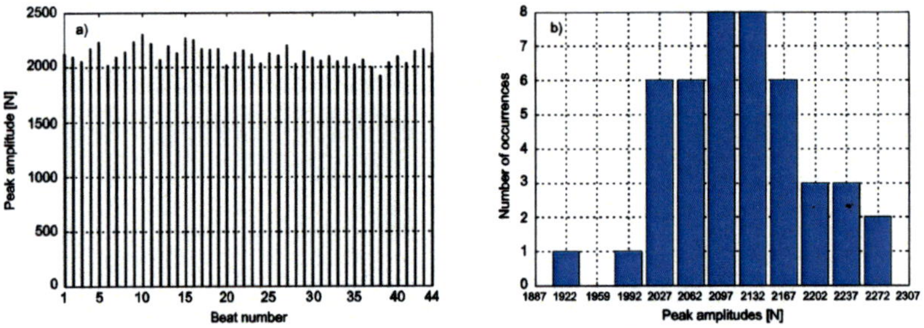

Figure 5: a) Peak amplitudes P_i and b) corresponding histogram.

4.1 PEAK-TO-PEAK INTERVALS

The histogram of the peak-to-peak intervals (Figure 4b) can be observed as a set of approximately normally distributed data in the narrow range of 0.475 s to 0.525 s. The standard parameters of normal distribution (mean value μ and standard deviation σ) can be extracted efficiently from the measurements, which make them very usable in the modelling process. A series of normally distributed random values x which corresponds to the realistic peak-to-peak intervals T_i given in Figure 4a, could be therefore generated artificially following a normal probability density function $\varphi(x)$ given by equation (Figure 4b):

$$\varphi(x) = \frac{1}{\sigma\sqrt{2\pi}} e^{-\frac{(x-\mu)^2}{2\sigma^2}}, \quad x \in R \tag{2}$$

where μ=0.5 s and σ=0.011 s. This will be demonstrated in Section 5.

4.2 JUMPING PULSE PEAK AMPLITUDES

Based on visual inspection [14] of the histogram in Figure 5b, it was found difficult to estimate if peak amplitudes P_i (i=1, ..., 44) follow any known probability distribution. Moreover, it should be noted that these values are not random sequence of numbers, but make a 'smooth' transition between amplitudes of consecutive jumping beats (Figure 1 and Figure 5a). This effect can be represented by a sequence of numbers $\overline{P}_i$ calculated as:

$$\overline{P}_i = \frac{P_i - \mu_P}{\mu_P},\ i = 1,\ldots,44 \tag{3}$$

where μ_P=mean(P_i). Therefore, $\overline{P}_i$ is a dimensionless array with zero mean value having the ability to reflect the variation of peak amplitudes P_i on a jump-by-jump basis (Figure 6). An auto-spectral density (ASD) of $\overline{P}_i$ could be obtained from their single-sided discrete Fourier amplitude spectra $A_{\overline{P}}(f)$ having spectral line spacing Δf=1/44, via the following equation [15]:

$$S_{\overline{P}}(f) = \frac{A^2_{\overline{P}}(f)}{2\Delta f},\quad f \geq 0 \tag{4}$$

where, loosely speaking, f is frequency (Figure 7). An important property of the ASD $S_{\overline{P}}(f)$ is that the variance of $\overline{P}_i$ is given by the area under the graph of $S_{\overline{P}}(f)$ and could be obtained via [16]:

$$\sigma^2_{\overline{P}} = \int_0^\infty S_{\overline{P}}(f)\,df \tag{5}$$

The ASD $S_{\overline{P}}(f)$ of 44 actual $\overline{P}_i$ values (Figure 7) can be fitted quite accurately by a Gaussian sum $S'_{\overline{P}}(f)$ using the non-linear least-square method [17]:

$$S'_{\overline{P}}(f) = \sum_{j=1}^{22} W_j e^{-\frac{(f-c_j)^2}{2\delta_j^2}} \tag{6}$$

where W_j are Gaussian peak amplitudes, c_j are the centres of individual Gaussian curves placed at 22 frequency centres of the real single-sided ASD and δ_j= Δf=1/44 are fixed Gaussian widths (Figure 7). The fitted Gaussian weights (amplitudes) W_j are given in Table 2.

Table 2: Fitting parameters of ASD $S'_{\overline{P}}(f)$:

j	c [Hz]	W [-]	j	c [Hz]	W [-]
1	0.0000	-0.0030	12	0.2500	-0.0003
2	0.0227	0.0221	13	0.2727	0.0004
3	0.0455	-0.0030	14	0.2955	0.0020
4	0.0682	0.0044	15	0.3182	-0.0000
5	0.0909	-0.0003	16	0.3409	0.0036
6	0.1136	0.0023	17	0.3636	0.0021
7	0.1364	-0.0006	18	0.3864	0.0002
8	0.1591	0.0052	19	0.4091	0.0009
9	0.1818	0.0031	20	0.4318	0.0004
10	0.2045	0.0001	21	0.4545	0.0004
11	0.2273	0.0034	22	0.4773	0.0015

A synthetic set of numbers $\overline{P}'_k$ (k=1, ..., N) having ASD $S'_{\overline{P}}(f)$ can be generated by taking the inverse Fourier transform of a sequence of complex numbers with amplitudes $\sqrt{2\Delta f' S'_{\overline{P}}(f)}$ (where $\Delta f'$=1/N) and phases which are uniformly distributed between $-\pi$ and π. Different realisations of the random phases may be specified by varying the seed of the random number generator, hence many different series $\overline{P}'_k$ may be generated with the same spectral properties (Figure 6). Because of the property of the ASD $S_{\overline{P}}(f)$ defined by Equation (5) (and therefore the property of its fit $S'_{\overline{P}}(f)$), all series $\overline{P}'_k$ generated in this way have the same variance $\sigma^2_{\overline{P}}$ for any specified number of jumping pulses N. Therefore, it is possible to generate a synthetic jumping force signal of arbitrary length (e.g. when N>>44) which will have statistically the same dynamics of variations of the peak amplitudes as the 44 real jumping beats. According to Equation (3), by multiplying such series by the scaling constant μ_P and adding the offset value μ_P calculated from the measured data, the resulting series can be used to artificially create different peak amplitudes a single individual generates during nominally similar jumping exercises. Finally, series $\overline{P}'_k$ can reflect the smooth modulation of measured beat-to-beat amplitudes in a synthetic jumping force signal, which conventional Fourier transform approaches cannot represent. All this will be demonstrated in Section 5.

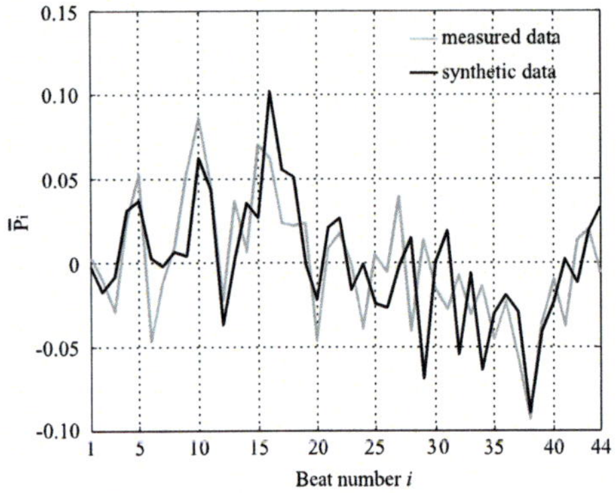

Figure 6: Variation of peak amplitudes on a beat-by-beat basis.

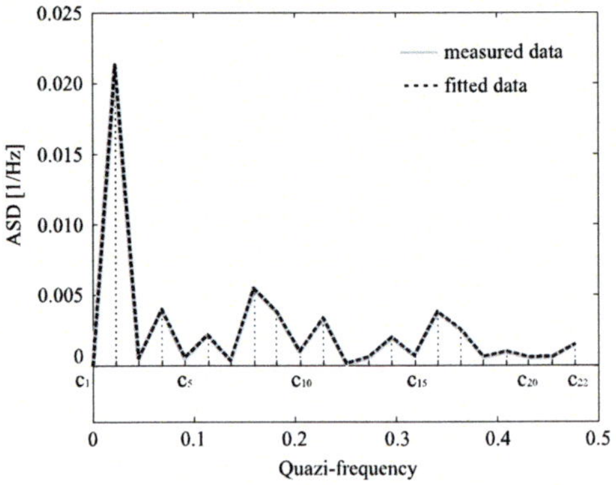

Figure 7: ASD of 44 peak beat amplitudes given in Figure 5a.

The strong linear correlation between peak amplitudes of the ECG waveform on the beat-by-beat basis and the corresponding peak-to-peak intervals imply that the heart generates higher ECG peaks when the peak-to-peak timing is increased [11]. However, based on the measured jumping force signal, this effect is not reflected in dynamics of human jumping. In the case of jumping, there is a very poor correlation between the peak-to-peak intervals T_i and associated peak amplitudes P_i, as shown in Figure 8. Hence, both jumping force parameters will be treated as independent random variables in the next section.

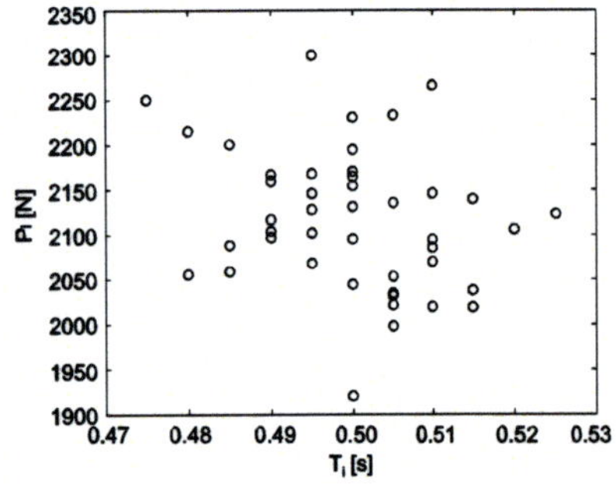

Figure 8: Peak amplitudes P_i vs. peak-to-peak intervals T_i.

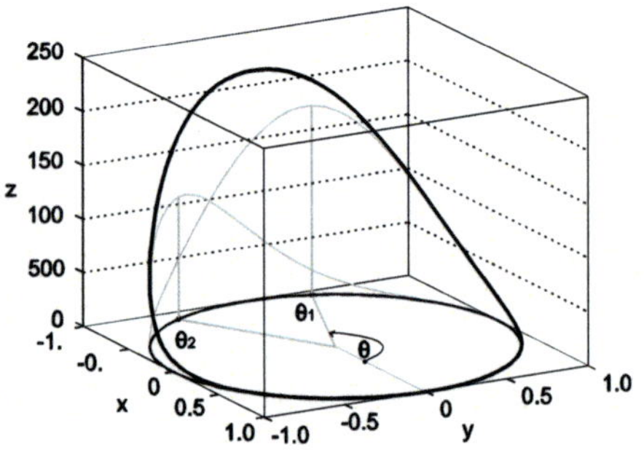

Figure 9: Three-dimensional trajectory generated by Equation (9) due to motion around the unit circle in the horizontal (x, y) plane.

5 DYNAMIC MODEL

The model generates a trajectory in a three-dimensional (3D) space with coordinates (x, y, z). Periodicity of the jumping force is obtained by retracing the Gaussian fit z(t), given by Equation (1), around a circle of unit radius in the (x, y) plane (Figure 9). The time required to complete one revolution on this circle corresponds to the period of one jumping cycle and lasts T seconds. The dynamic equations describing motion on the circle are therefore given by a set of two simple harmonic equations (Figure 9):

$$x(t) = \cos(\omega t)$$
$$y(t) = \sin(\omega t) \qquad (7)$$

where $\omega=2\pi/T$ is the angular frequency given in rad/s.

When transformed into the polar coordinate system, the oscillator given by Equation (7) becomes:

$$r = 1$$

$$\theta = \theta(t) = \omega t = \arctan\left(\frac{y(t)}{x(t)}\right), \quad \theta \in [0, 2\pi] \tag{8}$$

Further, Equation (1) can be rewritten as a function of the angular coordinate θ instead of time t:

$$z(\theta) = \sum_{i=1}^{2} (\alpha_i mg)\, e^{-\frac{(\theta-\theta_i)^2}{2\beta_i^2}} \tag{9}$$

where $\beta_i=4\pi b_i$ rad. The time positions of the two Gaussian peaks t_1 and t_2 now correspond to fixed angles θ_1 and θ_2 along the unit circle, respectively, as illustrated in Figures 8 and 9 and also given in Table 1. The trajectory $z(\theta)$ given by Equation (9) peaks away from the plane of the unit circle when θ approaches θ_1 and θ_2, whereas it is pulled back toward the circle and low ordinates as it moves away from these points (Figure 9).

Variations of the peak-to-peak intervals can be incorporated in the coupled system of equations (7)-(9) by varying the angular velocity ω on a cycle-by-cycle basis. Let T_k' be a series of peak-to-peak intervals calculated using Equation (2). The time-dependent angular frequency ω_k of circular motion is then given by the equation:

$$\omega_k' = \frac{2\pi}{T_k'}, \quad k = 1, \ldots, N \tag{10}$$

where N is the number of jumps. As a consequence of running the dynamic model given by a coupled system of equations (7)-(9) with variable angular frequency ω_k', the synthetic force signal is near-periodic only in terms of the peak-to-peak intervals, whereas the amplitudes of pulses remain equal to the average measured jumping pulse. To reflect the changes in both the amplitude and timing of the real jumping force from one pulse to another, amplitudes of each synthetic jumping pulse need to be multiplied by the corresponding element in series $\left(1+\overline{P}_k'\right)$ generated as explained in Section 4.1. This becomes evident when Equation (3) is rewritten as:

$$P_k' = \left(1+\overline{P}_k'\right)\mu_P, \quad k = 1, \ldots, N \tag{11}$$

where P_k' are synthetic peak amplitudes. Figure 10 illustrates examples of 20 s and 60 s long synthetic GRF signals when total number of jumps N is 40 and 120, respectively.

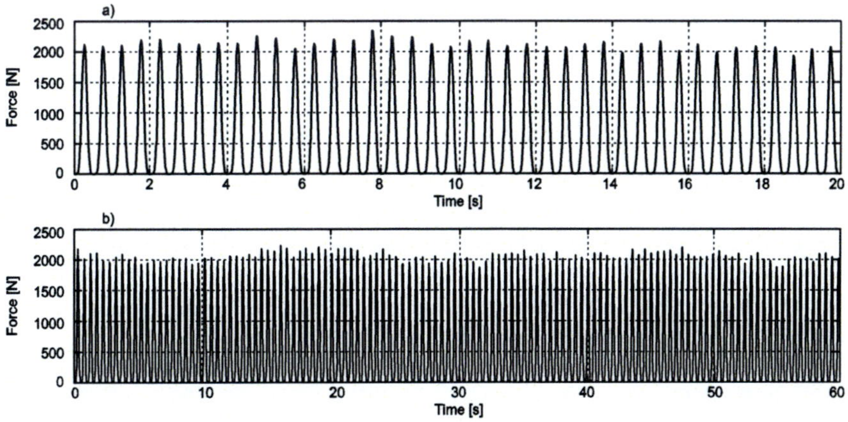

Figure 10: Synthetic jumping force signals a) 20 seconds and b) 60 seconds.

The similarity between the real measured and synthetic near-periodic vertical jumping force signals may be seen by comparison of Figure 1 to Figure 10. Moreover, the corresponding standard Fourier amplitude spectra are compared in Figure 11. For the first four dominant harmonics the relative errors are within the range of ±5%. Moreover, relative error in the area under the graph of the spectra is less than 10%. All this indicates good match in the frequency content between the measured and the synthetic GRF signals.

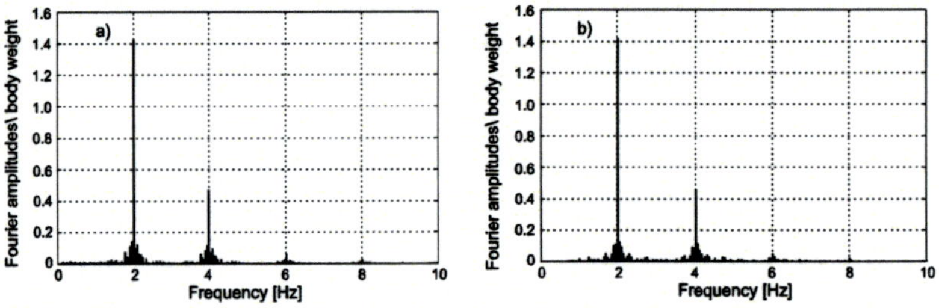

Figure 11: Fourier amplitude spectrum of a) measured and b) synthetic force signal due to jumping at average rate of 2 Hz.

6 SUMMARY AND CONCLUSIONS

This paper presents a new mathematical model to generate near-periodic synthetic jumping force signals with specified jumping rate and morphology. The near-periodic nature of the jumping force is modelled using a closed-loop trajectory throughout 3D space (x, y, z) around a circle of unit radius in (x, y) plane. The trajectory replicates size and shape of the average measured jumping pulse via a sum of two Gaussian exponentials. This modelling strategy can represent temporal and spectral features of the real human vertical jumping loading more effectively than the conventional half-sine pulses and Fourier series approach yielding more reliable predictions of dynamic structural response due to people jumping. The proposed Gaussian fit, coupled with equations of circular motion, has the following considerable advantages:

1) Two Gaussian exponentials, peaking just before and just after the peak of a realistic jumping beat, can reflect a lack of symmetry of the beat pattern quite accurately, as opposed to the symmetric half-sine pulses.

2) Variations of the cycle-by-cycle intervals can be included by varying the angular frequency for consecutive revolutions around the unit circle. As a result, the amplitude Fourier spectrum of corresponding synthetic jumping signal becomes narrow-band random phenomenon showing the leaking of energy in the vicinity of dominant Fourier harmonics.

3) Amplitudes of the synthetic jumping force signal can be changed on a cycle-by-cycle basis in a manner which allows the model to generate 'smooth' transition between consecutive beats, as it is measured in reality.

Furthermore, a number of extensions of the proposed model are possible:

1) Fitting the model to the jumping force patterns generated by different individuals under a range of jumping frequencies can yield a database of jumping force parameters for the general human population. The database can then be used to develop and calibrate a new generation of probability based stochastic models of jumping loading for individuals.

2) This framework can be extended further to probability based jumping loading models of groups and crowds. This presents an opportunity to enhance the vibration serviceability assessment of civil engineering structures occupied and dynamically excited by humans such as grandstands, footbridges, floors and staircases.

This framework can be extended further to probability based jumping loading models of groups and crowds. This presents an opportunity to enhance the vibration serviceability assessment of civil engineering structures occupied and dynamically excited by humans such as grandstands, footbridges, floors and staircases.

ACKNOWLEDGEMENTS

The authors would like to thank Mr Christopher Middleton for his help in collecting the data. Also, the authors would like to acknowledge the financial support provided by the UK Engineering and Physical Sciences Research Council (EPSRC) for grant reference EP/E018734/1 ('Human Walking and Running Forces: Novel Experimental Characterisation and Application in Civil Engineering Dynamics') and GR/T03017/01 ('Stochastic Approach to Human-Structure Dynamic Interaction').

REFERENCES

[1] H. Bachmann, W. Ammann, Vibrations in Structures Induced by Man and machines, IABSE - AIPC - IVBH, Zürich, Switzerland, 1987.

[2] M. Glackin, Stadia design rethink prompted by Cardiff fiasco. Building, January (2000) 11.

[3] P. Reynolds, A. Pavic, Vibration Performance of a Large Cantilever Grandstand during an International Football Match. ASCE Journal of Performance of Constructed Facilities 20 (2006) 202-212.

[4] H.Bachmann, A.Petlove, H.Rainer, Dynamic forces from rhythmical human body motions. In: Vibration Problems in Structures: Practical Guidelines, Appendix G, Birkhauser, Basel, Switzerland, 1995.

[5] B.R.Ellis, T.Ji, The response of structures to dynamic crowd loads, Digest 426. British Standards Institution, London, UK, 2004.

[6] J.H.H.Sim, A.Blakeborough, M.Williams, Statistical model of crowd jumping loads. ASCE Journal of Structural Engineering 134 (12) (2008) 1852-1861.

[7] BS 6399-1: 1996 incorporating amendment No. 1. Code of practice for dead and imposed loads. British Standards Institution, London, UK, 2002.

[8] Canadian Commission on Building and Fire Codes, User's guide - NBC 2005: structural commentaries (Part 4 of Division B). National Research Council Canada, Institute for Research in Construction, Ottawa, Canada, 2006.

[9] J.M.W.Brownjohn, A.Pavic, P.Omenzetter, A spectral density approach for modelling continuous vertical forces on pedestrian structures due to walking. Canadian Journal of Civil Engineering 31 (2004) 65-77.

[10] V.Racic, A.Pavic, Mathematical model to generate near-periodic human jumping force signals. Mechanical Systems and Signal Processing 24 (1) (2010) 138-152.

[11] P.E.McSharry, G.D.Clifford, L.Tarassenko, L.A.Smith, A dynamical model for generating synthetic electrocardiogram signals. IEEE Transactions of Biomedical Engineering 50 (2003) 289-294.

[12] AMTI product manuals, Advanced Mechanical Technology, Watertown, USA, 2007.

[13] J.G.Parkhouse, D.J.Ewins, Crowd-induced rhythmic loading. Structures & Buildings 159 (2006) 247-259.

[14] D.C.Montgomery, G.C.Runger, Applied statistics and probability for engineers, 2nd edition. John Wiley & Sons, Inc., New York, NY, 1999.

[15] K. McConnell, Vibration testing - theory and practice. John Wiley & Sons, Inc., New York, NY, 1995.

[16] D.E.Newland, An Introduction to Random Vibrations, Spectral and Wavelet Analysis, 3rd edition, Pearson Education Limited, Harlow, Essex, UK, 1993.

[17] D.M.Bates, D.G.Watts, Nonlinear Regression and Its Applications. Wiley, New York, 1998.

Proceedings of the IMAC-XXVIII
February 1–4, 2010, Jacksonville, Florida USA

LQR controller for an in-service floor

Donald S Nyawako, Paul Reynolds

Department of Structural and Civil Engineering, The University of Sheffield,
Sir Frederick Mappin Building, Mappin Street, Sheffield, S1 3JD, United Kingdom

E-mail: p.reynolds@sheffield.ac.uk

ABSTRACT

Of the numerous vibration mitigation technologies available, active vibration control is emerging as a realistic option for mitigation of human-induced vibrations in problem floors. This paper presents the design and implementation of an LQR controller that realizes a single input multiple output (SIMO) control strategy for suppression of human-induced vibrations in an in-service floor at the University of Sheffield. From results of modal testing of the floor structure, a suitable reduced-order model of the structure is derived. This is used to determine the optimal proportional and derivative gains that are implemented in both the analytical and experimental studies for SIMO control. Vibration mitigation performances attained in both the analytical and experimental implementation are presented. The analytical studies predicted a 58.7% reduction in vibration level for SIMO control for a single pedestrian walking across the floor. Experimental implementation yielded a 55% reduction in vibration level for the same.

Key words: Active vibration control, floor structures, human-induced vibration

1. INTRODUCTION

Human activity is one of the major causes of undesirable vibrations in civil engineering structures, primarily in floors, footbridges, grandstands and staircases. The increasing severity of human actions on some of these structures, coupled with advancements in structural design allowing for longer span and more lightweight structures often results in significant resonant, transient, steady-state or impulsive responses.

Passive, active and semi-active vibration mitigation techniques have been implemented in the above structures to mitigate the effects of human-induced vibrations [1], and active vibration control (AVC) is now emerging as a realistic option for mitigation of human induced vibrations in problem floors. Various AVC approaches for mitigation of human induced vibrations in floors, for example, direct velocity feedback DVF [2, 3], on-off non-linear velocity feedback control [4] direct and compensated acceleration feedback [5] and response dependent velocity feedback [6] have been explored. More recently, an approach that makes use of a linear quadratic regulator (LQR) controller has been investigated for mitigation of human-induced floor vibrations [7].

LQR controllers have been widely studied and implemented in the mechanical engineering and automatic control communities. They have been shown to have maximum effect on the systems they are implemented in while using a minimum amount of energy to achieve it. Comparisons of LQR control to classical methods of velocity feedback and on-off control have shown that this optimal control technique is able to bring the systems they are implemented in to rest much faster and is less prone to spillover [8, 9].

DOI 10.1007/978-1-4419-9831-6_24, © The Society for Experimental Mechanics, Inc. 2011

The design of an LQR controller makes use of optimal control principles and is achieved by minimization of a specified performance index. A suitable reduced order model (ROM) is required to formulate this optimal control problem. The ROM is derived based on uncontrollability/unobservability concepts of neglected modes of vibration and the placement of actuators and sensors at node points of the neglected higher modes [7, 10]. Based on pre-determined weighting matrices, suitable proportional and derivative gains of the LQR controller can then be derived from this ROM and implemented in both analytical studies and experimental implementation of the AVC system.

This work presents the design and practical implementation of an LQR controller for an in-service floor. An electrodynamic proof-mass actuator supplies the control force. Section 2 introduces the floor structure and section 3 describes the experimental modal analysis (EMA) carried out prior to the installation of the AVC and provides dynamic properties of the floor structure. Based on the results from modal testing, the locations of the AVC system (for the SIMO control structure) are selected and a ROM of the floor structure is developed in section 4. The LQR controller is designed in section 5, which also presents results from analytical studies of the AVC system on this floor. Results from implementation of the AVC system on the real floor are presented in section 6. Some conclusions drawn are presented in section 7.

2. DESCRIPTION OF FLOOR

The test structure is an open-plan reinforced concrete floor with timber decking, measuring 9.5m x 21.5m (Fig. 1a). It lies above three rooms: a Committee Room and Library (7.5m x 9.5m), the Entrance Hall (6.5m x 9.5m) and a Cloak Room (7.5m x 9.5m), as shown in Fig. 1b. This test structure rests on load bearing walls that are partitions of the three rooms described.

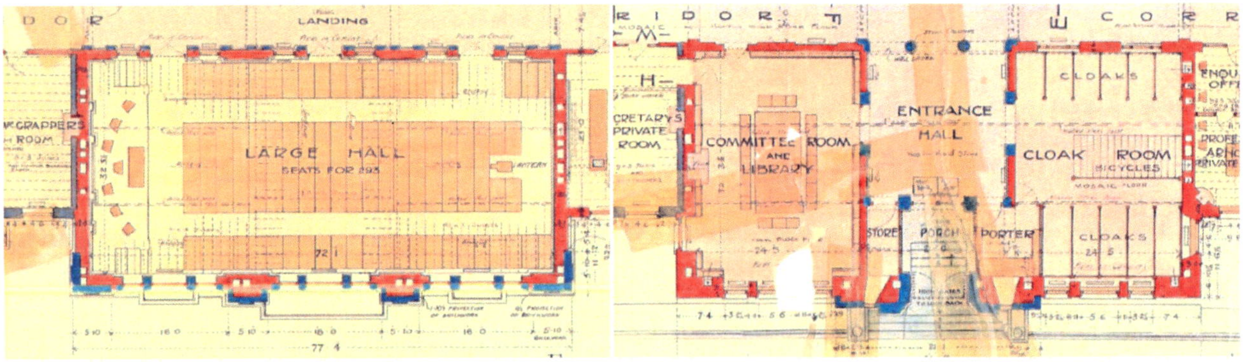

(a) (b)

Fig. 1 (a) Plan view of test structure. (b) Plan view underneath test structure.

From a preliminary inspection of the floor using heel drop excitation, the floor areas directly above the three rooms in Fig. 1b were found to be quite flexible. These were the initial locations proposed for installation of the AVC system.

3. MODAL TESTING

Modal testing of the uncontrolled structure was carried out using artificial excitation supplied by three electrodynamic shakers operated in inertial mode (2 APS Dynamics Model 400 mechanical shakers and 1 APS Dynamics model 113-HF). Three excitation points were used (Test Points (TPs) 113, 208 and 304) and responses were measured at all TPs shown in Fig. 2. The shakers were driven by a continuous uncorrelated random excitation at the three excitation points. 80 s time domain data blocks were utilized resulting in a frequency resolution of 0.0125 Hz. 20 averages were used with 75% overlap and a Hanning window was applied to all data blocks.

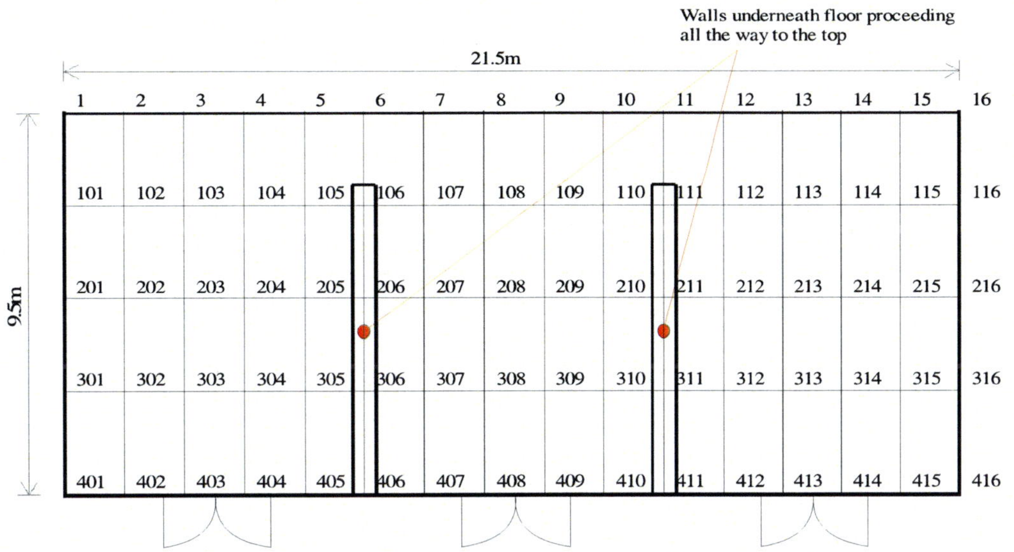

Fig. 2 Test grid for EMA.

Point accelerance frequency response functions (FRFs) acquired from EMA are shown in Fig. 3. The lowest mode occurs at 10.6 Hz and the highest amplitude within the range 10 Hz to 20 Hz occurs at 13.1 Hz. The lowest modes of vibration of this floor structure are quite high in frequency and are typical of a so-called high frequency floor [11]. Modal parameter estimation was carried out using the multiple reference orthogonal polynomial algorithm in ME'scope software. Tables 1, 2 and 3 show the estimated modes of vibration in the frequency bands 4 – 20 Hz from each of the shaker locations and Figs. 4, 5 and 6 show the associated mode shapes.

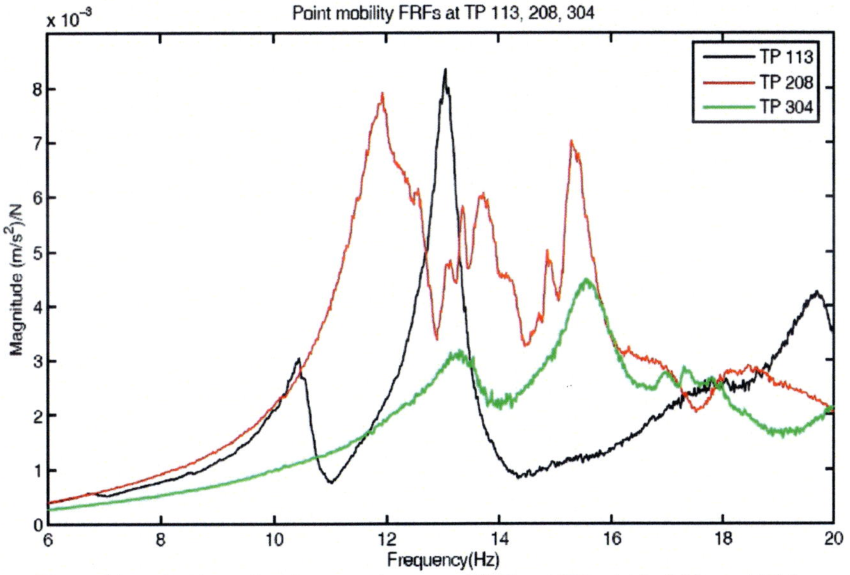

Fig. 3 Magnitudes of point accelerance FRFs at TPs 113, 208 and 304.

Table 1 Estimated vibration modes for shaker at TP113.

Frequency (Hz)	Damping (%)	Description
10.6	2.40	Bending mode over Committee Room
13.1	1.98	Bending mode over Committee Room
19.5	1.78	Bending mode over Committee Room

Table 2 Estimated vibration modes for shaker at TP208.

Frequency (Hz)	Damping (%)	Description
12.0	4.44	Bending mode over Entrance Hall
13.5	1.24	Bending mode over Entrance Hall
15.3	0.48	Bending mode over Entrance Hall
18.7	1.13	Bending mode over Entrance Hall

Table 3 Estimated vibration modes for shaker at TP304.

Frequency (Hz)	Damping (%)	Description
13.0	2.81	Bending mode over Cloak Room
15.3	4.09	Bending mode over Cloak Room
17.8	1.21	Bending mode over Cloak Room
19.2	0.641	Coupled mode over Cloak Room and Entrance Hall

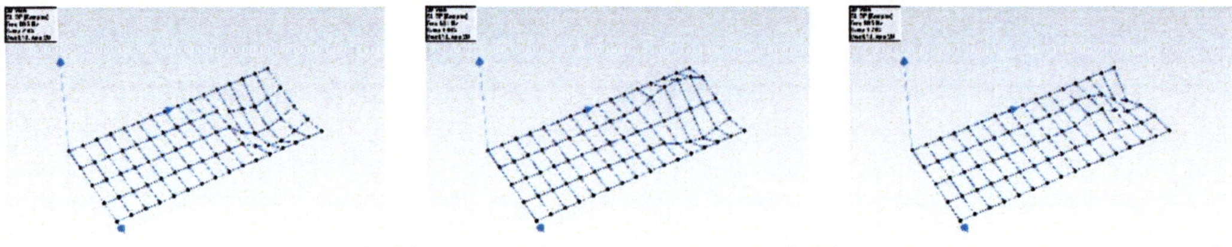

(a) (b) (c)

Fig. 4. Estimated vibration modes at TP113 (at 10.6 Hz, 13.1 Hz and 19.5 Hz).

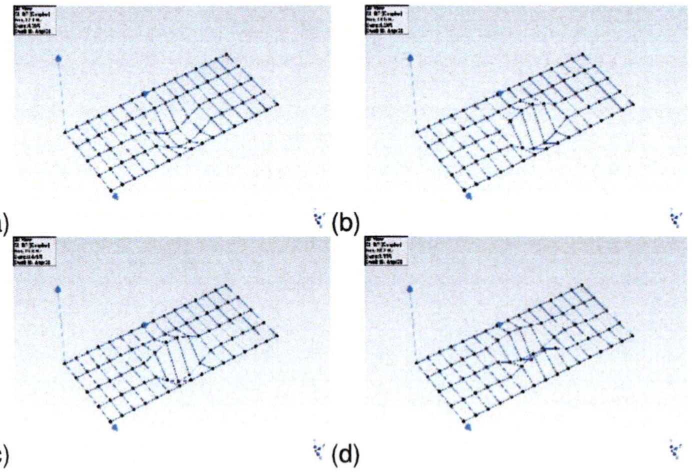

(a) (b) (c) (d)

Fig. 5. Estimated vibration modes at TP208 (at 12.1 Hz, 13.5 Hz, 15.3 Hz and 18.7 Hz).

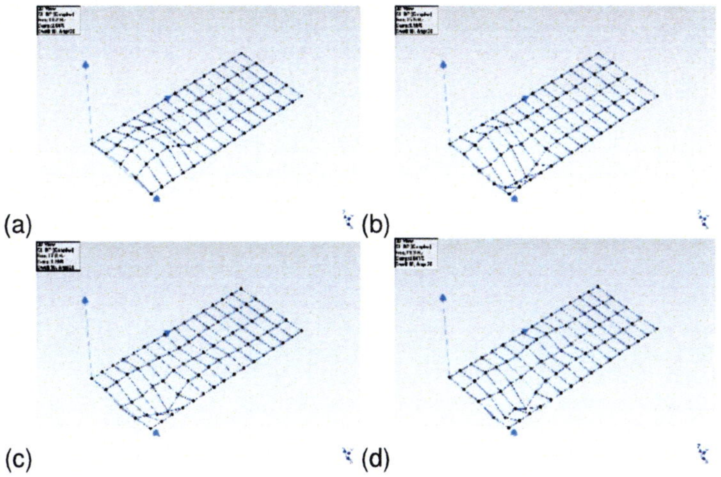

Fig. 6. Estimated vibration modes at TP304 (at 13.2 Hz, 15.5 Hz, 17.8 Hz and 19.3 Hz).

From modal testing, it can be seen that the lowest frequency modes of this floor (natural frequencies below 19.2 Hz) are localized within each of the floor sections above the Committee Room and Library, Entrance Hall, and the Cloak Room. The floor section above the Committee Room and Library was chosen to study the effectiveness of the LQR controller. This is primarily because TP 113 within this floor section had the highest point accelerance FRF peak at 13.1 Hz.

4. REDUCED ORDER MODEL

A detailed procedure of derivation of the ROM for any floor structure is presented in [7]. The mode order for producing the ROM in these studies has been chosen as 2 and the node points of mode 3 within this section of the floor (TPs 114 and 314) have been chosen as the locations of masses for the lumped parameter system, as shown in Fig. 7.

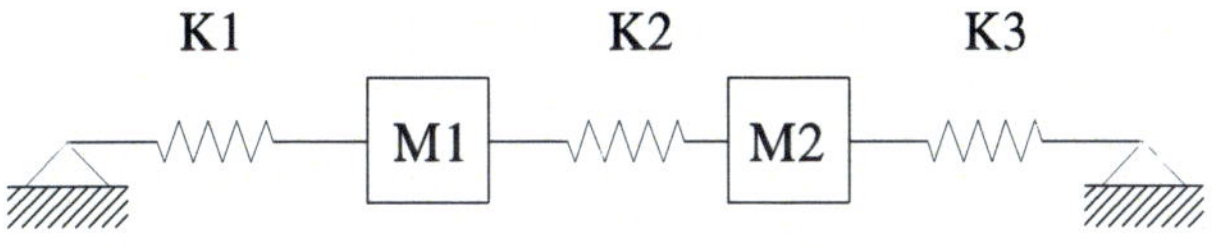

Fig. 7. ROM for LQR controller design.

From the mass normalized scaling of the mode shapes at TPs 114 and 314 for the first two modes of vibration of this section of the floor, the mass and stiffness matrices of the physical co-ordinates of Fig. 7 can be determined by: $M = \left(\Phi^T\right)^{-1}\Phi^{-1} = \left(\Phi\Phi^T\right)^{-1}$ and $K = \left(\Phi^T\right)^{-1}\Omega^2\Phi^{-1}$. For the first iteration, an error typically arises in diagonalization of the mass matrix. A generalized matrix for obtaining convergence can be derived by $[\delta\phi_{11} \quad \delta\phi_{21} \quad \delta\phi_{12} \quad \delta\phi_{22}]^T = \left[\frac{\partial \varepsilon_r}{\partial \phi}\right]^T \left[\left[\frac{\partial \varepsilon_r}{\partial \phi}\right]\left[\frac{\partial \varepsilon_r}{\partial \phi}\right]^T\right]^{-1} (-\varepsilon_r)$ [10], where ε_r is the error arising from diagonalization of the mass matrix in the first iteration. The mass normalized modal matrix at TPs 114 and 314 and spectral matrix for the first two modes of vibration are shown in Eqs. 1 and 2.

Mass normalized modal matrix:
$$\Phi = \begin{bmatrix} 0.0134 & -0.0209 \\ 0.0156 & 0.0178 \end{bmatrix} \tag{1}$$

Spectral matrix:
$$\Omega = \begin{bmatrix} 66.6 & 0 \\ 0 & 82.31 \end{bmatrix} \tag{2}$$

The mass matrix derived from the first iteration does not yield the required diagonalization. By applying the generalized matrix for obtaining convergence, the adjusted mass and stiffness matrices are now presented in Eqs. 3 and 4.

Mass matrix $$M = \begin{bmatrix} 1718.4 & 10.2 \\ 10.2 & 1898.0 \end{bmatrix} \quad (3)$$

Stiffness matrix $$K = \begin{bmatrix} 0.9947e+7 & -0.2027e+7 \\ -0.2027e+7 & 1.0267e+7 \end{bmatrix} \quad (4)$$

For the 2-DOF lumped mass model in Fig. 7, $M_1 = 1718.4\,\text{Kg}$, $M_2 = 1898.0\,\text{Kg}$ and the off-diagonal terms can now be neglected; $K_1 = 0.7920e+7\,\text{N/m}$; $K_3 = 0.8240e+7\,\text{N/m}$; $K_2 = 0.2027e+7\,\text{N/m}$. Proportional damping terms are: $C_1 = 5532.8\,\text{Ns/m}$; $C_3 = 6094.5\,\text{Ns/m}$; $C_2 = 22.8\,\text{Ns/m}$. These are determined from the expression $C = (\Phi^T)^{-1}(2\zeta\Omega)\Phi^{-1}$ where ζ term is obtained from Table 1. The derived state-space representation $\{\dot{x} = Ax + Bu;\ y = Cx + Du\}$ of the 2-DOF ROM used in the analytical studies and for determining optimal control gains is presented in Eq. 5.

$$A = \begin{bmatrix} 0 & 0 & 1 & 0 \\ 0 & 0 & 0 & 1 \\ -\frac{K_1+K_2}{M1} & -\frac{C_1+C_2}{M1} & \frac{K_2}{M1} & \frac{C_2}{M1} \\ \frac{K_2}{M2} & \frac{C_2}{M2} & -\frac{K_2+K_3}{M2} & -\frac{C_2+C_3}{M2} \end{bmatrix} \quad B = \begin{bmatrix} 0 & 0 \\ 0 & 0 \\ \frac{1}{M1} & 0 \\ 0 & \frac{1}{M2} \end{bmatrix}$$

$$C = \begin{bmatrix} 1 & 0 & 0 & 0 \\ 0 & 1 & 0 & 0 \\ 0 & 0 & 1 & 0 \\ 0 & 0 & 0 & 1 \end{bmatrix} \quad x = \begin{bmatrix} x_1 \\ x_2 \\ x_3 \\ x_4 \end{bmatrix} \quad \dot{x} = \begin{bmatrix} \dot{x}_1 \\ \dot{x}_2 \\ \dot{x}_3 \\ \dot{x}_4 \end{bmatrix} \quad x_3 = \dot{x}_1; x_4 = \dot{x}_2 \quad (5)$$

For the SIMO control setup, a single actuator is sited at TP114 (corresponding to location M1 on the ROM).

5. ACTIVE CONTROL DESIGN

This section presents the design of the LQR controller. Weights of the performance index are determined and proportional (k_p) and derivative (k_d) gains of the LQR controller are derived. Actuator and sensor dynamics (where dynamics of sensor refer to the integrator circuit) are presented. Performance of the AVC system with implementation of the LQR controller is assessed in simulations by using real human walking time histories obtained from treadmill walking tests.

5.1. Weights of performance index

Control design in optimal control is achieved by minimization of the performance index, J, in Eq. (6). The weights, Q and R, are determined by making use of Bryson's rule [12]. Q is based on the threshold of human perception for continuous vertical vibration. This is given in Eq. (7). For output constraint control (OCC), R = 1 in these studies.

$$J = \frac{1}{2}x^T(N)Sx(N) + \frac{1}{2}\int_{t_o}^{t_f}\left(x^T(t)Qx(t) + u^T(t)Ru(t)\right)dt \quad (6)$$

Where: x – n x 1 state vector; u – r x 1 control vector; Q, R – weights; A – n x n state matrix; B – n x r input matrix

$$Q = \begin{bmatrix} 2.56e+10 & 0 & 0 & 0 \\ 0 & 2.56e+10 & 0 & 0 \\ 0 & 0 & 3.19e+7 & 0 \\ 0 & 0 & 0 & 3.19e+7 \end{bmatrix} \quad (7)$$

5.2. Proportional and derivative gains

From the state-space matrix of the ROM in Eq. (5), and the weighting matrices Q and R, the optimal proportional (k_{p1} and k_{p2}) and derivative (k_{d1} and k_{d2}) gains can be evaluated using MATLAB software. These gains for the SIMO control structure are presented in Table 4. Fig. 8 shows a block diagram of a typical feedback control scheme with the LQR controller. The ideal integrator expressions are replaced by the sensor dynamics in Eq. (9).

Table 4. Optimum proportional and derivative gains for SIMO control for Q with R = 1.

R = 1	k_{p1}	k_{p2}	k_{d1}	k_{d2}
Q	1920.4	-2671.6	2765.6	103.6

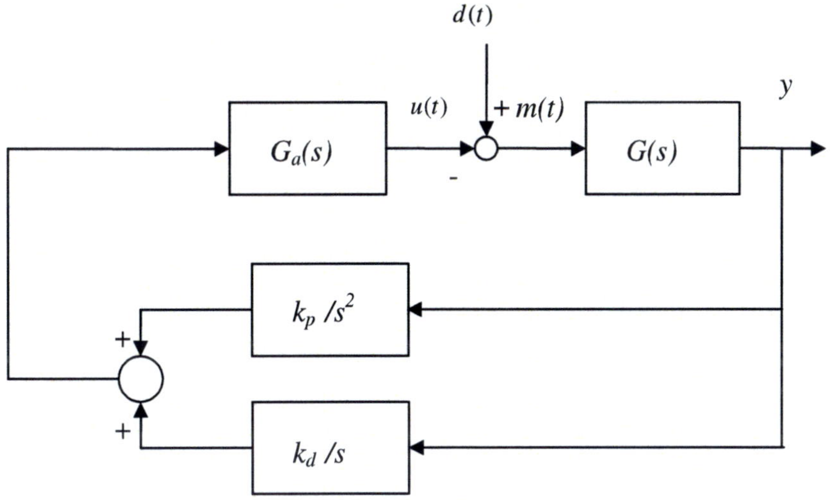

k_p proportional gain k_d derivative gain

$u(t)$ actuator output signal $y(t)$ acceleration response

$d(t)$ input disturbance signal

Fig. 8. Block diagram of typical feedback control scheme ($m(t) = d(t) - u(t)$).

5.3. Actuator and sensor dynamics

The actuator is an APS Dynamics model 113-HF. The sensors are piezoelectric accelerometers (Endevco Model 7754A-1000). The saturation voltage is taken as V_s = 0.5 V. The controller hardware comprises of a PC with a National Instruments PCI-6036E DAQ Card installed. Velocity is calculated digitally by integration of the measured acceleration signal together with the computation of the control law. The actuator, having inertial mass of 13 kg each, can be described closely by the third order model in Eq. (8). ς_A and ω_A are the damping ratio and natural frequency of the actuator. K_A and ε are positive numbers. The natural frequency of the shaker is estimated as $\omega_A = 11.30$ rad/s (1.80 Hz). Velocity and displacement quantities are obtained digitally from the acceleration and DC corrections are applied. The integrator circuit can be closely modeled by the filter in Eq. (9).

$$G_A(s) = \frac{K_A s^2}{s^2 + 2\zeta\omega_A s + \omega_A^2} \cdot \frac{1}{s+\varepsilon} = \frac{13200s^2}{s^3 + 46.75s^2 + 469s + 4822} \quad (8)$$

$$G_S(s) = \frac{s}{s^2 + 2\zeta_S\omega_S s + \omega_S^2} = \frac{s}{s^2 + 8s + 100} \quad (9)$$

5.4. Stability analysis

The closed-loop stability of the system is investigated using the Nyquist stability theorem by plotting the contour of $K_c(j\omega)G(j\omega)$ where $s = j\omega$ [13, 14] as shown in Fig. 9. $K_c(s)$ incorporates k_{p1}, k_{p2} *and* k_{d1}, k_{d2} (Table 4) and actuator and sensor dynamics in Eqs. (8) and (9). The closed-loop system is stable if the contour of $K_c(j\omega)G(j\omega)$ does not encircle the $-1+j0$ point when the number of poles of $K_c(s)G(s)$ in the right-hand s-plane is zero [14].

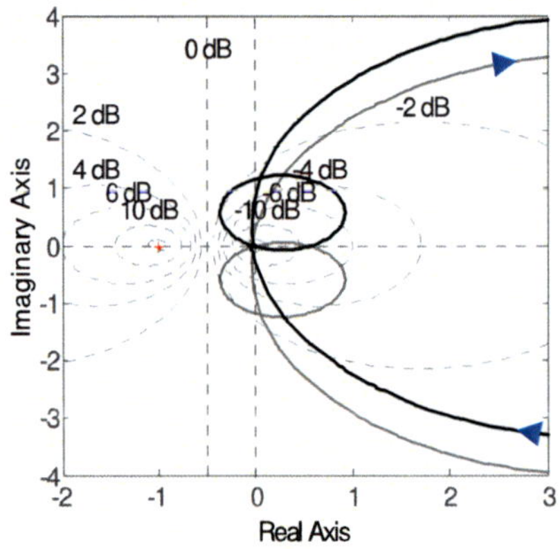

Fig. 9. Nyquist contour of $K_c(j\omega)G(j\omega)$ for SIMO control for Q - GM = 15.9, PM = 48.2^0.

5.5. Performance assessment

Fig. 10 shows the uncontrolled and controlled responses from analytical studies at M1 (corresponding to TP114 on the test structure). Table 5 shows the results of the analytical simulations in Fig. 10 for the uncontrolled and controlled (for SIMO control) responses carried out using the gains established in section 4.2. Real walking excitation obtained from treadmill tests is used to provide the excitation [15]. This comprises of a single pacing frequency at 2.12 Hz in such a way that the first floor vibration mode might be excited by the fifth harmonic. A 58.7 % reduction in the vibration level was predicted using SIMO control.

Table 5: Simulation performance assessment for SIMO control to walking excitation

	Analytical studies		Experimental implementation	
	Uncontrolled	SIMO	Uncontrolled	SIMO
Walking at 2.12 Hz				
MTVV [(1)] (m/s^2)	0.179	0.074	0.219	0.0987
Reduction (%)	—	58.7 %	—	55.0 %

[(1)] Maximum Transient Vibration Value defined as the maximum value of the 1s running RMS

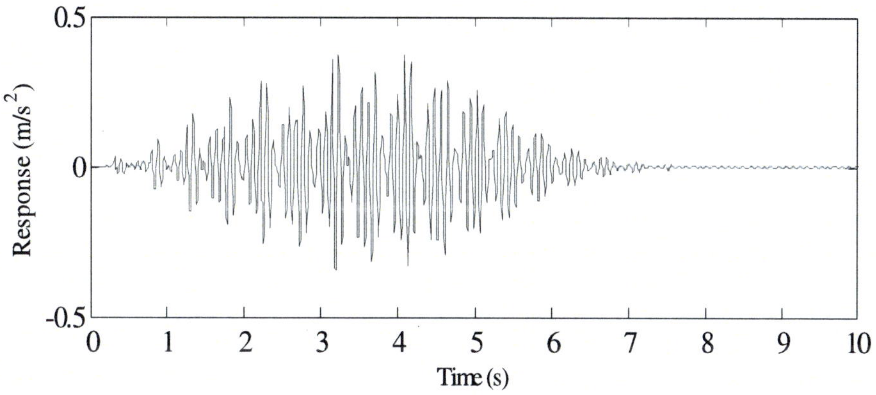

(a) Uncontrolled

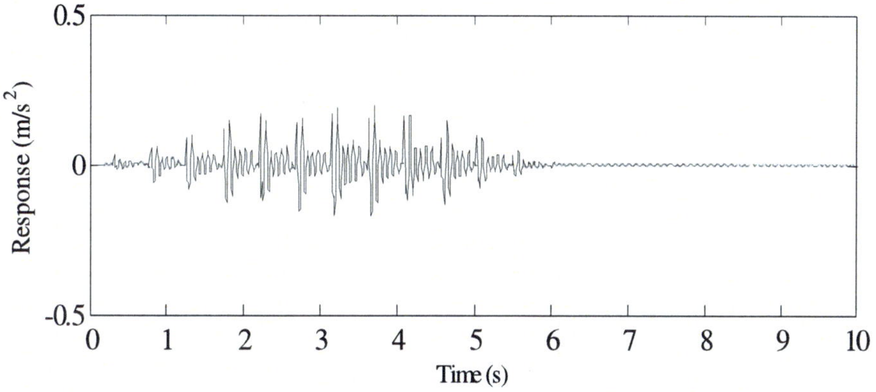

(b) Controlled – SIMO control

Fig. 10. Uncontrolled and controlled (for SIMO control) responses at M1

6. RESULTS FROM EXPERIMENTAL IMPLEMENTATION OF ACTIVE VIBRATION CONTROL

Walking tests were carried out using the same walking excitation frequencies noted in Section 5.5. Fig. 11 shows a photograph of the experimental setup.

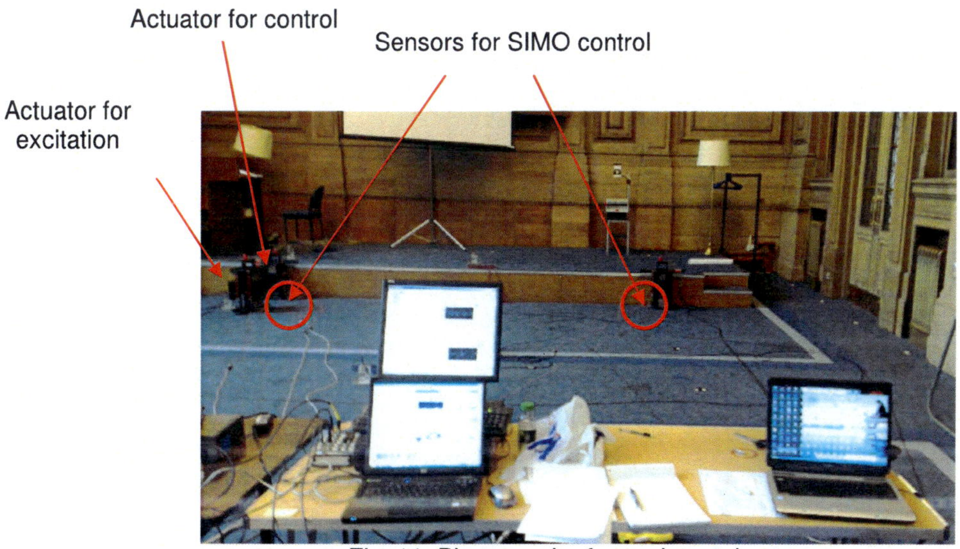

Fig. 11. Photograph of experimental setup

The walking path comprised of walking from TP14 to TP414. Fig. 12 shows the uncontrolled and controlled responses. The reductions in controlled responses with SIMO control structure was 55.0 %. The experimental results agreed well with the analytical predictions.

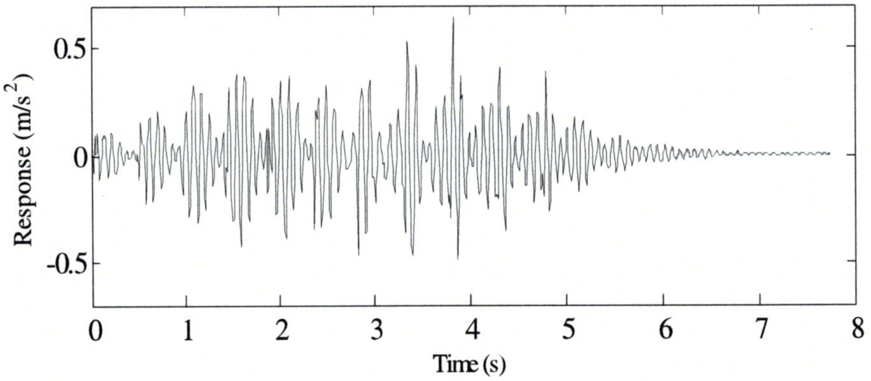

(a) Uncontrolled

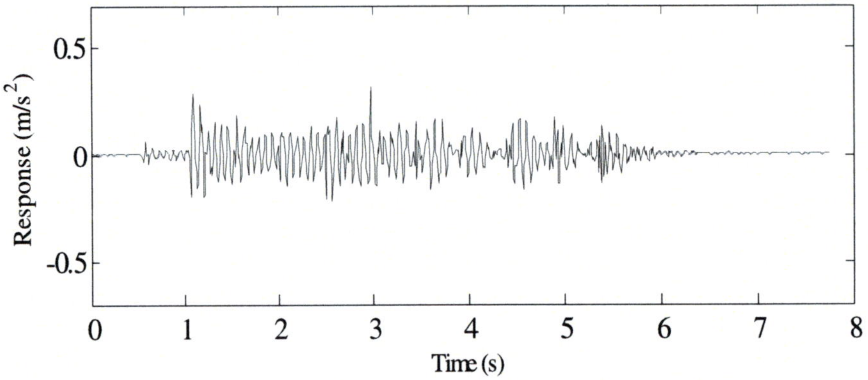

(b) Controlled – SIMO control

Fig. 12. Uncontrolled and controlled (for SIMO control) responses at M1

7. CONCLUSIONS

With a suitable ROM, an LQR controller can be designed for mitigation of human induced vibrations in floors. This paper demonstrates the successful development and implementation of an LQR controller for suppression of human-induced vibrations in an in-service floor at the University of Sheffield. 55-60% reductions in vibration levels with the AVC on were noted in both the analytical studies and experimental implementation of the AVC system by making use of SIMO control. There was a good agreement between the analytical studies and experimental implementation of the AVC system.

ACKNOWLEDGMENTS

The authors would like to acknowledge the financial assistance provided by the UK Engineering and Physical Sciences Research Council (EPSRC) through a platform grant entitled “Dynamic Performance of Large Civil Engineering Structures: An Integrated Approach to Management, Design and Assessment” (Ref: EP/G061130/1).

REFERENCES

[1] Nyawako D. and Reynolds P. (2007), Technologies for mitigation of human-induced vibrations in civil engineering structures, The Shock and Vibration Digest, 39 (6), 465-493.

[2] Hanagan, L.M. (1994) Active control of floor vibrations, PhD dissertation, Virginia.

[3] Hanagan, L.M. and Murray, T.M.(1997), Active control for reducing floor vibrations, *Journal of Structural Engineering*, Vol. 123, pp. 1497–1505, 1997.

[4] Diaz. I.M., Nyawako D.S. and Reynolds P. (2008), On-off nonlinear velocity feedback control for cancelling floor vibrations, Proceedings of *Fourth European Conference on Structural Control*, St. Petersburg, vol. 1, pp. 175 182, 2008.

[5] Diaz. I.M., and Reynolds, P. (2009), Acceleration feedback control of human-induced floor vibrations, Accepted for publication in *Engineering Structures,* September 2009.

[6] Nyawako, D. and Reynolds, P. (2009a), Response-dependent velocity feedback control for mitigation of human-induced floor vibrations, *Smart Materials and Structures,* 18, 075002, 1-14.

[7] Nyawako, D and P. Reynolds (2009b), LQR control for mitigation of human-induced floor vibrations', under review in the Journal of Sound and Vibration.

[8] Han, J.H. and Lee, I. (1997), Optimal placement of piezoelectric sensors and actuators for vibration control of a composite plate using genetic algorithms, Smart Materials and Structures 8, 257 – 267.

[9] Bai, M.R. and Lin, G.M. (1996), The development of a DSP-based active small amplitude vibration control system for flexible beams by using the LQG algorithms and intelligent materials, *Journal of Sound and Vibration*, 198, 411-427.

[10] Seto, K. and Mitsuta, S. (1992), A new method for making a reduced order model flexible structures using unobservability and uncontrollability and its application in vibration control, In: *Proc. 1st Int. Conf. Motion and Vibration Control (MOVIC)*, 152–158.

[11] Smith, A.L., Hicks, S.J., and Devine, P.J. (2007), Design of floors for vibration, a new approach, SCI Publication P354, Published by The Steel Construction Institute, Silwood Park, Ascot, Berkshire.

[12] Hespanha, J.P. (2007), Lecture notes on LQG/LQR controller design, University of California Santa Barbara.

[13] Maciejowski, J.M. (1989), Mutlivariable feedback design, Addison-Wesley Publishing Company, Inc. ISBN 0-201-18243-2

[14] Skogestad, S. and Postlethwaite, I. (1996), Multivariable Feedback Control – Analysis and Design, John Wiley and Sons Ltd. ISBN 0 471 94277.

[15] Brownjohn, J.M.W., Pavic, A. and Omenzetter, P. (2004), A spectral density approach for modelling continuous vertical forces on pedestrian structures due to walking, *Canadian Journal of Civil Engineering*, 31(1), 65–77.

Proceedings of the IMAC-XXVIII
February 1–4, 2010, Jacksonville, Florida USA

Development of a Wireless Bridge Monitoring System With Energy Harvesting Modules

Prof. Dr.-Ing. J.H.T. Schmidt,
University of Applied Sciences Magdeburg-Stendal
Breitscheidstrasse 2, 39114 Magdeburg
Thomas.Schmidt@hs-magdeburg.de

NOMENCLATURE

m	*Mass*
d	*Damping constant*
k	*Spring stiffness*
ξ_0	*Damping ratio*
$p(t)$	*Acting force on geophone*
f_0 ,ω_0	*Natural (angular) frequency of the geophone*
$U(\omega)$, $u(t)$	*Voltage output of geophone in the frequency and time domain*
G	*Geophone sensitivity (generator constant)*
$x(t)$	*Relative displacement between geophone housing and geophone base*
x_b	*Displacement of the base*
R_c	*Coil resistance*
R	*Equivalent resistance*
C	*Equivalent capacitance*
L	*Equivalent inductivity*

ABSTRACT

The paper presents the first results obtained in a research project started in 2009. Goal of the project is the development of a multi-sensor wireless bridge monitoring system with sensors that are supplied by energy harvesting modules. Parallel to the hardware development one group of our research team develops the software for the online data handling and evaluation. Our research team consists of four groups: Structural Engineering Group, Sensor Group, Wireless Group and Algorithm Group. Each group contributes to the project. Currently a standard monitoring system with external power supply is running on a pre-stressed concrete bridge. Vibrations, cracks, temperature, and moisture of the five span bridges are constantly monitored for at least one year. This system shall be replaced by the new system. The paper presents the concept of the energy harvesting modules and the online algorithms used to evaluate the measurements. The results obtained so far from the currently running monitoring system will also be presented.

INTRODUCTION

Besides railway bridges there are actually over 120.000 highway and road bridges in Germany. According to the German Standard DIN 1076 [1] each of these bridges has to be visually maintained in pre-defined intervals between two and six years. Many of the bridges, especially pre-stressed concrete bridges, are currently in a condition which makes reconstruction urgently necessary. The costs for maintenance and reconstruction steadily increase and many local communes can hardly bear them. This is where the research project presented in this paper deals with. The idea is to develop a monitoring system that can reduce the costs for maintenance on the

T. Proulx (ed.), *Dynamics of Civil Structures, Volume 4*, Conference Proceedings of the Society for Experimental Mechanics Series 13,
DOI 10.1007/978-1-4419-9831-6_25, © The Society for Experimental Mechanics, Inc. 2011

one side and gain more information of the current health state of a bridge or any other structure on the other side. What we want to have is what we call a “nerve system” of a structure as it is meanwhile common in each modern care or plane. This nerve system shall also help to better understand the behavior of the structure. The development of new and the improvement of already available sensors alongside the frantic development of low power electronics and devices in last years seems to make such “nerve systems” possible. But it is and will be a great challenge for researchers all over the world to establish such systems. A major challenge is how we can easily gather data of a structure and how to store the data most effectively. The power consumption of such a nerve system should be as low as possible. Further issues are: which data should be monitored, how should the data be evaluated to get the relevant information, and how is this information to be interpreted so that the goal to simplify and improve maintenance and to reduce its costs can be achieved. The project presented here is a step in this direction.

PROJECT CONCEPT

The whole project/research team consists of four groups, each group contributes to the project in a special sub-project. The four main sub-projects are.

1. The Structural Engineering Group (the author of the paper and his co-workers) is responsible for the structures (the bridges) that should be examined. The group decides what kind of structural information has to be monitored, where and with which data acquisition schedule. It helps to find the sensors and to define its specification. The group defines the data evaluation process and helps in the developing the appropriate algorithms to “get most out of the data”. The group shall model the structure with an appropriate method (e.g. FEM) to prepare a guideline for the correct maintenance decisions. It is responsible for the data interpretation and has to propose the procedures to improve the maintenance of the structure.
2. The Sensor Group [2] is responsible for the development of a self-sufficient sensor. That means, the group shall develop the sensor and the energy harvesting module to power the sensor attached to an independent data logger.
3. The Wireless Group [3] has to develop a low power heterogeneous network system so that the data from the sensors and its loggers can be transmitted to a central data base. The group has to choose the appropriate method for the wireless communication. It is responsible for the whole communication concept within the project.
4. The Algorithm Group [4] is responsible for the appropriate algorithms to evaluate the data obtained from the monitoring system. It has to develop the best possible data compression methods for local and global data storing and for the data transmission in the network. It helps the Structural Engineering Group the “get most out of the data” with the appropriate data presentation.

An overview of the wireless monitoring concept that shall be developed together with the responsibilities of each group is shown in Figure 1. It consists of several layers:

1. the sensors with its own energy supply through energy harvesting modules.
2. the data loggers with its own energy supply by energy harvesting modules and special “intelligence”, e.g. data compression; additionally the loggers may have a wireless module connected
3. the wireless network module – consisting of single modules for each sensor/logger
4. the analysis layer where the data is prepared and evaluated by special algorithms
5. the common data base for the whole system with its visual interface
6. the decision layer where the state of the structure is determined and the appropriate reaction to that state.

In the following paragraphs some of the first results of group 1 and 2 are presented. The algorithms currently used to evaluate the data are shortly described.

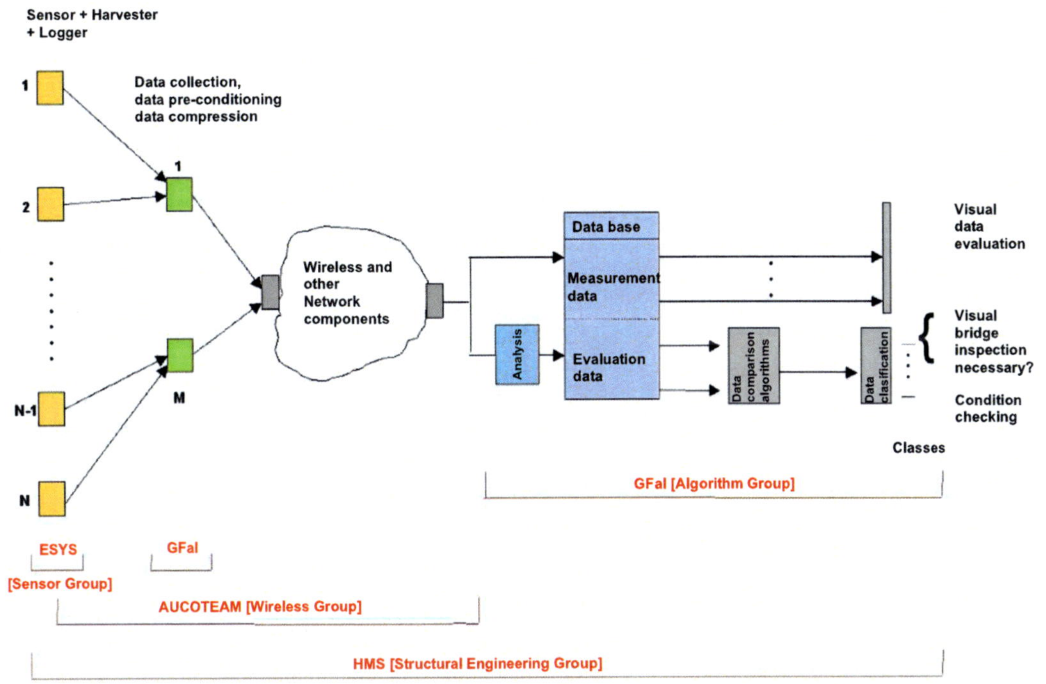

Figure 1. *Overall concept of the research project*

SENSORS AND HARVESTING MODULES

Since the project is the first step in the development of a multi-sensor wireless sensor system, at first a very simple vibration sensor was chosen that has several advantages. As sensor we (group 1 and 2) choose a geophone.
This rather old fashioned sensor has the advantage that it needs no power supply at all to gather vibration data. The functional principal of the sensor is shown in Figure 2.

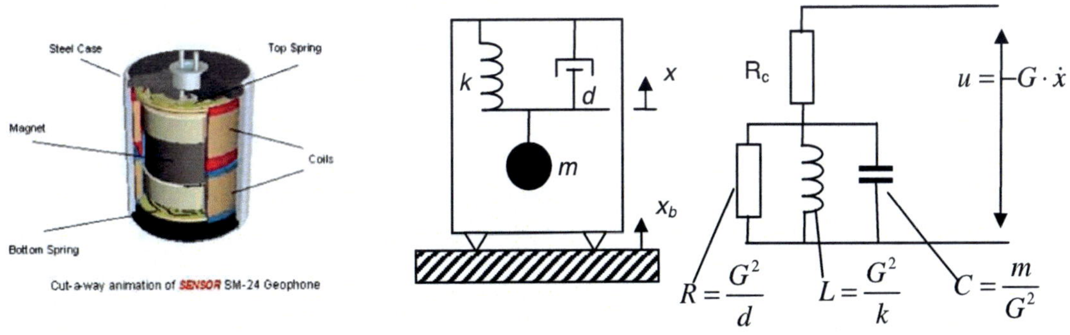

Figure 2. *Components of a Geophone [5] and its principal – mechanical pendulum – electrical equivalent*

Actually a geophone is an electro dynamical device, very similar to a microphone, where a coil moves in the field of a permanent magnet. The structural system of the sensor represents a simple mechanical pendulum which can be very well described by a single degree of freedom system. The mathematical description of the pendulum can be found in many textbooks. Here only the governing differential equation in the time domain is given.

(1) $$m \cdot \ddot{x} + d \cdot \dot{x} + k \cdot x = p(t)$$

with

(2) $$p(t) = -m \cdot \ddot{x}_b$$

that represent the force on the geophone mass due to the movement of the base. There are several other forces that may have an impact on the geophone mass, e.g. air pressure that are neglected here.
Since the permanent magnet is surrounded by a coil, suspended by springs that are connected to the case, a voltage is induced in the coil when the geophone base moves. The voltage can be calculated by Faraday's Law

(3) $$u(t) = -G \cdot \dot{x}$$

The transfer function for the voltage output can be derived in the frequency domain as

(4) $$U(\omega) = \frac{-\omega^2 G \cdot \dot{X}}{-\omega^2 + 2i\xi_0 \omega\omega_0 + \omega_0^2}$$

with

(5) $$\omega_0 = 2\pi f_0, \quad \sqrt{\frac{k}{m}}, \quad \xi_0 = \frac{d}{2\sqrt{km}}$$

The sensor works like a generator. So the sensor itself could be used as an energy harvester which will be described below. The electrical equivalent of the whole system (Figure 2) can be easily derived [5].
The electrical equivalent can be used in the development and optimization of the harvester circuits in a simulation program like [6].

A literature revue reveals that currently man attempt are made by researchers all over the world to develop energy harvesting modules. A lot of different physical effects are used, e.g. the piezoelectric effect, the electrostatic effect, the thermoelectric effect, the magnetostriction effect and the electrodynamic effect. We decided to use the electrodynamic effect because of its simplicity and because of the experiences we already have with different types of electrodynamic transducers.
A first prototype of our energy harvester was build according to the principal of a geophone. Additionally, different geophones were used as harvesters, e.g. an SM-6, 4.5 Hz Geophone [6] and an LC 4, 1 Hz Geophone from [7]. Since this first prototype showed some disadvantages it was re-designed by the Sensor Group. Figure 3 shows the current development model.

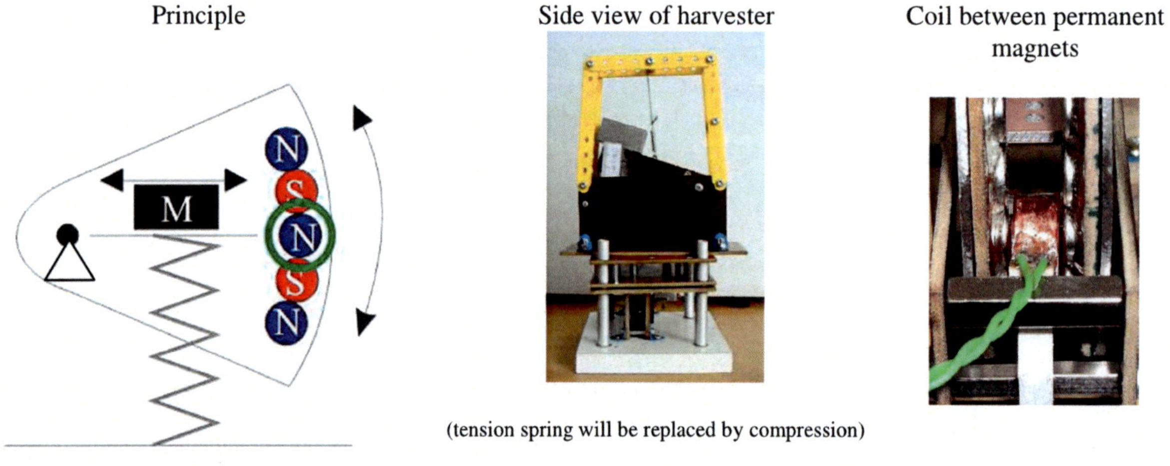

Figure 3: *Re-designed harvester (generator) module*

Here the initial translational system was removed by a rotational system. The harvester = generator consists of two rotary arms that are fixed with a ball bearing to reduce friction. On the rotary arms a magnet and an additional moveable mass is attached. Two arms are used in order to produce a homogeneous magnetic field in which the coil is moved when the system is excided. The arms are supported by springs. The main harvesting principal is not changed. Even the differential equation (1) is equivalently assumed that the rotations are small compared to the length of the rotary arms. The re-design was made in order to improve the damping behavior of the harvester and to make it easier to adopt the resonance frequency of the harvester to predefined values. By this adoption the amount of energy that can be harvested on a particular bridge can be optimized which is obvious from the transfer function equation (4) for the output voltage. Figure 4 shows the transformed voltage output of a harvester with 35 gram additional mass by which the system is tuned to a resonance frequency of 3.1 Hz. The system is excided by sinusoidal input signal with +/-0.5 mm displacement amplitude of the harvester base. In order to use this output, a special circuit is currently developed.

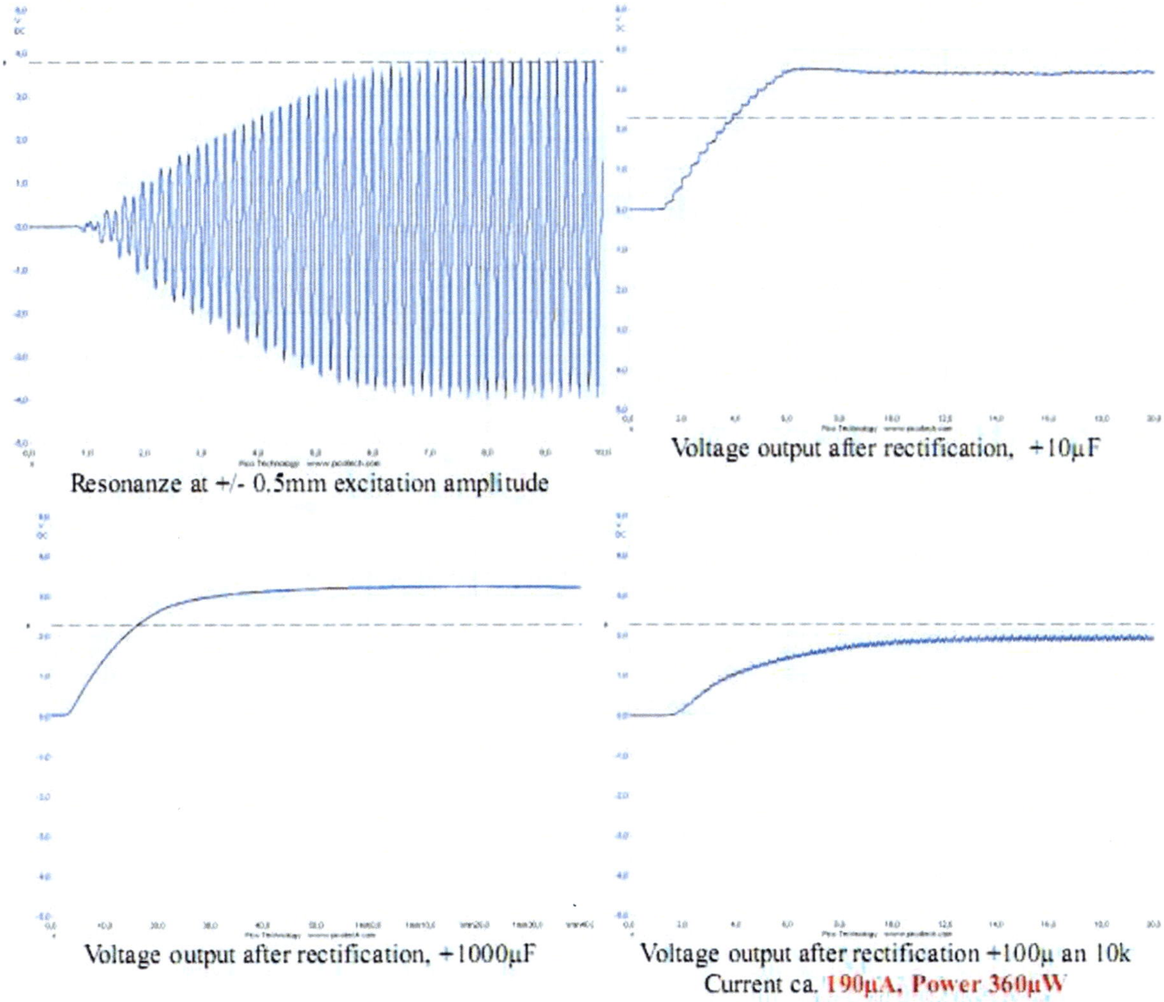

Figure 4: *Results obtained with harvester prototype*

LONG-TERM MONITORING PROJECT

Parallel to the development of the sensor and harvester module a standard monitoring system was installed at a pre-stressed concrete bridge. The five span bridge, finished in 1996, is shown in Figure 6. It has a total length of 230 m and the maximum width of the mid-span is 64 m. This bridge was chosen since it was recently checked (2008) according to [1] and it was found, that after the preceding major bridge check in 2002 large shear cracks have developed. The reason for the cracks could not be clearly determined so far. It was decided in 2009 that aided by a permanent measurement system especially the shear cracks should be monitored. The monitoring system that is currently installed in the bridge (Figure 7) consists of ten velocity transducers (geophones, SM-6, 4.5 Hz [7]), six temperature sensors, six moisture sensors, and five crack width sensors. Additionally the air temperatures are measured by a special station. Via an UMTS module the whole data acquisition system can be controlled were ever an internet connection is available. In order to determine final transducer positions and the natural frequencies, damping values, and mode shapes that existed when the long term measurements were started a two day measuring campaign was executed. The measured and evaluated mode shapes, natural frequencies, and damping values are summarized in Figure 7. The installation of the first harvesting prototype on the bridge will be at the end of 10-2009. In the next weeks additional sensors will be integrated in the current measurement system to measure the pillar settlements.

Figure 6: *The Friedensbrücke-Nord in Magdeburg on which the long term monitoring system is installed*

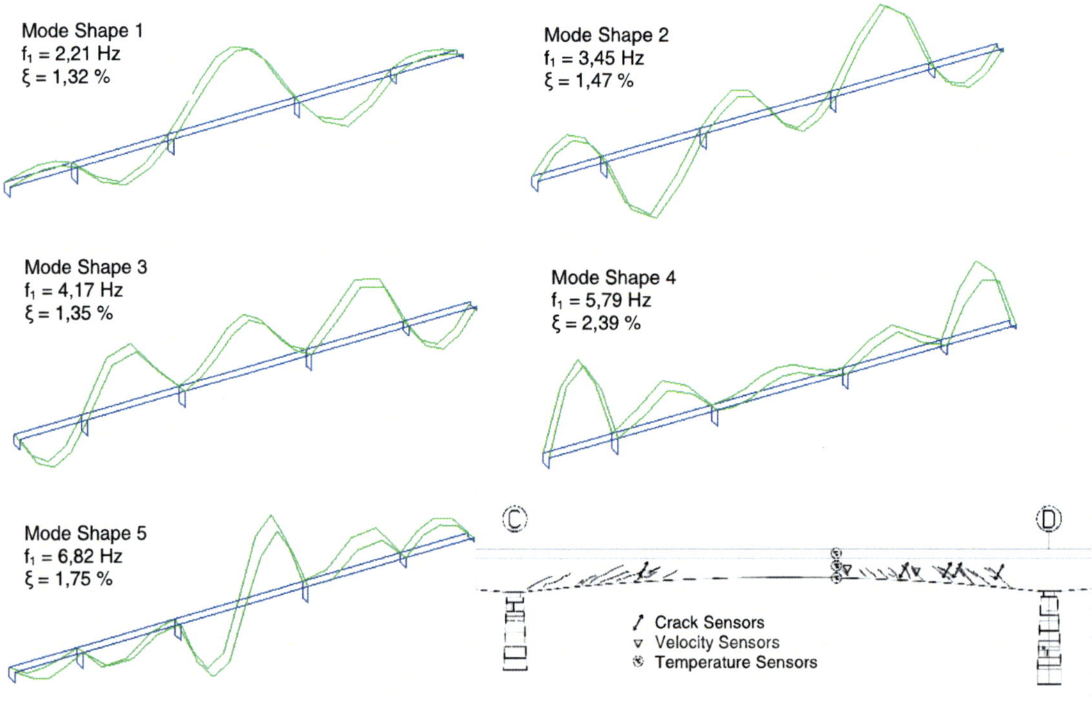

Figure 7: *Measured mode shapes of the bridge when the long-term measuring program started*

DATA EVALUTATION

Meanwhile the bridge monitoring system is running for about two month. The evaluated changes of the natural frequencies and temperatures of a time segment of two weeks can be seen in Figure 8. The evaluation is performed online during the measurements. The measurements and online evaluations were visually programmed in DASYLab [9]. The natural frequencies are evaluated with the help of a standard power spectral density analysis. The pikes of the natural frequencies are identified with an algorithm which is coupled to a modal analysis module. The frequency bands where the PSD pikes have to be identified come from this modal analysis module. The modal analysis module, which is a stand-alone-application is started during the measurement process from within DASYLab after each hour. This module has very sophisticated algorithms implemented to automatically identify the natural frequencies, corresponding damping values, and mode shapes. At the moment we use the commercial software product Artemis Extractor [10] as modal analysis module. Within Artemis the EFFD and SSI methods are use to find the frequencies and mode shapes. After Artemis calculated the frequencies form the data of the last hour the PSD Algorithm in DASYLab is adopted to search for the PSD pikes within the appropriate frequency bands. Several other online statistical evaluations are performed in DASYLab, e.g. maximum amplitudes of temperature, velocity, crack width. The crack width development is constantly checked within DASYLab. An alert is send via e-mail when a certain maximum value is exceeded.

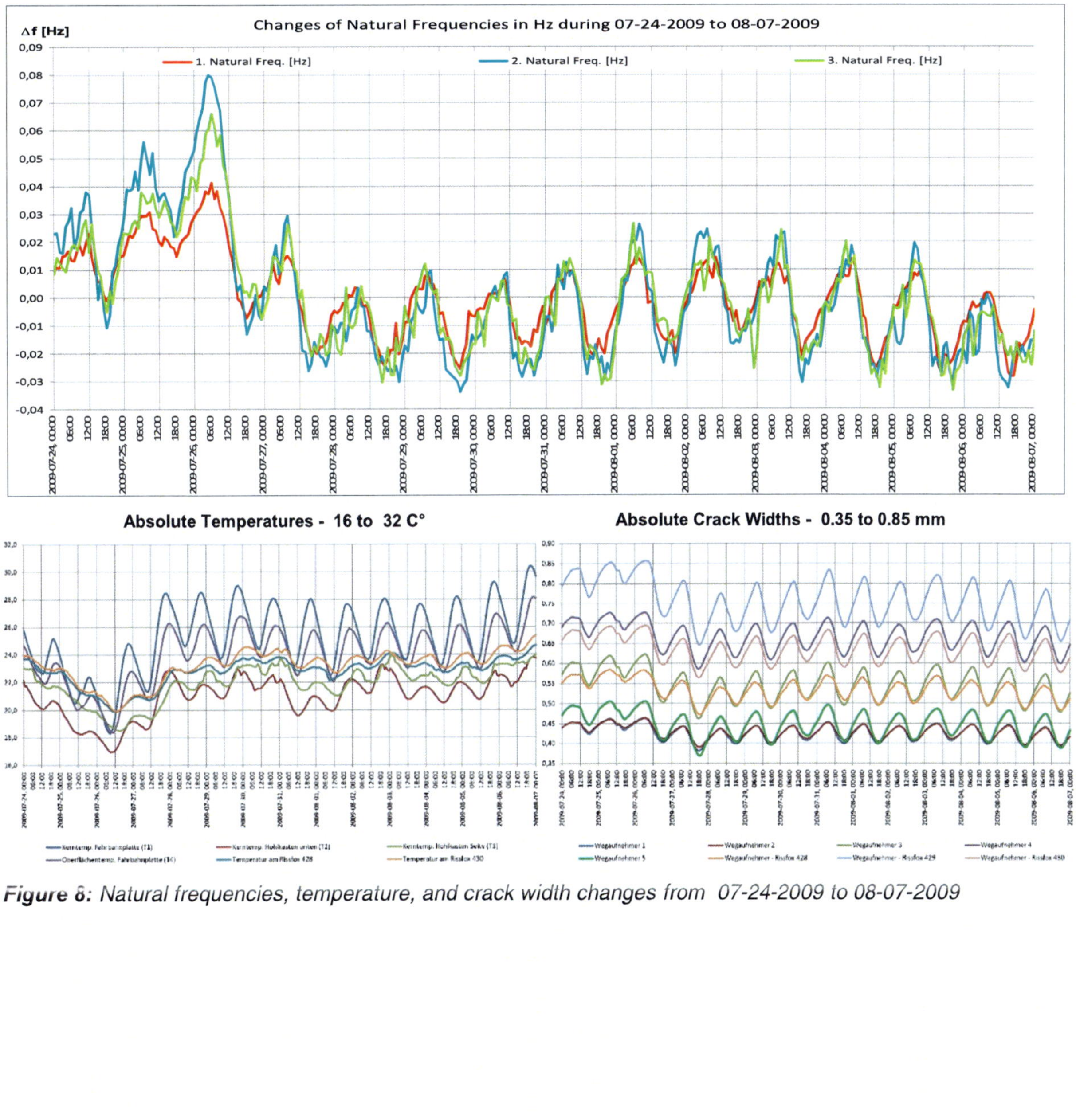

Figure 8: Natural frequencies, temperature, and crack width changes from 07-24-2009 to 08-07-2009

FURTHER STEPS

Currently, an own modal analysis module is implemented by the Algorithm Group. Additionally, other algorithms shall be implemented to count the cars that pass the bridge and an algorithm to identify the car type (e.g. automobile, truck) by pattern recognition.
The Wireless Group is testing several commercially available wireless modules. Based on the results obtained from the test and in coordination with the Sensor/Harvester Group own modules will be developed.
In the near future the currently running long term monitoring will be enhanced by sensors that measure the settling of the bridge pillars. Additionally, other commercially available harvesting modules (solar based and others) will be tested on the bridge.
At the time of the writing of the paper the shear reinforcement of the bridge that was actually built in the bridge is detected by different non-destructive methods (e.g. pulse induction method).
Furthermore, an additional long-term monitoring project will be started at the end of 2009. Besides the sensors mentioned above strain gages, force transducers, and video cameras will be used.
The statistics toolbox OptiSlang [7] with its state-of-the-art statistical algorithms will be included into the online data evaluation process to identify correlations between crack width, natural frequencies, damping, temperature, etc.
The harvester prototype described above will be tested on the Friedensbrücke-Nord (Figure 6) at the end of October 2009.
The Friedensbrücke-Nord has been modeled with ANSYS Workbench/Classic 12. The new elements for modeling pre-stressed concrete structures are used. This model is going to be optimized by the integration of the already mentioned statistic tool OptiSlang.

ACKNOWLEDGEMENTS

The author gratefully thanks the Arbeitsgemeinschaft industrieller Forschungsvereinigungen "Otto von Guericke" e.V. (AiF) [12], for granting the above presented research project. A particular acknowledgement is for all members of our research team for their help in preparing the material for this paper.

REFERENCES

[1] DIN 1076, Edition 1999-11, Engineering structures in connection with roads - inspection and test
[2] ESYS GmbH, http://www.esys.de/
[3] Aucoteam, http://www.aucoteam.de/
[4] Gesellschaft für angewandte Informatik, GfaI, http://www.gfai.de/
[5] A. Stienstra, P.W. Maxwell, A.S. Badger, Intrinsic noise in geophones, Presented at the 55th Meeting of the EAEG, Stavanger, June 1993
[6] Tina Design Suite, Analog, Digital, Symbolic, RF, VHDL, MCU and Mixed-Mode Circuit Simulation & PCB Design, http://www.tina.com
[7] Sensor Netherlands, http://www.geophone.com/index.asp
[8] L-4 Geophone Seismometer from Mark Product sold by Sercel Inc., http://www.sercel.com
[9] DASYLab, http://www.measx.com, http://www.dasylab.com
[10] Artemis Extractor, http://www.svibs.com/
[11] OptiSlang, http://www.dynardo.de/
[12] Arbeitsgemeinschaft industrieller Forschungsvereinigungen "Otto von Guericke" e.V.(AiF) http://www.aif.de

Proceedings of the IMAC-XXVIII
February 1–4, 2010, Jacksonville, Florida USA

Operational Modal Monitoring of Ancient Structures using Wireless Technology

Rafael Aguilar, Civil Engineering PhD student, Department of Civil Engineering – ISISE, University of Minho, Campus de Azurém 4800 - 058 Guimarães, Portugal

Luis F. Ramos, PhD in Civil Engineering, Assistant Professor, Department of Civil Engineering – ISISE, University of Minho, Campus de Azurém 4800 - 058 Guimarães, Portugal

Paulo B. Lourenço, PhD in Civil Engineering, Full Professor, Department of Civil Engineering – ISISE, University of Minho, Campus de Azurém 4800 - 058 Guimarães, Portugal

Ricardo Severino, Electrical and Computer Engineering PhD Student, CISTER/IPP-HURRAY! Research Unit - School of Engineering Polytechnic Institute of Porto, 4200-072 Porto, Portugal

Ricardo Gomes, Junior Researcher, CISTER/IPP-HURRAY! Research Unit - School of Engineering Polytechnic Institute of Porto, 4200-072 Porto, Portugal

Paulo Gandra, PhD in Electrical and Computer Engineering, Researcher, CISTER/IPP-HURRAY! Research Unit - School of Engineering Polytechnic Institute of Porto, 4200-072 Porto, Portugal

Mario Alves, PhD in Electrical and Computer Engineering, Researcher, CISTER/IPP-HURRAY! Research Unit - School of Engineering Polytechnic Institute of Porto, 4200-072 Porto, Portugal

Eduardo Tovar, PhD in Electrical and Computer Engineering, Research Leader, CISTER/IPP-HURRAY! Research Unit - School of Engineering Polytechnic Institute of Porto, 4200-072 Porto, Portugal

NOMENCLATURE

f	Natural Frequency of the system
ξ	Damping coefficient
ω_k	Process noise
v_k	Measured noise
x_k	Discrete-time space state vector
y_k	Observation vector
A	Discrete-time state matrix
B	Discrete-time input matrix
C	Discrete-time output matrix
D	Discrete-time direct transmission matrix
h	Hankel matrix
j	Number of columns in the Hankel Matrix
Y_f	Future information of the Hankel matrix
Y_f^{ref}	Past information of the Hankel matrix
L,Q	Factors from a LQ factorization
Λ	Eigenvalues matrix
S	Observability matrix
Ψ	Complex modes shape matrix
Φ	Eigenvectors matrix
λ	Real eigenvalues or real eigen frequencies

DOI 10.1007/978-1-4419-9831-6_26, © The Society for Experimental Mechanics, Inc. 2011

ABSTRACT

Operational Modal Analysis is currently applied in structural dynamic monitoring studies using conventional wired based sensors and data acquisition platforms. This approach, however, becomes inadequate in cases where the tests are performed in ancient structures with esthetic concerns or in others, where the use of wires greatly impacts the monitoring system cost and creates difficulties in the maintenance and deployment of data acquisition platforms. In these cases, the use of sensor platforms based on wireless and MEMS would clearly benefit these applications. This work presents a first attempt to apply this wireless technology to the structural monitoring of historical masonry constructions in the context of operational modal analysis. Commercial WSN platforms were used to study one laboratory specimen and one of the structural elements of a XV century building in Portugal. Results showed that in comparison to the conventional wired sensors, wireless platforms have poor performance in respect to the acceleration time series recorded and the detection of modal shapes. However, for frequency detection issues, reliable results were obtained, especially when random excitation was used as noise source.

1. INTRODUCTION

For most advanced economies, buildings represent one of the largest single investments of national resources. While recently built buildings are very important, there are other existent structures that require even more attention. The permanent study of ancient buildings, which represents the architectural heritage of a region, is a key aspect to modern societies since the impact of a possible loss is not only economical but also cultural. This work intends to contribute on the process of preservation of the architectural heritage.

In this line, it is intended to evaluate possible damages and to perform structural monitoring according to the modern strategies of minimum repair and the use of non intrusive methods. On the latter, the advancements on Micro-Electro-Mechanical Systems (MEMS) and the wide range of alternatives on wireless communications are turning Wireless Sensor Network (WSN) technology into a promising candidate to develop new structural monitoring solutions for this research area. With this respect, the aim of this work was to rely on commercially available wireless and MEMS technology in real operational modal analysis case studies to evaluate their capabilities and state future research lines.

The paper starts by describing the conventional wired based and wireless based sensors and data acquisition equipments (DAQ) used for structural dynamic monitoring works. Operational modal analysis is then overviewed by describing the SSI method as it was used as data processing tool in the experimental tests. Furthermore, two case studies where the WSN platforms were used to perform operational modal analysis are shown and discussed. Finally, a new WSN platform, which is being built as part of future developments, is presented as well as the conclusions.

2. DYNAMIC MONITORING SYSTEMS FOR CIVIL ENGINEERING STRUCTURES

During the past years significant hardware developments occurred for structural monitoring purposes. Conventional monitoring sensors used for these applications, involve a large number of wires (fiber-optic cables or other wired communication medium) and centralized data acquisition systems with remote connections. The fact that the conventional sensor platforms use wires increases the cost of the monitoring systems, creates difficulties in the maintenance and deployment in the field. In this line, the recent years have witnessed an increasing interest in a new technology based on WSN platforms as a low-cost alternative for being applied in civil engineering structures [1].

2.1 Wired Based Monitoring Systems

The wired based systems (also called in this work as conventional systems) used for structural monitoring are composed by three parts: measurement sensors, data acquisition equipments and, in some cases, remote connection systems. The measurement sensors are connected with cables to the data acquisition systems which can also be remotely connected to a central station.

The conventional measurement sensors used for dynamic identification tests are piezoelectric, piezoresistive, capacitive or force-balanced accelerometers. For data acquisition purposes, platforms with capabilities of medium sampling rates (around 2000 Hz) and Analog Digital Converters (ADCs) with resolutions over than 16 bits are chosen. In the case of the remote connection systems the most popular approaches include the use of IEEE

802.11a, b, and g standards or cellular data (such as CDMA, GSM/GPRS, EDGE, and so on) for communication purposes [2]. Figure 1 shows an example of piezoelectric sensors and USB data acquisition systems.

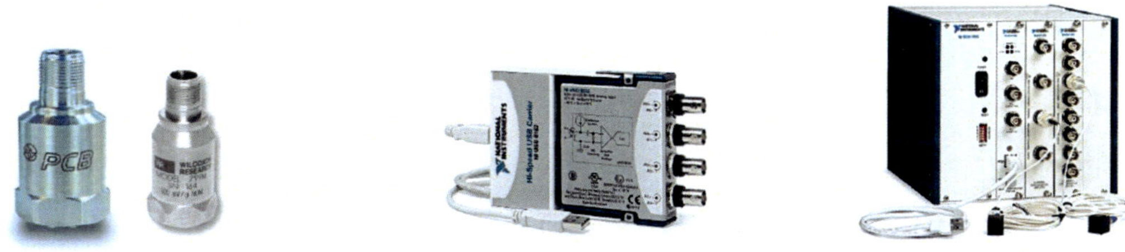

(a) (b) (c)

Figure 1 - Conventional equipments used for dynamic identification. (a) Accelerometers models PCB 393B12 and WR 799M [3], [4]; (b) and (c) USB data acquisition equipment models NI USB-9233 with 24 bits and NI SCXI-1531 with 16 bits [5].

2.2 Wireless Based Monitoring Systems

The use of wireless technology with embedded MEMS for structural monitoring was first proposed by [6], [7], [8] and, [9]. In those works, the authors proposed the integration of wireless communications with sensors to develop a near real-time monitoring system of structures. Since then, a lot of academic efforts to improve the technology have been made.

The first WSN platform was developed by the University of California-Berkeley and subsequently commercialized by Crossbow [10] since 1999. Since then, this commercial technology has been constantly advancing. In Ref. [11] a compilation that highlights the state of the art of this type of sensors up to 2005 is presented and is resumed in Figure 2.

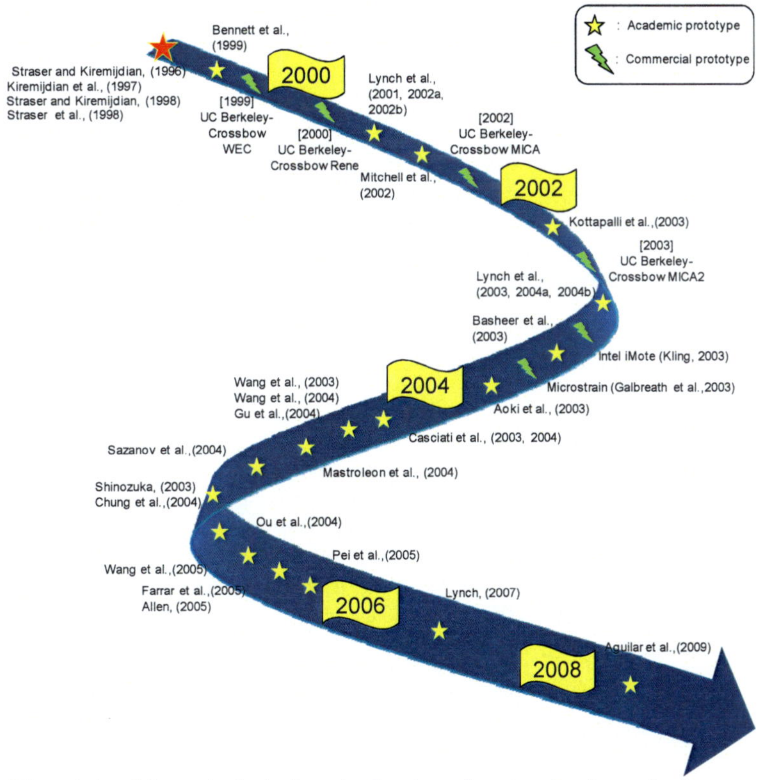

Figure 2 - Overview of the state of the art of wireless technology for structural monitoring

A monitoring system based on WSN platforms with embedded MEMS is composed by three parts: 1) Measurement units, 2) base station and, in some cases, 3) remote connection system. The equipments and the technology used for the third part of the system (remote connection) are, most of the times, the same as the ones used in the conventional systems.

The measurement units are composed of several sensing interfaces with embedded MEMS. MEMS are an emerging technology through which miniature mechanical systems are built using the standard Integrated Circuits technologies on the same chip as the electronic circuitry. The main advantage of MEMS is that, because of the high effectiveness used in their fabrication process, these can perform measurements with relatively low cost and low power consumption [12]. There are several commercially available MEMS-based transducers like micro-accelerometers, temperature, humidity and, pressure micro-sensors.

Besides the obvious issues related to energy consumption, there are other aspects related to the measurement sensors and DAQ characteristics to consider when these platforms are being evaluated as an alternative of conventional systems. Commercial micro-accelerometers are primarily made based on the capacitive principle, are able to measure in 1, 2 or 3 axis configuration with a range of measurement from ±2 g to ±400 g and low resolutions (in comparison to conventional platforms). For operational modal analysis, platforms which have embedded microaccelerometers able to measure at least in a range of ±2 g with the higher resolution available should be chosen. Another important aspect to take into account are the DAQ systems embedded in the platforms. As it was previously shown, conventional systems have high resolutions over 16 bits; unfortunately, commercial platforms are only equipped with ADCs with low resolutions varying from 10 bits to 12 bits.

3. OPERATIONAL MODAL ANALYSIS OF CIVIL ENGINEERING STRUCTURES

For dynamic monitoring, the first and natural tendency of civil engineering researchers was to take profit from important previous developments made in the area of system identification and experimental modal analysis in the electrical and mechanical engineering. Depending on the excitation source, two different groups of techniques are currently used for performing structural dynamic monitoring, the Input-Output and the Output-Only techniques.

Nowadays Output-Only techniques (also called operational modal analysis) in which the structures are excited just by ambient noise are desirable and commonly used due to the higher costs and difficulties involved in the Input-Output Tests [13]. Output-Only methods are based on the premise that wind, traffic and human activities can adequately excite structures.

The main assumption of the Output-Only identification methods is that the ambient excitation input is as a Gaussian white noise stochastic process in a frequency range of interest. Due to the random nature of the excitation, the response includes not just the modal contributions of the ambient forces and the structural system but also the contribution of the noise signals from undesired sources. For this reason, the identification techniques must have the ability to separate the noise modes.

Considering previous experiences, when dynamic monitoring works are performed, there is no perfect method that can be generally applied. However, the Stochastic Subspace Identification method (SSI) or its variants is the method that gives more reliable results in most of the cases and so, the one that is used in this work.

The stochastic subspace identification (SSI) method was originally proposed by [14] and then modified (SSI-Data method) by [15]. The Data-Driven Stochastic Subspace Identification method (SSI-Data) is based on the stochastic space model theory from output-only measurement and has as focus the identification of the state matrix A and the output matrix C, which contains the modal information of the studied system. The SSI method uses robust numerical techniques such as QR-factorization, singular value decomposition (SVD) and least squares. The QR-factorization results in a significant data reduction, whereas the SVD is used to reject system noise. The method is summarized in Figure 3.

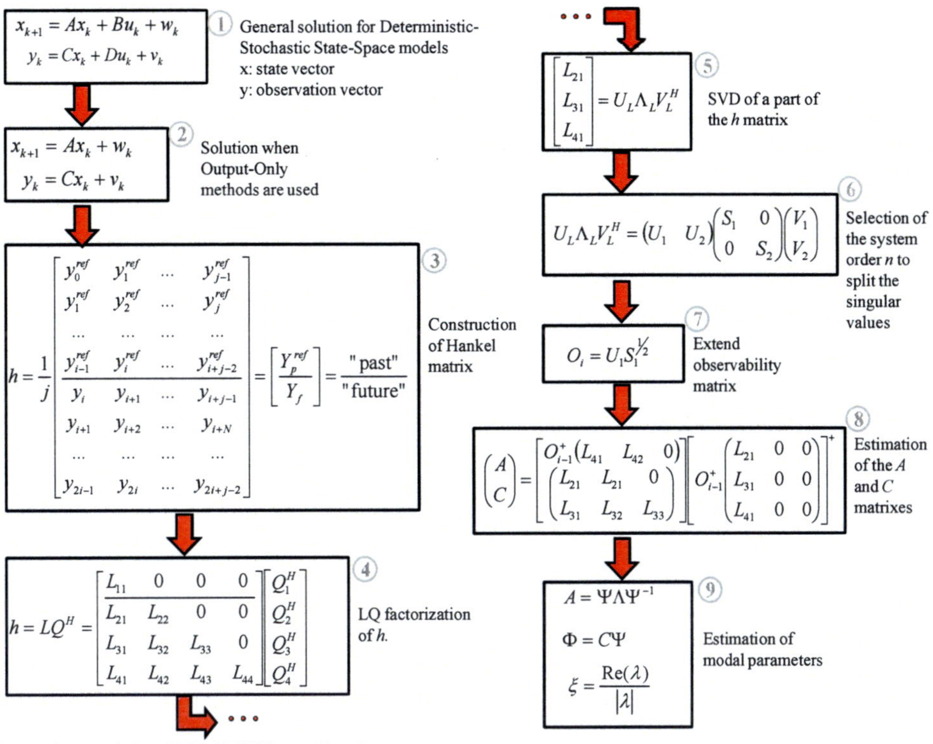

Figure 3 - Flow chart of the SSI-DATA method.

4. OPERATIONAL MODAL ANALYSIS OF ANCIENT STRUCTURES USING WIRELESS PLATFORMS

4.1 Sensors and Data Acquisition Equipments

In this work, conventional wired monitoring systems with accelerometers, data acquisition equipments and commercial WSN platforms were used for performing the experimental studies.

The conventional wired monitoring systems used the high sensitivity piezoelectric accelerometers model PCB 393B12 [3]. For the conventional data acquisition equipment the NI-USB9233 platform [5] which offers ADC resolution of 24 bits was selected.

For the WSN platforms, Crossbow technology [10] was chosen as they offer a low-cost solution and low power boards with embedded micro-accelerometers.

Two case studies were carried out with the aim of comparing the performance of the WSN platforms with respect to the conventional wired based systems and assess the possibility of their use to perform operational modal analysis of civil structures. Details, results and comments of the tests are following shown.

4.2 Case study I: Dynamic response of an inverted Pendulum

A single degree of freedom structure represented by an inverted pendulum is one of the simplest examples that are used by civil engineers to explain the fundamentals of the dynamics of structures. In this work, this structure is also used as a tool to evaluate and understand the behavior of the commercial wireless based platforms and their utility for engineering works.

The specimen consists in an inverted pendulum with 1.70 m high built by wood with steel plates in its top and base (Figure 4a). To perform a complete dynamic characterization of the pendulum, three conventional wired based one-axis accelerometers and three wireless + MEMS platforms working with their accelerometers only in one direction were used. These sensors were arranged in two setups where the node 1 was kept as common for both measurements as shown in Figure 4b, Figure 4c and Figure 4d. For comparison purposes the wired and wireless systems were set for running in parallel. The data acquisition was done at a sampling rate of 128 Hz.

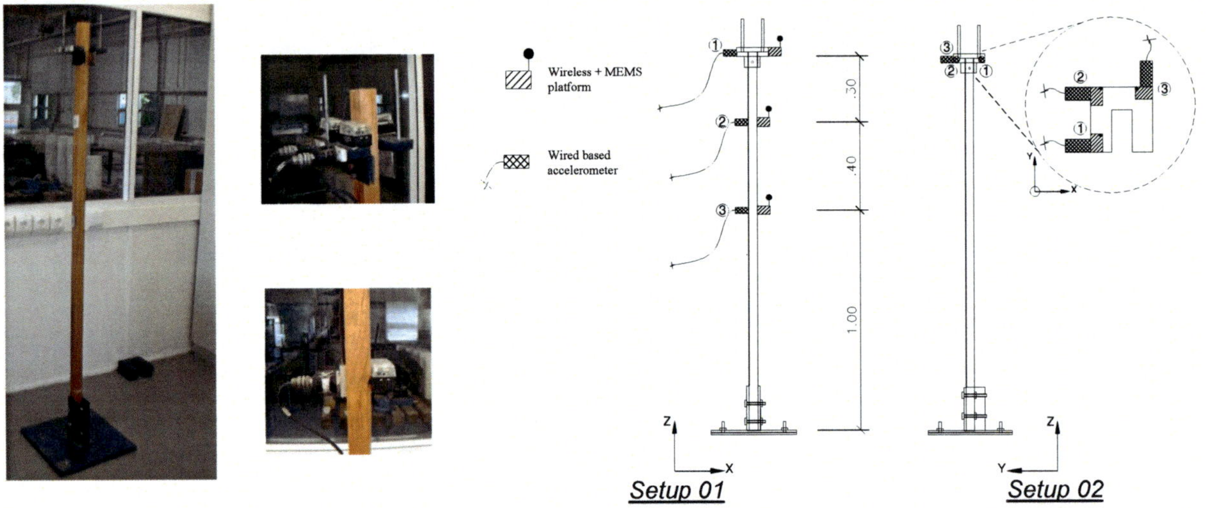

(a) (b) (c) (d)

Figure 4 - Description of the case study I. (a) Inverted Pendulum in the laboratory; (b) Close up of the wired and wireless based sensors arrangement; (c) Setup 01 description and; (d) Setup 02 description.

The first analysis consists on the study of the performance of the wireless platforms with respect to the conventional sensors by means of the acceleration time series collected. With this purpose, tests under random excitation and under ambient noise were performed. The Figure 5 shows the recorded signal by mote 3 and by accelerometer 3 in both scenarios.

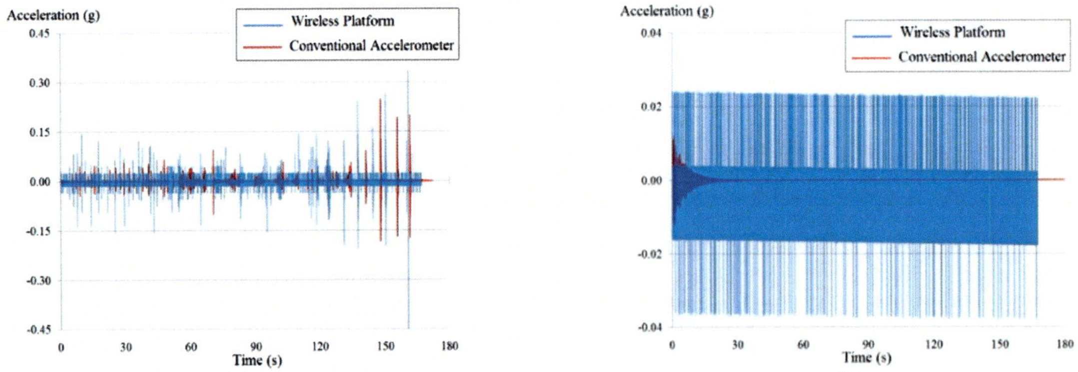

(a) (b)

Figure 5 - Time series collected by mote and accelerometer # 03 in the tests of the Pendulum. (a) Response under random excitation in Setup 01 and; (b) Response under ambient noise in Setup 02

The results evidence the poor performance of the MEMS micro-accelerometers embedded in the wireless platforms for measuring vibrations with low amplitude. In the case when random noise was exciting the structure the maximum peaks detected by the wireless platforms vary in an range of 2 to 9 times while when ambient noise is used as source of excitation, the maximum peaks detected vary from 3 to 6 times with respect to the peaks detected by the conventional wired based accelerometers.

Using again ambient and random noise as source of excitation, the second analysis performed in the specimen consists on the study of its dynamic properties.

For determining the dynamic properties of the system the SSI method, which is implemented in the ARTeMIS extractor software [16], was used. In Figure 6 the stabilization diagram corresponding to the random excitation system is shown. The Table 1 shows a resume of the results of the wireless and conventional accelerometers.

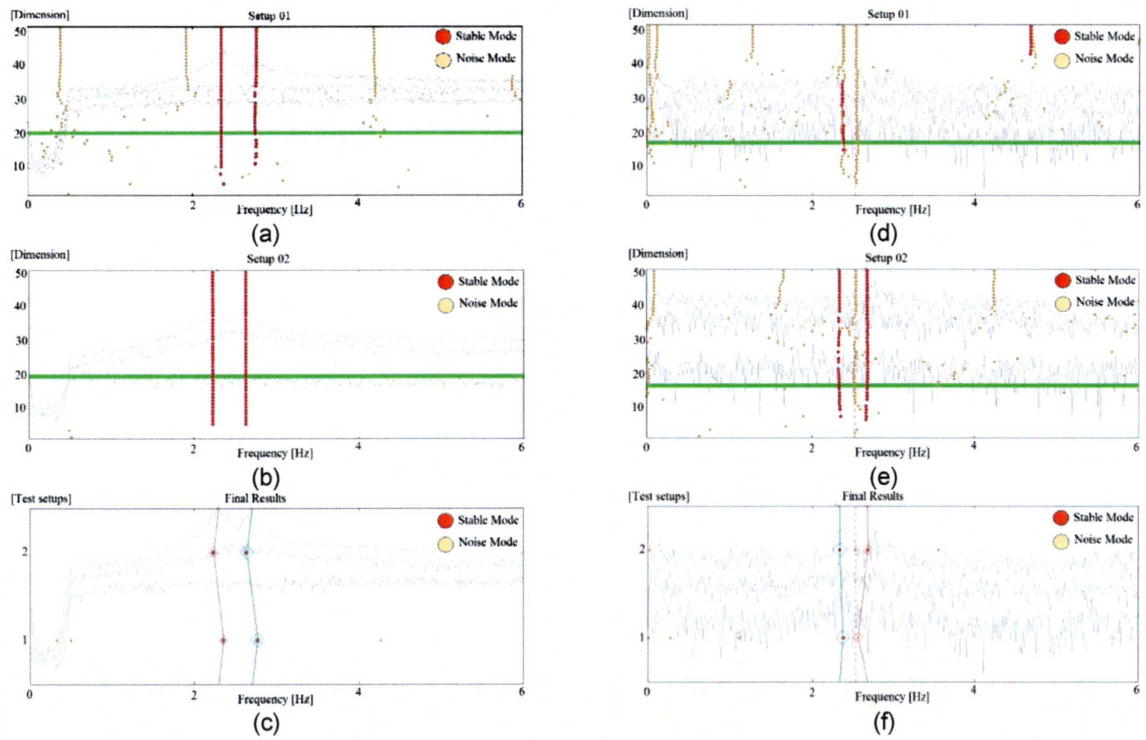

Figure 6 - Stabilization diagrams of the analysis of the time series recorded under random excitation in the Pendulum tests. (a) (b) and (c) Results of the conventional wired based accelerometers; (d) (e) and (f) Results of the wireless platforms.

Table 1: Results of the dynamic response of the Pendulum

		Conv. Accelerometers		Wireless Platforms			
	Mode	f (Hz)	ξ (%)	f (Hz)	ξ (%)	f_{ERROR} (%)	ξ_{ERROR} (%)
Excited	1	2.30	1.45	2.35	3.57	2.13	59.39
	2	2.71	1.57	2.68	2.94	1.12	46.60
Ambient	1	2.26	0.82	2.41	9.82	6.22	--
	2	2.63	2.12	2.83	10.42	7.07	--

According to the frequency content, the results show that the wireless platforms give accurate results, even under environmental noise. When the structure is lightly and randomly excited the modal identification is easy because the stable poles are properly aligned in the natural frequencies. In the case of ambient noise the dynamic identification becomes more complicated for the appearance of noise modes in the stabilization diagrams (stabilization diagrams are not shown in this work); however interesting results were also achieved. Due to the lack of synchronization algorithms implemented on the motes, information about the mode shapes is useless.

4.3 Case study II: Dynamic Response of Monuments - The Chimneys at Paço dos Duques Building

The Paço dos Duques Building was built between 1422 and 1433 by D. Afonso (bastard son of the king of Portugal D. Joao I) in Guimarães located at the north of Portugal. In the beginning the building was used as a residence of the "Duques de Bragança" and was inhabited from 1480 to 1807 [17]. In 1807 the building was used as barracks and in 1888 the Architects and Archeologist Portuguese Society considered it as a historical monument [18]. In 1937 the building was re-built with the base of the available information and also new structural elements were introduced which gave the recent impressive character to the building. The Figure 7 shows the view of the actual situation of the building and its original condition before the intervention in 1937.

(a) (b)

Figure 7 - Paço dos Duques Building. (a) Actual situation [19] and; (b) Front view of the palace in 1935 [18].

One of the most important changes in the structure was the addition of chimneys. The original building had only 4 chimneys and later in the intervention of 1937, 34 more chimneys were added. Since 2002 the building is under study and under restoration as well as preservation works. As part of those works, the structural analysis of the chimneys was considered due to the fact that those elements are the most damaged in the building.

Based on the previous reliable results of the experiments in the pendulum (case study I), the use of the commercial wireless platforms for structural dynamic monitoring was explored.

The dynamic response of one of the 4 original chimneys was studied using conventional and wireless platforms. The Figure 8 shows a general view of the restoration works that are being carried out and the wireless platforms placed in the studied chimney.

(a) (b)

Figure 8 - Chimneys at Paço dos Duques Building. (a) Current restoration works and, (b) Sensors location.

The advantages of using wireless platforms were evidenced in this case study. In addition to the economical aspects, the fact that the wireless platforms are lighter in comparison to the conventional accelerometers allowed to place them in an easier way on the surface of the chimneys. The data acquisition process was also easier allowing a safer work in difficult zones such as the roof of the building.

The Figure 9 shows the stabilization diagrams of the analysis of the time series recorded under random excitation. Table 2 shows the results of the identified frequencies using the conventional accelerometers and wireless platforms.

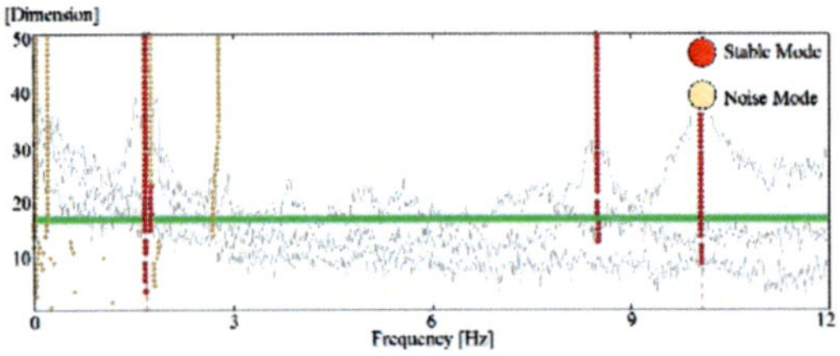

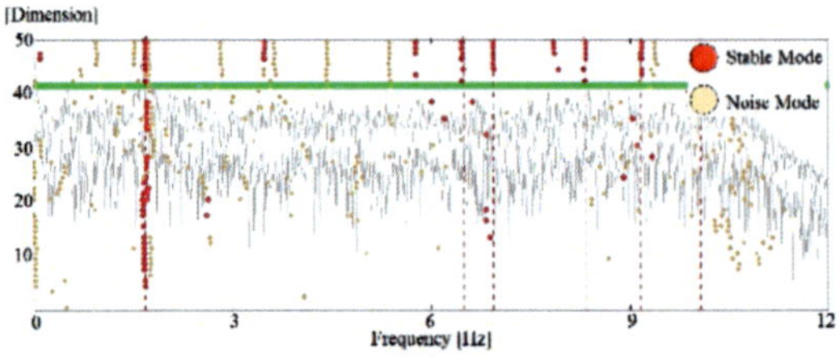

(a) (b)

Figure 9 - Stabilization diagrams of the analysis of the time series recorded under random excitation in the studied chimney at Paço dos Duques Building. (a) Results of the conventional wired based accelerometers and; (b) Results of the wireless platforms.

Table 2: Dynamic response of the studied Chimney at Paço dos Duques building

Mode	Conv. Accelerometers f (Hz)	ξ (%)	Wireless Platforms f (Hz)	ξ (%)	f_{ERROR} (%)	ξ_{ERROR} (%)
1	1.69	1.34	1.68	1.61	0.60	16.77
2	1.77	4.22	1.71	0.72	3.51	--

The results show small differences in the identified frequencies obtained by using the conventional and the wireless platforms. When the tests are performed with ambient noise (results are not shown in this work) the same frequencies could be identified but, as in the previous experience on the pendulum, with more difficulties as more noise modes appear in the stabilization diagrams.

5. NEW WIRELESS PLATFORM TO PERFORM STRUCTURAL DYNAMIC MONITORING WORKS

Since the strengths and weakness of the commercial wireless platforms were identified, current works are now oriented to develop a new platform to fulfill the requirements of the operation modal analysis of civil structures.

The limitations of the commercial WSN platform for performing operational modal analysis can be synthesized in the following aspects: not enough resolution of the microsensors and DAQ systems embedded in the platforms, lack of synchronization algorithms, data loss issues and, no solutions available related to energy consumption. As a first stage of the developing works it was intended to board the first 3 aspects as, in this stage, long term monitoring is out of the scope. The last aspect will be considered in future stages of the research works.

In this first stage, it was considered to specify, develop and produce a new sensing device capable of measuring low amplitude accelerations. It is also considered that the new device will be compatible with current commercial-of-the-shelf WSN platforms capable of supporting standard communication protocols IEEE 802.15.4 protocol, as well as the OpenZB communication stack [20].

At the moment, the new wireless platform is under construction. With this new tool it is scheduled to perform future validation tests in laboratory specimens and also in real monuments.

6. CONCLUSIONS

This paper explores a new tool based on wireless technology with embedded MEMS sensors for performing operational modal analysis of structures. With this respect, commercial WSN platforms available in the market were chosen for being studied by comparing their performance against well known and tested conventional wired based systems.

The results of the laboratory tests showed that these platforms have poor performance with respect to the acceleration time series recorded due to the low resolution of the microaccelerometers and DAQ systems embedded. The wireless platforms showed also poor performance for the detection of modal shapes. In the case of frequency detection, reliable results were obtained especially when the systems were randomly excited.

In order to study the performance of the wireless platforms in the field, a second series of tests was carried out in the chimneys of a historical XV century monument in Portugal. Interesting results were achieved once again in the frequency detection task of the modal analysis process. In this case, very small differences were detected between the measured frequencies with conventional and wireless platforms.

With the obtained results, the problems of the commercial wireless platforms and their application for civil engineering studies were identified. Future works are based on the development of a new wireless sensing board considering this time microaccelerometers and DAQ systems with higher resolutions and the implementation of standard communication protocols.

7. ACKNOWLEDGMENTS

The first author gratefully acknowledges to Alban, European Union Programme of High Level Scholarships for Latin America, for the financial support with the scholarship number E07D400374PE.

8. REFERENCES

[1] Ruiz-Sandoval, M.E. "Smart" sensors for civil infrastructure systems. PhD Thesis, University of Notre Dame (2004).

[2] NI. Overview of Structural Health Monitoring Solutions. www.ni.com (2009).

[3] PCB. Product Catalog, (2007).

[4] http://www.wilcoxon.com/. Wilcoxon Research, Inc. (2009).

[5] NI. User guide and specifications. www.ni.com (2009).

[6] Straser, E.G. and Kiremijdian, A.S. A modular visual approach to damage monitoring for civil structures. Proceedings of SPIE v2719, Smart Structures and Materials (1996).

[7] Kiremijdian, A.S.; Straser, E.G.; Meng, T.H.; Law, K. and Soon, H. Structural damage monitoring for civil structures. International Workshop - Structural Health Monitoring (1997).

[8] Straser, E.G. and Kiremijdian, A.S. A modular, wireless damage monitoring system for structures. The John A. Blume Earthquake Engineering Center (1998).

[9] Straser, E.G.; Kiremijdian, A.S.; Meng, T.H. and Redlefsen, L. A modular wireless network platform for monitoring structures. Proceedings – SPIE The International Society for Optical Engineering (1998).

[10] Crossbow. MoteWorks Getting Started Guide, Crossbow (2008).

[11] Lynch, J.P. and Loh, K.J. A Summary Review of Wireless Sensors and Sensor Networks for Structural Health Monitoring. The shock and vibration digest (2006).

[12] Stiharu, I. MEMS, general properties, Encyclopedia of vibration. Academic Press (2002).

[13] Doebling, S.W.; Farrar, C.R. and Cornwell, P. A Statistical Comparison of Impact and Ambient Testing Results from the Alamosa Canyon Bridge, International Modal Analysis Conference (1997).

[14] Van Overschee, P. and De Moor, B. Subspace algorithms for the stochastic identification problem. Proceedings of the 30th Conference on Decision and Control, Brighton, England, (1991).

[15] Peeters, B. and De Roeck, G. Reference-based stochastic subspace identification for output-only modal analysis. Mechanical Systems and Signal Processing. Academic Press, 855-878 pp, (1999).

[16] SVS. ARTEMIS extractor pro user manual, Release 4.5. Structural Vibration Solutions (2009).

[17] Azevedo. A. Paço dos Duques, Livraria Lemos (1964).

[18] Silva, J. H.P. Paço dos Duques em Guimaraes, Secretaria de Estado de Informacao e Turismo (1974).

[19] http://www.bing.com/maps/ (2009).

[20] http://www.open-zb.net. Open-ZB open-source toolset for the IEEE 802.15.4/ZigBee protocols website (2009).

Proceedings of the IMAC-XXVIII
February 1–4, 2010, Jacksonville, Florida USA

CONSIDERATION ON STRUCTURE VIBRATION ISOLATION ASYMMETRY. TWO FUNNY CASES.

Titus Gh. Cioara, T.C, Ionel Nicolae, I.N, Ioan Cireș, I.C, Dumitru Cristea, D.C, Dragoș Cireș, Cristiana Căplescu, C.C

., University "Politehnica" of Timisoara
Bv. Mihai Viteazul nr. 1 Timisoara -Romania
E-mail: cioara@mec.upt.ro

SUMMARY: Generally, in the vibration structure isolation the solutions take in consideration symmetry of the isolation elements placement in the structure the separation from adjacent surrounding structures. But there are situations when, accidentally, may appears hidden links bypassing one or more isolation elements by hard stiffness links, creating an asymmetry of the isolation system.
One of this isolation, accidentally asymmetryzed, usually creates large vibration levels of the whole structure, sometimes dramatically, as in the paper case study. A vertical axis ventilator for burning gases, serving a tall chimney in a power plant, being placed the roof of the aggregate building, is isolated from surrounding structure, supporting his metallic box frame structure on a string of the helical spring boxes, symmetrically placed along the metallic structure foot. A fanny hidden mistake in the time of the spring boxes placement, had created a mechanical asymmetry, occurring dramatically vibration levels, which was analyzed by our research team, the results being presented in the present paper.
A second, the same funny case, is presented: two air compressors fixed on massive concrete foundation, being mechanic identically present large different in vibration motion distribution, due to an asymmetric characteristic distribution of elastic pad from compacted gravel.

KEYWORDS: vibration, isolation, mechanical asymmetry

1. ON ASYMMETRY IN STRUCTURE VIBRATION ISOLATION

The importance of the symmetry in vibration isolation may be clearly confirmed by a simple structure (Fig. 1), a rigid beam B_e with a lumped mass m placed on its middle length of 2a, isolated by two springs placed in the both end of the beam, A and B, of stiffness k_1 in point A and of stiffness k_2 in point B. The two end point allow to have the linear motions of law $x_1(t)$ and $x_2(t)$, so the linear motion of the middle point of the beam Be, where is considerate placed a lamped mass m is of the law motion $x(t)=(x_1(t)+x_2(t))/2$, solution of the differential equation

$$m\ddot{x}(t)+c\dot{x}(t)+\left[(k_1+k_2)-\frac{(k_1-k_2)^2}{(k_1+k_2)}\right]x(t)=Q(t) \quad (1)$$

where $c\dot{x}(t)$ is the damping force acting on mass m, and Q(t). the excitation force.

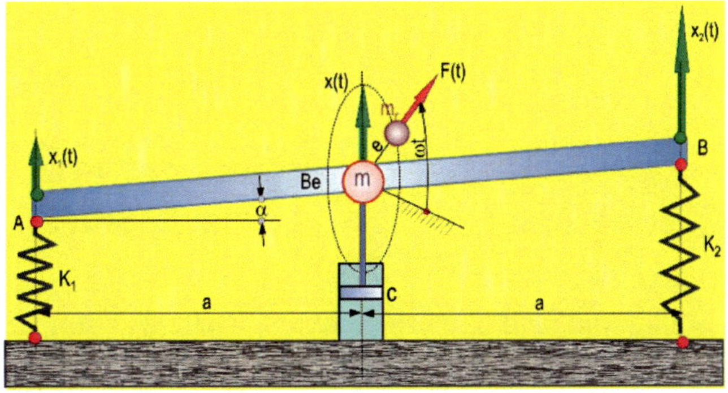

Figure. 1 Illustration of an isolation asymmetry to simple vibration structure

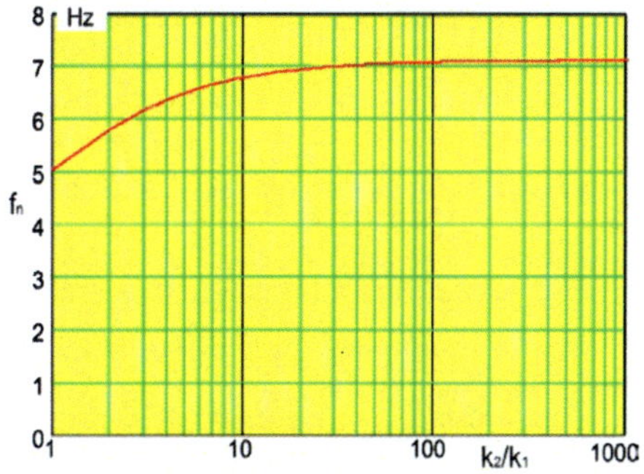

Figure. 2 Variation of the natural frequency f_n function of mechanical asymmetry ratio k_2/k_1

From cinematic conditions, the two laws of motion, $x_1(t)$ and $x_2(t)$ are gf forms,

$$x_1(t)=x(t)-\varphi(t)a; \quad x_2(t)=x(t)+\varphi(t)a \quad (2)$$

T. Proulx (ed.), *Dynamics of Civil Structures, Volume 4*, Conference Proceedings of the Society for Experimental Mechanics Series 13, DOI 10.1007/978-1-4419-9831-6_27, © The Society for Experimental Mechanics, Inc. 2011 where, from dynamic equilibrium, the angular rotation motion of the beam is

$$\varphi(t) = \frac{(k_1 - k_2)}{(k1 + k2)a} x(t) \tag{3}$$

For a mechanical symmetry of the vibration structure in Fig. 1 the two stiffnesses are equal, $k_1=k_2=k$. and from (2) and (3) results, $x_1(t)=x_2(t)= x(t)$ and $\varphi(t)=0$, the beam motion being a translation vibration

The natural frequency of the system (1) is of the form

$$f_n = \frac{1}{2\pi}\sqrt{\frac{k_2}{m}\left[(1+\frac{k_1}{k_2}) - \frac{\left(\frac{k_1}{k_2} - 1\right)^2}{(\frac{k_1}{k_2} + 1)}\right]} \tag{4}$$

which of mechanical symmetry become of form $f_n = \frac{1}{2\pi}\sqrt{\frac{2k}{m}}$. As example, considering, m=1kg and k=1000 N/m, the value of natural frequency is f_n=5.033 Hz.

Now, considering that, by accidental coupling with a hard stiffness link in point A, its stiffness increasing $k_1>k_2$ the natural frequency according to (4) increasing function of ratio k_1/k_2 asymptotic to f_n=7.1 Hz, at $k_1/k_2>10$, (Fig. 2).

In this condition, from (2) and (3) results

$$\varphi(t) = \frac{1}{a}x(t);\quad x_1(t) \approx 0;\quad x_2(t) = 2x(t) \tag{5}$$

For an excitation force *Q(t)* as it comes from a rotative unbalance rotor, of moment *($m_r e$)*,

$$Q(t) = m_r e (2\pi f)^2 \sin(2\pi f t) \tag{6}$$

and a transient starting speed, from the rotation frequency, of value *f=0,* to the working frequency of the constant value $f=f_w$ the sweeping low is considered of the form

$$f(t) = \frac{1}{2}\left(1 - \cos(\pi\frac{t}{T_s})\right) f_w \ to \to t \le T_s$$
$$f(t) = f_w \ to \to t > T_s \tag{7}$$

where T_s is the transient time to f_w.

Figures 3a and 3b present, graphically, the transient frequency f(t) and the corresponding excitation force Q(t) for a transient time T_s=2 seconds, during frequency it is rising from 0 to .working value f_w=8 Hz.

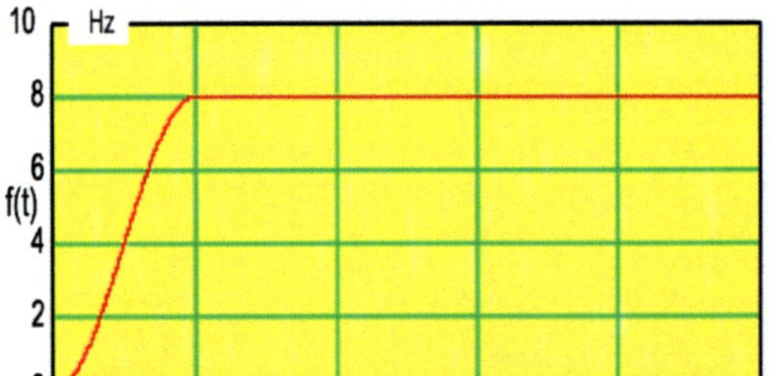

Figure 3a Time history of transient frequency variation from f=0 to $f=f_w$ =8 Hz

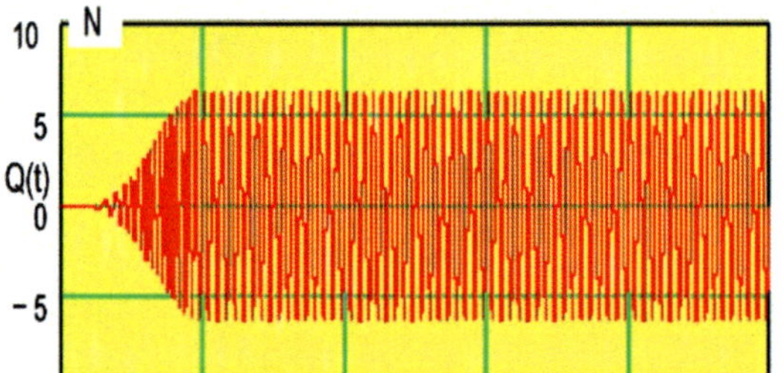

Figure 3b Time history of transient transient excitation force Q(t)'

For the two cases, of mechanical symmetry, ($k_1 = k_2$=1000N/m), and of mechanical asymmetry ($k_1 >> k_2$), in the figure 4 are graphic presentation the time history, $x_2(t)$ of the point B responses during 10 seconds , first two being considered of the transient speed. It is evident that, for the same excitation, in the asymmetry case, ($k_1 >> k_2$), the vibration levels are more highest, as in case of mechanical symmetry ($k_1 = k_2$)

So, for a structure optimal isolated by design, using elastic elements, placed symmetric along to separation planes with other surrounding structures present low vibration levels. But accidentally can appear hard elastic hidden coupling, which may change, dramatically, dynamic response of the isolated structure, usually with large vibration level, sometime dangerous for integrity of the structure and connected surrounding structures.

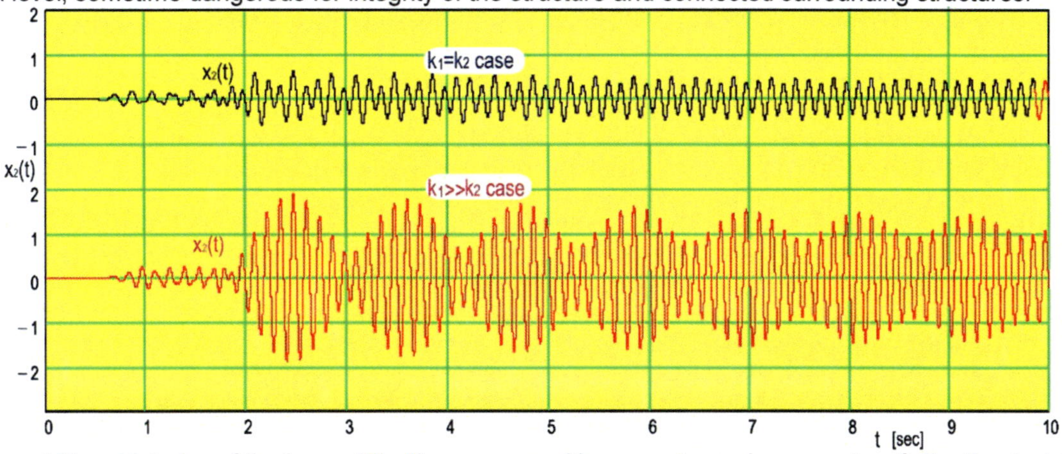

Figure 4 Time histories of the beam,(Fig.1), response with symmetry and asymmetry of vibration isolation

2. TWO CASES ANALYZED

During long time of our team in field vibration investigation we met a lot of professional interesting and even exciting problems, as two which are presented below, related to asymmetry accidentally arising in vibration isolation of a structure.

2.1. A ventilation aggregate for burn gases serving a tall chimney

An funny case was met one time when we was calling from a thermo power plant to investigate a ventilation aggregate which at the first start after his built, presented dangerous vibration levels, in contrast with other the same aggregate serving the six tall chimneys to eliminate the burning gases coming from steam boilers.

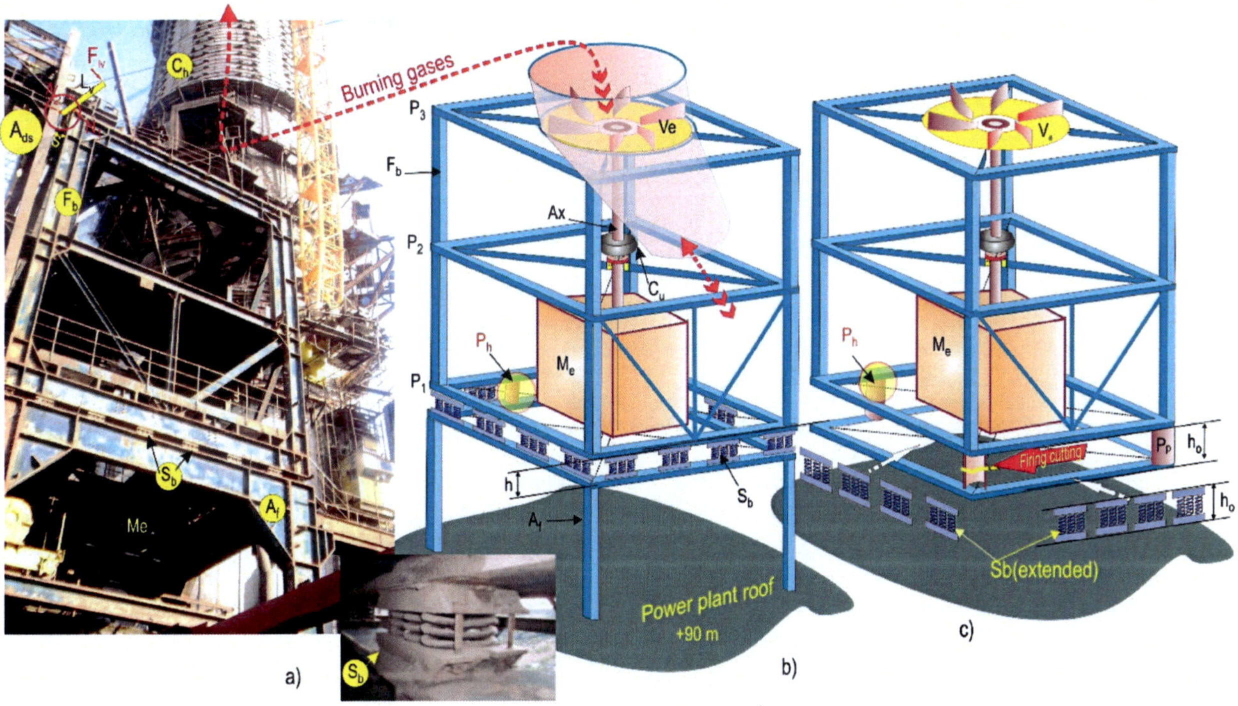

Figure 5 Vibration Isolation system for a vertical centrifugal ventilation aggregate serving a tall chimney in a thermo power plant: a) general view; b) system of vibration isolation and c) building technological aspects.

For this thermo power plant the general designer proposed a new solution for burn gases evacuation to atmosphere through the tall chimney: the ventilation aggregate to be placed on the roof of the turbo machinery building, of 90 m high. The motivation was related to efficiency of the burn gases elimination, but a lot of problem where arise during working time.

The aggregate, composed from, an electric motor M_e of 3.2 MW at 500rot/min acting, by a coupling C_u the ventilator V_e are vertical axed on a metallic frame box F_b, which is isolated from an auxiliary metallic frame A_f,. by a string of helical spring boxes S_b, placed symmetrically on perimeter of the contact boundary. The auxiliary metallic frame is fixed on the turbo machinery building roof and allows horizontal motion of the ensemble F_b and A_f elastic in horizontal plane, the vibration transmissibility being law.

The burn gases content high density of abrasive particle, which excavate non-uniform the blades of the ventilator rotor creating high level of mass unbalance, and after a short period of running had occurs a lot of major faults, the main being a premature failure of the superior slide bearing of electric motor M_e.

By vibration complex measurement, performed by the team of Vibration Testing and Research Laboratory of University " Polithnica" Timisoara" were determinate the vibration distribution along the ensemble metallic structure Af_s motor electric M_e and the house of the ventilator V_e.the vibration motion distribution being plotted in figure 6.

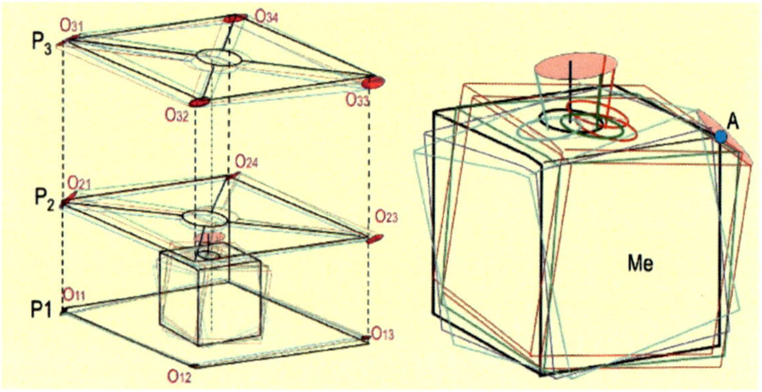

Figure 6 Vibration motion distribution along ensemble structure F_b and motor M_e , corresponding to first

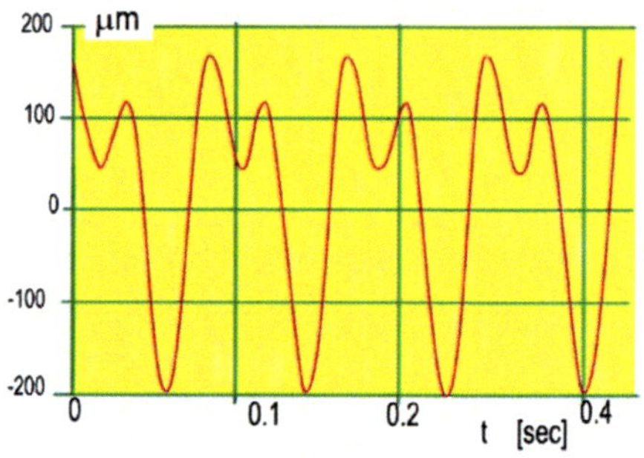

Figure 7 Time history of horizontal vibration motion for point A on the motor M_e

The time history of one vibration signal recorded on motor M_e (point A) structure look as in the figure 7, which present clearly two harmonical components. The second harmonical being correlated with the dynamic misalignment of the aggregate vertical axis, occurring from bolted rigid coupling, C_u. The technical solution to decrease the vibration levels along to aggregate ensemble was by replacing the bolted rigid coupling with a Periflex toroidal coupling more elastic as bolted coupling.

After this successful application solution for a next new built aggregate we were called to find out what is cause of more high level vibration arising again. When we come in the site to record vibration on the aggregate structure the operator was worry to put on the aggregate being convinced to be a disaster so large vibration occurring in the single time when they tried to put on. In this situation we tried to determine the natural frequencies in possibility of a resonant effect tuned by rotation frequency of the aggregate axle.

Existing a another structure A_{ds} ,(fig 5a), closed, by span **s, to** frame box F_b structure using a lever L_v we did a impulsive force to frame box, F_b , expected to record a law level free decay vibration signal , containing the modal components of vibration system, frame box F_b on the string of the helical spring blocks. But the signal recorded didn't look as a free vibration, it was as a impulsive one which explained that something rigid links is connected to elastic suspension. It was a funny surprise finding, in an hide space a pipe (Ph, Fig. 5 b and c) instead of one spring block (S_b), creating a fixed point connected between the two structures F_b and A_f so the suspension being strong asymmetrically, so, normally, the fear of operator was real.

Situation created occurs from technology applied during the frame box built direct to the site, which in built time is placed on a string of pipes, of length h_0, which after all the aggregate was made, the spring blocks extended at the same high h_0 are placed between two structure, then by firing all the pipe being cutting and eliminated, the entire weight of aggregate, (40 tone), is supported by symmetric placed of helical spring blocks, which are compressed to high h.

Of course, after cutting and elimination the hide pipe sector all was ok, remaining only the situation amusement and keeping in mind: don't hurry!

2.2. Different vibration distributions of two mechanical identically air compressor aggregate.

Another analyzed an interesting case was the vibration of the two air compressors aggregate in an oxygen station (fig.8) The two aggregates and foundations were the same, but one of them has presented more large vibration levels as the other.

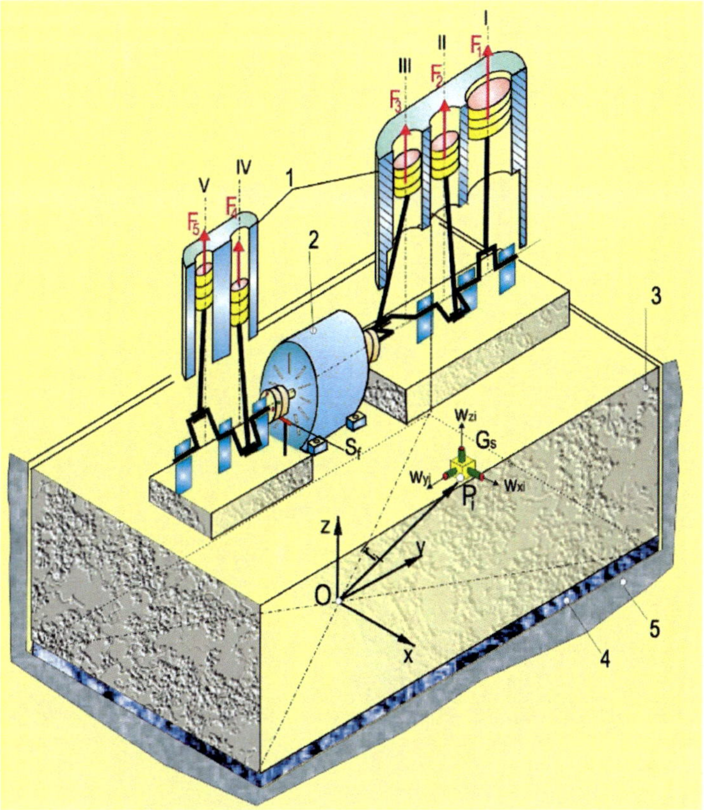

Fig. 8 Illustration of a multicilinder air compressor

A multi-cylinder compressor, as a five stage air compressor 1 (Figure 8) is composed from two parts, law pressure stage (cylinders I, II and III) and high pressure stage (cylinders IV and V), both being driven by an electric machine 2, and seating on a massive the concrete foundation 3..

The massive block foundation is used for two reasons: to machine axis line stiffening and to diminishing machine-fundation vibration excited by residual inertial unbalanced forces F_1.... F_5 acting on machine,. The levels of the vibration depending of the dynamical characteristics of the isolation of the supported pad of foundation 4 and of inertial mass and moments of the rigid block foundation 3

Considering the perturbation forces F_1.......F_5 for both ensemble are the same for both compressor aggregate, first supposition, to explain the large difference between vibration of the two aggregate, it was focus on the integrity of the concrete foundation verify which as a massive mass must to has vibration rigid body motions. In case of a possible hidden crack the significant deviation from the rigid body motions may occurs.

To determined the distribution of vibration was measured the vibration motions components, $w_{xi}(t)$ $w_{yi}(t)$ and $w_{zi}(t)$, in a lot of points P_i along the ensemble machine massive foundation. For a rigid body motion the vibration distribution is of law

$$\{w_i(t)\} = \{w_o(t)\} + [R]\{\varphi(t)\} \tag{8}$$

where the column matrix vector

$$\{w_i(t)\} = \left\{w_{xi}(t) \quad w_{yi}(t) \quad w_{zi}(t)\right\}^T \tag{9}$$

$\{w_o(t)\}$ being the same column matrix vector of a given point of the ensemble, in this case the foot geometric foot center O of the rigid foundation,

$$\{\varphi(t)\} = \{\varphi_x(t) \quad \varphi_y(t) \quad \varphi_z(t)\}^T \tag{10}$$

the column matrix vector of rotation laws around of reference frame Oxyz axes, and

$$[R_i] = \begin{bmatrix} 0 & z_i & -y_i \\ -z_i & 0 & x_i \\ y_i & -x_i & 0 \end{bmatrix} \tag{11}$$

the position matrix of point P_i

The vibration distribution law (8) can be extended to amplitude distribution

$$\{W_i\}_{c,s} = \{W_o\}_{c,s} + [R]\{\Phi\}_{c,s} \tag{12}$$

Index c, and s being related with the terms in cos and sin.

The values of foot amplitude vibration motion distribution, for first harmonic amplitude is given in the below table

Agregate no. 1				Agregate no. 2			
	cos(ωt)	sin(ωt)			cos(ωt)	sin(ωt)	
Wo_x	0,56	-0,84	μm	Wo_x	1,77	-0,10	μm
Wo_y	-1.70	0,87		Wo_y	-1,77	-0,68	
Wo_z	**-8,35**	**2,88**		Wo_z	**-12,13**	**-3,03**	
Φ_x	**-4,40**	**-2,00**	μm/m /	Φ_x	**-6,44**	**-6,38**	μm/m /
Φ_y	-0,35	-0,13		Φ_y	-0,52	-1,75	
Φ_z	0,04	-0,42		Φ_z	0,04	-0,42	

and the illustration of vibration motion distribution is geven in figure 9.

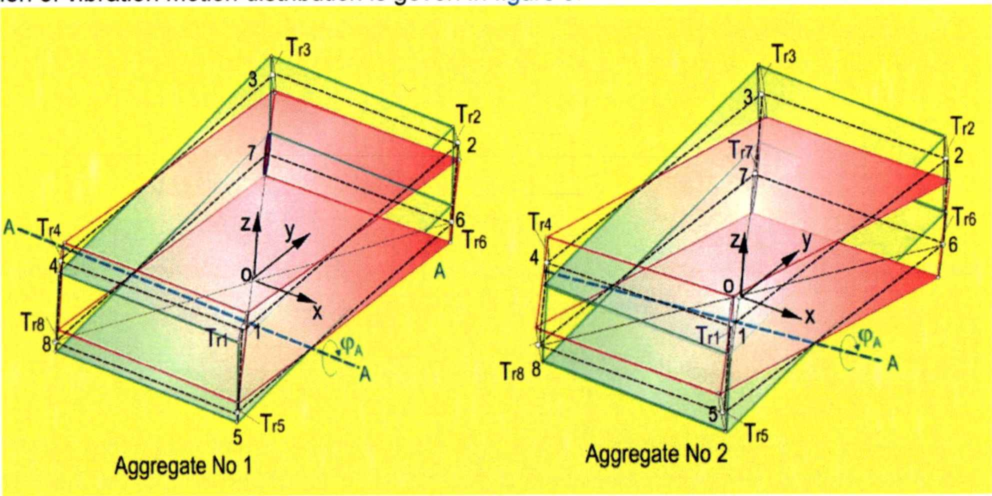

Figure 9 Illustration of massive foundation vibration motion distribution for two aggregate

The value difference between the experimental measured vibration distribution along the both aggregates and the vibration distributions identified by laws (8) were small, so that the idea of the cracked aggregate no 2 foundation was eliminated. It was assumed by the oxygen station staff that by harrying works in gravel compaction time they not respected exactly technological prescription of designer.conclusion: Ones again don't hurry!

REFERENCES

[1] Cioara T, Gh, "Vibration measurement", Lectures, electronic edition, (in Romanian)

[2] Vibration Testing and Research Laboratory Report No 10/ 1982(Univ.Politehnica of Timisoara)

[3] Vibration Testing and Research Laboratory Report No 8/ 1990(Univ.Politehnica of Timisoara)

Proceedings of the IMAC-XXVIII
February 1–4, 2010, Jacksonville, Florida USA

Evaluating Axle Loads of Moving Vehicle Using Bridge Deck Plate Responses

[1]Byeong Hwa Kim, [2]Min Seok Park, [3]Heui-Gu Park, [4]In Hwan Bae, [5]Soo Jin Kim, [5]Keum Soo Yeo,

[1] Assistant Professor, Department of Civil Engineering, Kyungnam University, 449 Wolyoung, Masan, Kyungnam, 631-701, Republic of Korea
[2] Chief Researcher, Highway & Transportation Technology Institute, Korea Expressway Corporation, 50-5, Sancheok-ri, Dontan-myeon, Hwaseoung-si, Kyronggi, Republic of Korea
[3] Researcher, Newconstech, 906 Byucksan Sunyoung Technopia, Ojeon, Uiwang, Kyeonggi,437-821, Republic of Korea,
[4] Deputy manager, New Airport Hiway Co.Ltd., 414-16 Keomam-dong, Seo-gu, Incheon, 404-160, Republic of Korea,
[5] Master Student, Department of Civil Engineering, Kyungnam University, 449 Wolyoung, Masan, Kyungnam, 631-701, Republic of Korea

ABSTRACT: A new algorithm to extract influence sequence has been introduced. Using the resulting influence sequences, a way to identify the axle loads of moving vehicles has been introduced. The approaches are based on an optimal linear filter design theory. The algorithms make use of the strain time histories of a bridge deck plate. The performance of the algorithm has been examined experimentally for a folded deck plate on Yongjong Grand suspension bridge. It turns out that the algorithms are feasible and economical while some further works for the accuracy improvement are expected.

INTRODUCTION

Information regarding the overweight trucks on a bridge could be used in many applications. The applications include planning, economic and enforcement of the bridge. For such purposes, many Bridge Weigh-In Motion (BWIM) systems [1-4] have been developed and instrumented in the field after the first pioneer work by Moses [5].

The Moses-based BWIM systems utilize a set of static moment influence sequence for the dynamic responses of a bridge. Thus, the system errors significantly rely on the accuracy of the influence sequence. However, to date, there is no systematic algorithm to extract the influence sequence in the field. In addition, the dynamic responses of the bridge girders are often affected by the global system loading conditions. Thus, the applications to long-span bridges are strictly limited because there are always many truck loads on the bridge that are able to affect the system accuracy. Furthermore, the BWIM system utilizing the response of the bride girders has poor accuracy for the estimation of single axle weight, while the good accuracy could be achieved for the gross weight estimation.

To resolve such drawbacks, this study attempts to develop a systematic algorithm to extract the influence sequences through a field test. In addition, this study investigates a way to use the responses of deck plates instead of girders. This is because the responses of deck plates are very local and sensitive to single axle loads.

EXTRACTION OF INFLUENCE SEQUENCE

Consider a truck that has three wheel loads as shown in Fig.1. As the truck moves over the deck plate of a bridge, the bending moment $M(t)$ of the deck plate at the location 'A' becomes a function of time. If the truck keeps a constant velocity of v in the considered region, the bending moment could be expressed as a function of

T. Proulx (ed.), *Dynamics of Civil Structures, Volume 4*, Conference Proceedings of the Society for Experimental Mechanics Series 13, DOI 10.1007/978-1-4419-9831-6_28, © The Society for Experimental Mechanics, Inc. 2011 space 'x',

$$M(x) = w_1 I(x - s_1) + w_2 I(x - s_2) + w_3 I(x - s_3) \quad (1)$$

where, $I(x)$ denotes the moment influence at the location x. The terms w_i and s_i denote the *i*th wheel load and location, respectively. It is noted that the moment function can be expressed as a linear combination of a set of the s_ith shifted basis of $I(x)$.

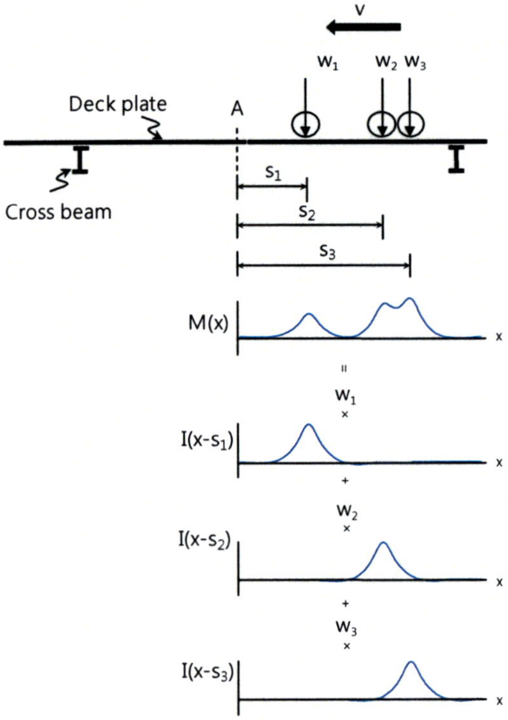

Fig.1 Moment influence sequence

In general, the *n*th discrete sample of the bending moment $M(n)$ among total N samples and total L wheel axles can be expressed by

$$M(n) = \sum_{k=0}^{N-1} w(k)\, I(n-k) \quad (2)$$

where, $w(k)$ denotes the wheel load at the sample k. Note that the number of nonzero element in the wheel load sequence $\{w(k)\}$ is only L. The mean-square value of the error between the measured moment $M(n)^*$ and $M(n)$ is

$$e = \sum_{n=0}^{N-1}\left[M(n)^* - \sum_{k=0}^{N-1} w(k)\, I(n-k)\right]^2 \quad (3)$$

Since this is a quadratic function of the moment influence, the minimization of the error yields a set of linear equations:

$$\sum_{k=0}^{N-1} \gamma_{ww}(p-k)\, I(k) = \gamma_{wM^*}(p) \quad \text{for } p = -N+1, \cdots, 0, \cdots, N-1 \quad (4)$$

where, $\gamma_{ww}(k)$ is the autocorrelation of the wheel load sequence $\{w(k)\}$ and $\gamma_{wM^*}(k)$ is the cross correlation between the measured moment sequence $\{M(n)^*\}$ and the wheel load sequence $\{w(n)\}$. Note that seeking $I(k)$ in Eq.(4) is an optimal linear filter design problem and the optimal linear filter, in the sense of minimum mean-square error, is called a Wiener filter [6]. In general, Eq. (4) can be expressed in the form as

$$\mathbf{\Gamma}_{ww}\mathbf{I}_N = \boldsymbol{\gamma}_{wM^*} \quad (5)$$

where, $\boldsymbol{\Gamma}_{ww}$ is a (2N-1)×(2N-1) Toeplitz matrix with elements $\gamma_{ww}(p-k)$ and $\boldsymbol{\gamma}_{wM^*}$ is a (2N-1)×1 cross correlation vector with elements $\gamma_{wM^*}(p)$. The solution for the optimal coefficients of the moment influence sequence is

$$\mathbf{I}_N = \boldsymbol{\Gamma}_{ww}^{-1}\,\boldsymbol{\gamma}_{wM^*} \tag{6}$$

EXTRACTION OF AXLE LOADS

Assume that the moment influence sequences $I(n)$ have been identified from Eq.(6) and the moment sequences $M^*(n)$ have been measured. Then Eq.(3) is a quadratic function of the wheel load sequence $w(n)$, the minimization of the error yields the set of linear equations:

$$\sum_{k=0}^{N-1}\gamma_{II}(p-k)\,w(k) = \gamma_{IM^*}(p) \quad \text{for } p = -N+1,\cdots,0,\cdots,N-1 \tag{7}$$

where, $\gamma_{II}(k)$ is the autocorrelation of the moment influence sequence $\{I(k)\}$ and $\gamma_{IM^*}(k)$ is the cross correlation between the measured moment sequence $\{M(n)^*\}$ and the moment influence sequence $\{I(n)\}$. In general, Eq. (7) can be expressed in the form as

$$\boldsymbol{\Gamma}_{II}\,\mathbf{w}_N = \boldsymbol{\gamma}_{IM^*} \tag{8}$$

where, $\boldsymbol{\Gamma}_{II}$ is a (2N-1)×L matrix with elements $\gamma_{II}(p-k)$ and $\boldsymbol{\gamma}_{IM^*}$ is a (2N-1)×1 cross correlation vector with elements $\gamma_{IM^*}(p)$ while $\mathbf{w}_N$ is a L×1 desired wheel load vector. Since the number of sample N is much larger than the number of wheel axle L, Eq.(8) becomes over-determined matrix equation. Then the optimal solution for the wheel load vector can be achieved by

$$\mathbf{w}_N = \left[\boldsymbol{\Gamma}_{II}^T\,\boldsymbol{\Gamma}_{II}\right]^{-1}\boldsymbol{\Gamma}_{II}^T\,\boldsymbol{\gamma}_{IM^*} \tag{9}$$

It is can be shown that the Mose's algorithm [5] is a special case of Eq. (8), in the sense of the collocation method, because a set of L equations at the wheel load locations agrees with the Mose's system equation.

FIELD APPLICATIONS

Field Test Setup

The proposed algorithms has been examined through Yongjong Grand suspension bridge that lies on the 40.2 km long expressway linking Seoul metropolitan and the Incheon international airport in Korea. The 550 m long bridge has a center span of 300 m and side spans of 125 m. The self-anchored suspension bridge has two 107 m high towers and the 35 m wide double decks to carry a six-lane highway on the upper deck and four-lane highway and double-track railway on the lower decks. The lower decks consist of two independent deck plate sections. Each lower deck plate section carries two-lane highway as shown in Fig. 2.

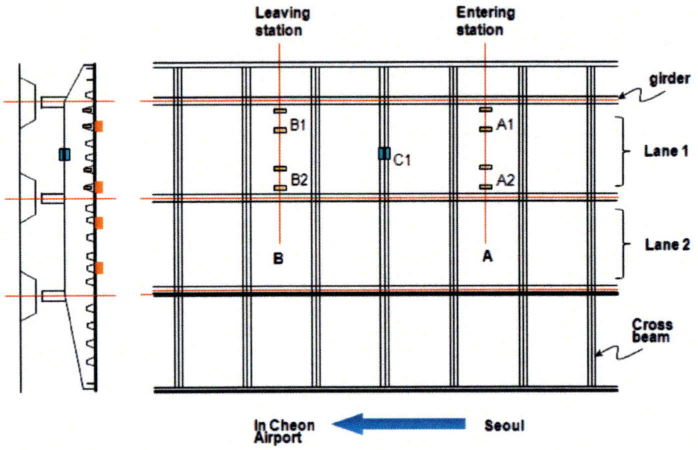

Fig.2 Sensor map

The considered test bed consists of a set of the rectangular regions that are surrounded by two girders and two cross beams. The sensors are placed at two stations on Lane 1 as depicted in Fig.2. One is entering station 'A' and the other is leaving station 'B'. For each sensor station, the four strain gauges are placed at the center span of the rectangular deck plate. Using a half-bridge connection to improve the signal to noise ratio, only four strain gauge signals, A1, A2, B1 and B2, are recorded. In addition, a signal C1 of the cross beam using two strain gauges are recorded when a truck passes by.

As shown in Fig.3, the deck plate consists of folded steel plate, and the strain gages are attached on the convex parts. As shown in Fig.4, two trucks are used for the verification task. One is unloaded three-axle truck and the other is fully loaded four-axle truck. To record the strain gauge time history, the NI Compact-RIO data logger has been used with the sampling rate of 2000Hz.

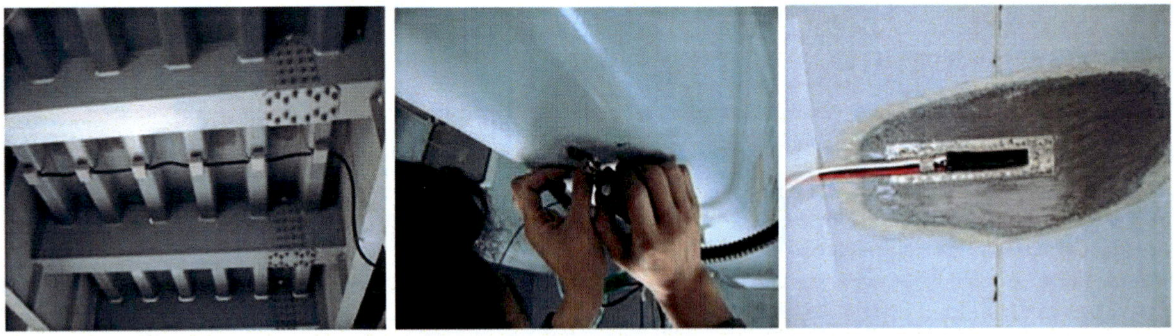

Fig.3 Sensor attachment on the deck plate of bridge

Fig.4 Field tests

Table 1. Measured static axle weight of calibration trucks

	three axle truck (kN)	Four axle truck (kN)
First axle	48.608	89.572
Second axle	73.206	80.654
Third axle	70.756	90.846
Fourth axle		90.356
Gross weight	192.570	351.428

Evaluation of Influence Sequence

To apply this proposed identification algorithm for moment influence sequence, a strain time history has been captured for an arbitrary two wheels truck with a wide wheelbase. The speed of the target truck is obtained by inspecting the strain time history. It turns out 93.78km/h. Since the signal energy of the measured strain time history lies below 40Hz, the measured signals have been filtered with a low pass digital filter of 60Hz.

For total number of sample N=2401 and total number of axles L=2, the 4801×4801 Toeplitz matrix $\mathbf{\Gamma}_{ww}$ and the 4801×1 cross correlation vector $\boldsymbol{\gamma}_{wM^*}$ in Eq. (5) have been constructed. The desired moment influence sequence vector $\mathbf{I}_N$ has been obtained through Eq. (6). The resulting moment influence sequence is shown in Fig.5. As depicted in Fig.6, the measured moment sequence agrees very well with that predicted by the estimated influence sequence. It can be confirmed that the proposed algorithm works very well in the field condition.

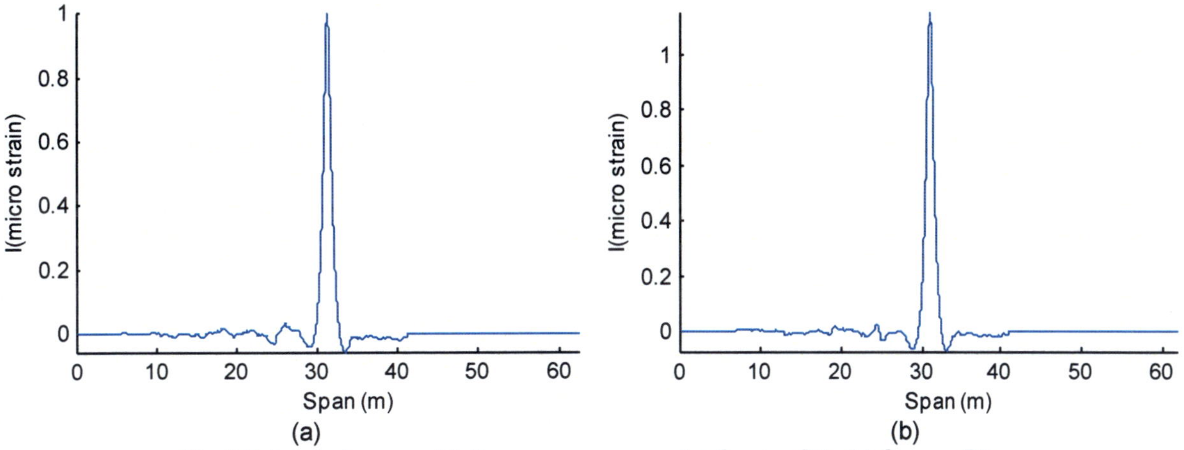

Fig.5 Extracted moment Influence sequence: (a) Sensor B1; (b) Sensor B2.

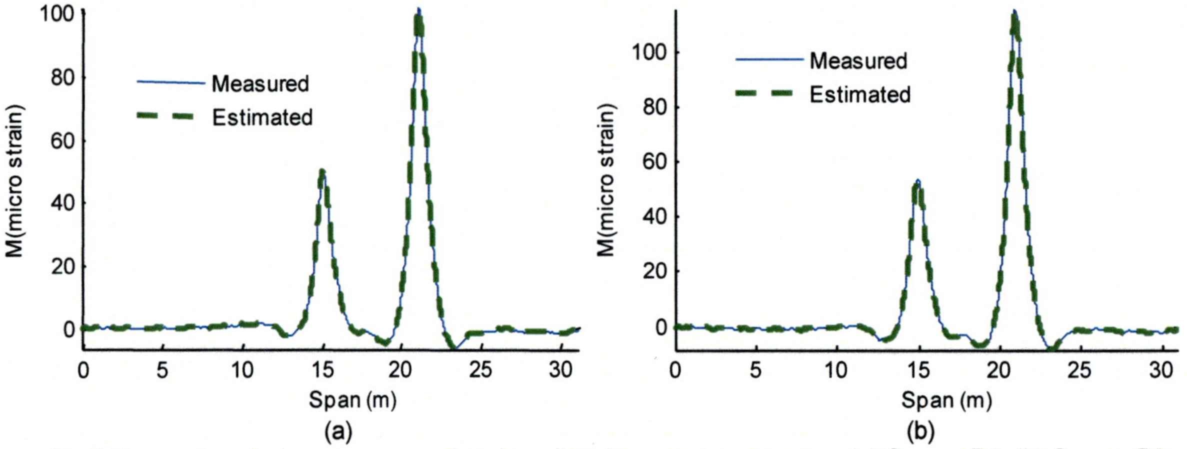

Fig.6 Comparison between measured and predicted moment sequences: (a) Sensor B1; (b) Sensor B2.

Evaluation of axle loads

To examine the accuracy of the proposed algorithm for the single axle load identification, two calibration trucks shown in Table 1 pass by the test stations. For the strain time samples of N=1123, the over-determined matrix $\mathbf{\Gamma}_{\mathrm{II}}$ and the cross correlation vector $\boldsymbol{\gamma}_{\mathrm{IM}^*}$ are constructed and the desired axle load vector $\mathbf{w}_{\mathrm{N}}$ is obtained by Eq. (9). The estimated moment sequence with the identified individual axle loads has been compared in Fig.7. Total eight tests are repeated. The first four cases are for the passage of the three-axle truck and the other four cases are for the four-axle truck.

The estimated results and their errors with respect to the static axle load in Table 1 are compared in Table 2. The variation of errors in the gross weight is much smaller than those in the single axle weight. The maximum errors for the three-axle truck and the four-axle truck have occurred at Case 2 and Case 8, respectively. One possible explanation is that the constant velocity condition has not been satisfied because the strain time histories between the entering station and the leaving station are different. The truck driver might put the brake pedal at the leaving station to keep the ordered constant velocity. Also, it should be reported that the time shift of the influence sequence s_i has high sensitivity to the system accuracy.

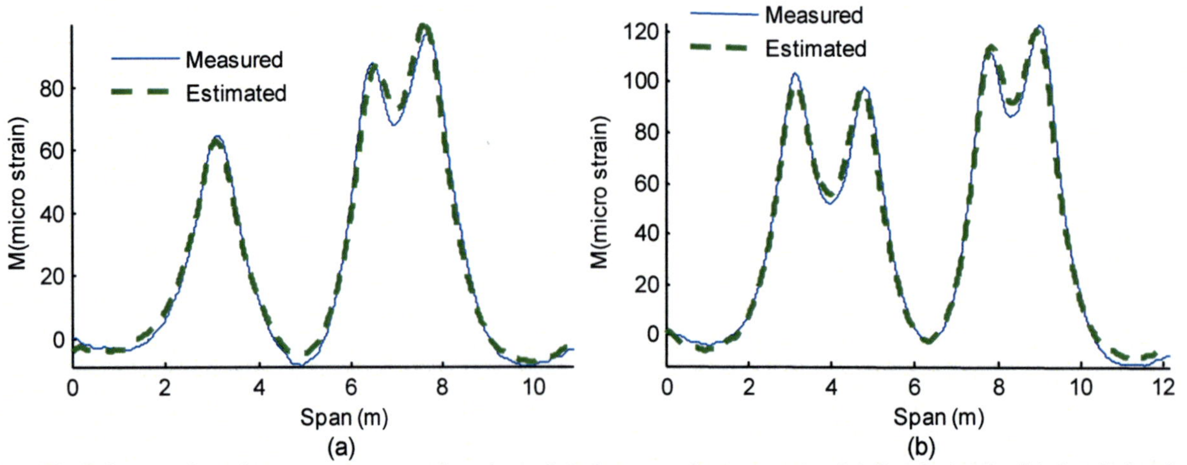

Fig.7 Comparison between measured and predicted moment sequences: (a) 3-axle truck; (b) 4-axle truck.

Table 2. Estimated axle weight of calibration trucks

Case	Velocity	Wheelbase	Axle Weight (kN)				Gross Weight
No.	(Km/h)	(m)	First	Second	Third	Fourth	(kN)
1	69.36	4.57	50.339 (-3.56)	68.588 (6.31)	73.568 (-4.02)	-	192.525 (-0.31)
2	73.67	4.53	46.054 (5.26)	63.746 (12.92)	67.283 (4.91)	-	177.083 (7.73)
3	71.17	4.55	52.213 (-7.42)	66.804 (8.75)	75.665 (-6.94)	-	194.681 (-1.44)
4	70.89	4.62	51.636 (-6.23)	70.891 (3.16)	74.530 (-5.33)	-	197.057 (-2.67)
5	71.45	5.87	82.626 (7.75)	81.173 (-0.64)	86.071 (5.26)	92.370 (-2.23)	342.240 (2.29)
6	78.49	5.84	86.980 (2.89)	85.532 (-6.05)	80.253 (11.66)	93.615 (-3.61)	346.380 (1.11)
7	77.48	5.86	90.246 (-0.75)	87.247 (-8.17)	82.274 (9.44)	94.507 (-4.59)	354.275 (-1.15)
8	75.66	5.81	91.545 (-2.2)	88.698 (-9.97)	91.301 (-0.50)	102.563 (-13.51)	374.107 (-6.81)

* the value in brace denotes a percentage error.

SUMMARY AND CONCLUSIONS

Unlike the girder responses of a bridge, the deck plate response has very local and sensitive to single axle load. To use these advantages, the identification algorithm of the moment influence sequence has been newly proposed. With the similar concept, the new algorithm for the axle load identification has been proposed. To examine the feasibility and accuracy of the proposed algorithms, a set of field tests has been conducted.

Based on the experimental results, the following three conclusions could be made. First, the proposed algorithm for the influence sequences is very systematic and it works very well in a field condition as confirmed in Fig. 6. Second, the proposed algorithm for the identification of single axle load using the deck plate responses seems to be feasible and economic. Third, the accuracy improvement of the proposed axle load identification algorithm is expected for the commercial applications. To improve the accuracy of the proposed approach, the wondering effect of truck passage should be considered. In addition, the axle location parameter also simultaneously identified with the axle loads. This is necessary because the sensitivity of the axle locations is considerable if the deck plate responses are used.

ACKNOWLEDGEMENTS

This work was supported by National Research Foundation of Korea Grant funded by the Korean Government (2009-0072212)

REFERENCES

[1] Peters, R.J. (1986), "CULWAY - an Unmanned and Undetectable Highway speed Vehicle Weighing System", in Proceedings of the 13th AARB Conference, Australian Road Research Board.

[2] Znidaric, A., Moses, F. (1995), "Bridge Weigh-In-Motion Testing of Vehicle Gross Weights in Slovenia", the First European WIM Conference, Zurich, March 1995.

[3] Dempsey, A.T., Jacob, B. and Carracilli, J. (1999). "Orthotropic Bridge WIM for determining Axle and Gross Vehicle Weights", in Proceedings of the Final symposium of the project WAVE, Ed. B. Jacob, Hermes Science Publications, Paris, France, pp.227~238.

[4] COST 323 (1999), "Weigh-in-Motion of Road Vehicles", Final Report, August 1999.

[5] Moses, F. (1979), "Weigh-in-Motion System Using Instrumented Bridges", transportation Engineering journal, ASCE, 105, TE3, pp.233~249.

[6] Proakis, J.G., Manolakis, D.G. “Digital Signal Processing : Principles, Algorithms, and Applications”, 3rd Ed., Prentice Hall, New Jersey, 1996.

Proceedings of the IMAC-XXVIII
February 1–4, 2010, Jacksonville, Florida USA
©2010 Society for Experimental Mechanics Inc.

SELF EXCITED VIBRATION OF A CYLINDRICAL BODY WIND BIDIRECTIONAL FLOWMETER.

Dorin Simoiu, D.S., Titus Gh. Cioara, T.C.

University "Politehnica" of Timisoara
Bv. Mihai Viteazul nr. 1 Timisoara -Romania
E-mail: cioara@mec.upt.ro

SUMMARY: In the paper is analyzed the dynamical interaction between air flow and a two components wind flowmeter, consisting from a rigid cylinder clamped to a double direction elastic element, with two strain gage sensors mountings. The output sensors signals are proportional with induced wind force components on the two sensors orthogonal sensitivity axes, components which allow to be determined both, wind speed and direction. Interaction between wind and elastic coupled rigid structure induce self excited vibration of structure at critical speeds of the flow around the cylinder, occurring high level perturbation components in the output flowmeter signals. Based on a coupled flow - structure system of equation are simulated the self excited vibration in the frequency range of interest, 0 to 30 Hz, domain of wind flow. Finally are analyzed by experimental data the response of the flowmeter to different wind flow including the strong gusts.

KEYWORDS: wind flowmeter, self excited, wind gust

1. WIND PARAMETERS MEASUREMENT METHODS

1.1. General considerations

The wind induces complex loads to the structures, placed in the free air conditions, as: buildings, industrial installations, communication antennas, and many others. These structures have to be designed to support wind loads of maximum intensity of placed area, including, strong gale, at mean speed 22 to 25 m/sec, and storm at 30 to 35 m/sec. But in the majority of time it is necessary to record wind, to know the frequency spectrum which is necessary to define dynamic excitation on structure. In this case is necessary to be used a wind sensor of which modal frequency domain to be more high as the wind spectrum. A wind spectrum is usually located in the low frequency band, of 0-to 30 Hz, so, the value of first modal frequency of the wind sensor has to be located two or three more times high.

A type of this sensor is presented in the figure 1, [1], a bidirectional flowmeter composed from a light cylinder 1. of aluminum, connected, by a rod 2 to a sensor body 3, on which are profiled by milling two elastic lamellas a_x and a_y of which planes are rectangular axed between. Placing the flowmeter transversal on direction of the wind stream, given by velocity vector $\vec{V}$, the force vector $\vec{F}$, induced by wind to cylinder, is parallel to vector $\vec{V}$, and the magnitude can be approximate

$$F \approx C_F h D \rho \frac{V^2}{2} \tag{1}$$

where C_F is a coefficient of proportionality called the shape factor. For cylinder the factor C_F is not a constant for all wind speeds, but depends of Reynolds number, and cylinder geometry *D* and *h*.

The two components F_x and F_y, of the force $\vec{F}$, parallel to orthogonal axes $O_x x$ and $O_y y$ give two bending moments around this two axes

$$M_{ix} = L_x F \sin\alpha; \quad M_{iy} = L_y F \cos\alpha \tag{2}$$

and corresponding to bending stresses σ_{ix} and σ_{iy} on the lamellas a_x and a_y faces, where are applied two strain gages,

$$\sigma_{ix} = \frac{M_{ix}}{W}; \quad \sigma_{iy} = \frac{M_{iy}}{W} \tag{3}$$

where W is the geometric moment of inertia of the transversal sections of lamellas a_x and a_y.

Having recorded the stress signals σ_{ix} and σ_{iy} the value of wind velocity result of the form,

$$V = \sqrt{\frac{2}{C_F h D \rho}\left(W\sqrt{\left(\frac{\sigma_{ix}}{L_x}\right)^2 + \left(\frac{\sigma_{iy}}{L_y}\right)^2} \right)} \tag{4}$$

T. Proulx (ed.), *Dynamics of Civil Structures, Volume 4*, Conference Proceedings of the Society for Experimental Mechanics Series 13,
DOI 10.1007/978-1-4419-9831-6_29, © The Society for Experimental Mechanics, Inc. 2011

and direction of the wind, relative to sensor frame oxy, by position angle

$$\alpha = tan^{-1}\sqrt{\left(\frac{L_y\sigma_{ix}}{L_x\sigma_{iy}}\right)} \tag{5}$$

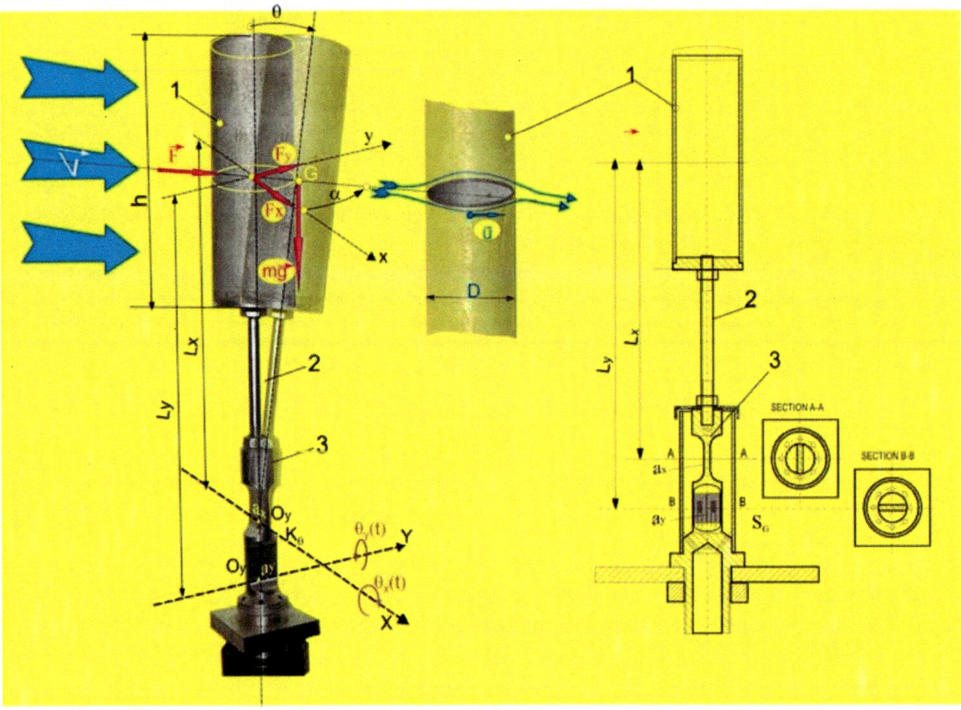

Figure 1 Illustration of bidirectional strain gage flowmeter, [1]

1.2. Perturbation in interaction flow elastic clamped flowmeter body

In the immediate vicinity of a bounding surface between fluid and structure (cylinder, figure 2a), due to viscosity of fluid the distribution of velocity is special one. Flow in this layer, called boundary layer, can be either laminar or turbulent. There is a point P_{ds} of flow separation where the velocity vector $\vec{u}$ falls almost to zero. The fluid flow becomes detached from the surface of the object, and instead takes the forms of eddies and vortices. Also the distribution of the pressure in the boundary layer is the special one, resulting drag force $\vec{F}$

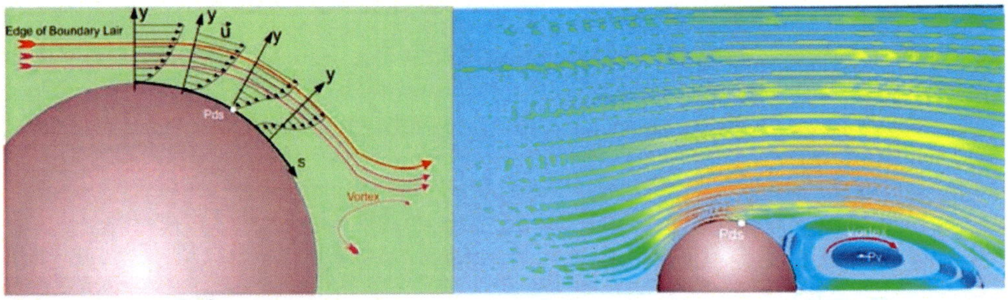

a) b)

Figure 2 Illustration of 2D flow in boundary layer of a cylinder

Unsteady separation of the flow can developed repeating pattern of swirling vortices, (illustrated in figure 3), the frequency of vortices being of the form

$$f_v = S_t \frac{V}{D}. \tag{6}$$

S_t is the Strouhal number, for a cylinder, S_t=0.2.

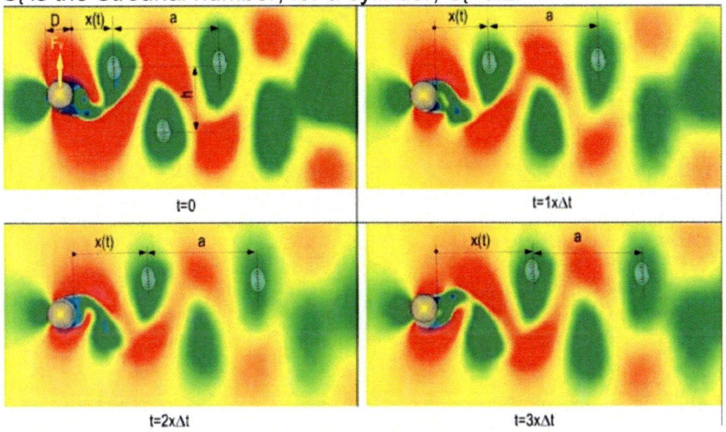

Figure 3 Illustration of Karman vortex street periodicity (Δt=0.1 sec)

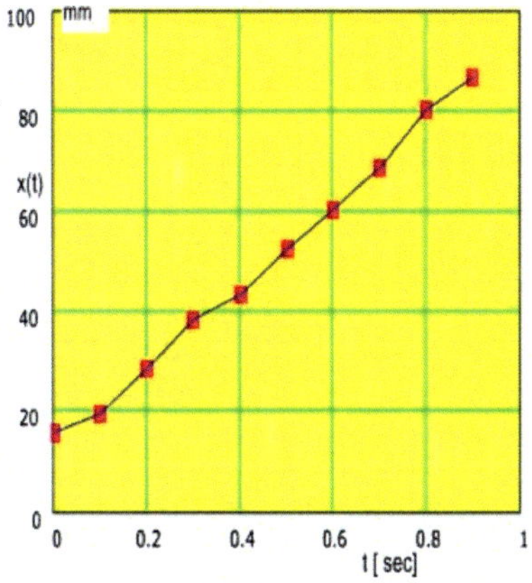

Figure 4 Constant speed displacement x of the vortex

The displacement law x(t) of the vortex street is linear in time, (figure 4) of constant speed, different of stream speed V. Periodicity of vortex create a periodicity of pressure distribution at surface of the structure (in this case the cylinder), resulting a transversal periodic force, approximated by an empirical harmonic form,

$$F_v(t) = C_v \frac{1}{2} \rho V^2 A \sin(2\pi f_v t) \tag{7}$$

When the structure is elastic one, or it is elastic connected to a fixed structure, as in case of the flowmeter, (sensor body 3, elastic rotations by laws $\theta_x(t)$ and $\theta_y(t)$), the dynamical equations of the structure are coupled with the equations of the flow control volume, a space defined relatively around the structure space, where the influence of the structure shadow. By deformation of the structure a quantity of fluid, in flow, energy is transferred to structure, which coming into self vibration, and for a critical value of fluid velocity V_c the energetic transfer is maxim accepted by, vibration motion being "*locking-in*" by one of natural frequency of the structure, as, illustrated in figure 5

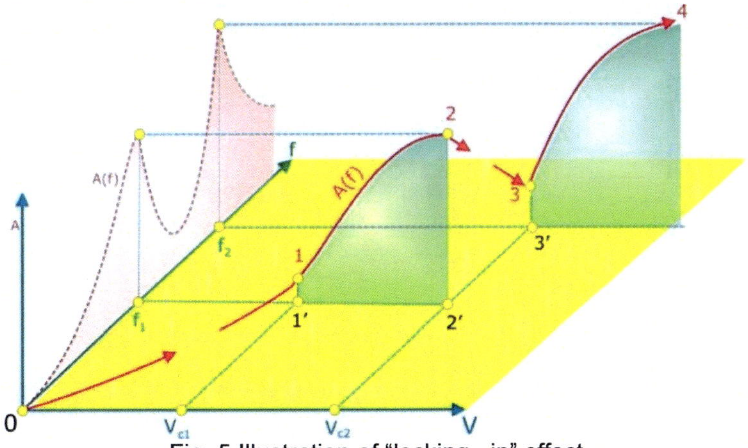

Fig. 5 Illustration of "locking –in" effect

Considering a rectangular reference frame *OVfA, OV* being axis of velocity. *Of* axis of frequency and *OA* axis of amplitude, sweeping the velocity along it axis at a special value called critical speed

$$V_{c1} \approx f_1 \frac{D}{S_t} \tag{8}$$

where f_1 is the value of structure first natural frequency. Raising more the flow velocity V, the vibration frequency *f* is keeping strictly the same value f_1 and amplitude is raising from point 1 to 2, point 2 corresponding to a second critical speed V_{c2} when the vibration steady state jump suddenly from point 2 to point 3, corresponding to a harmonical vibration, tuned by frequency f_2 of the second mod, which is as an *attractor*.

2. SIMULATIONS

2.1. Simulation problems

In view to design an optimal bidirectional flowmeter, by above technical solution it was simulated the interaction of the wind at different speed resolved numerically the Navier Stokes equation in a Reynolds transport form, coupled with the elastic connected structure.

As software was used ANSYS CFX, in the Reynolds-stress turbulence model. Is a simulation which involving two-way Fluid-Structure Interaction, where the fluid is solved in CFX and the solid is solved in the ANSYS. The two solvers are coupled on the entire solution process in order to model the interaction between fluid and solid. The ANSYS Multi-field solver, using the MFX setup is used for coupling the two solvers. CFX uses a unique hybrid finite-element/finite-volume approach to discretizing Navier-Stokes equations.

The spatial domain is discretized into finite control volumes using a mesh. The governing equations are integrated over each control volume. Fluid domain was discretized in 129797 elements and structure domain was discretized in 3971 elements.

In the figure 7a are plotted the time histories of the stress $\sigma_x(t)$, corresponding to bending deformation of the elastic lamella a_x, due to load F_y, induced by wind, of constant speed at V=5,10,20, and 30 m/sec at direction $\alpha=0^0$. $\sigma_x(t)$ represents the bending stress of the mid point Ox of lamella a_x, on which the strain gage T_{1x}, T_{2x}, T_{3x} and T_{4x} montage is glued. In the figure 7b are plotted the time histories of the stress $\sigma_y(t)$, corresponding to bending deformation of the elastic lamella a_y, due to load F_y, induced by wind, of constant speed at V= 0.5,10,20, and 30 m/sec, at direction $\alpha=90^0$. In both figures, during first 0.2 sec, it can be observes a transient processing, of two stress components, one of free decay, by natural frequencies of the flowmeter, and second component, a constant one corresponding to constant value of wind speed.

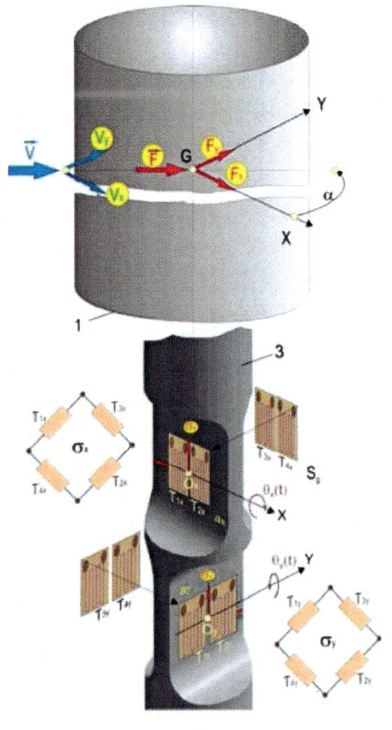

Figure 6 Strain gage montages for bidirectional flowmeter measurement

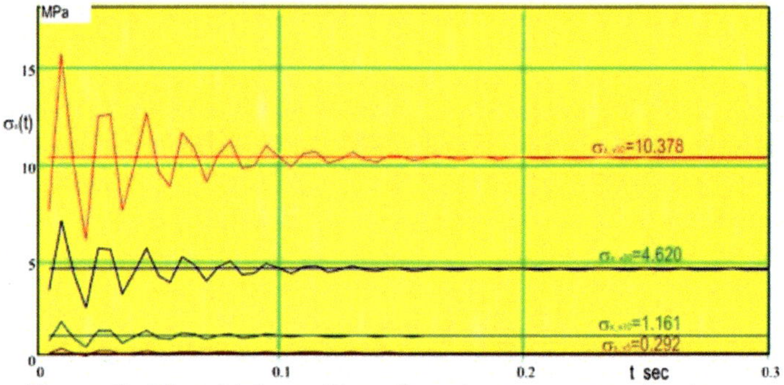

Figure 7a Time history of bending stress σ_x around axis $o_x x$ (figure 6), for four simulated wind speed V=5,10,20 and 30 m/sec, at direction $\alpha=0^0$

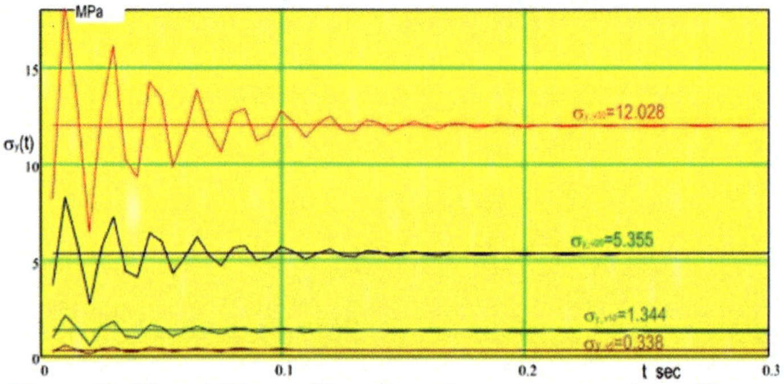

Figure 7b Time history of bending stress σ_y around $o_x y$ (figure 6), for four simulated wind speed V=5,10,20 and 30 m/sec, at direction $\alpha=90^0$

From plot of strain $\sigma_{x,y}$ against speed V (figures 8a and 8b), as expected, it is observes a nonlinear dependent of two variable, as,

$$\sigma_x = 0.0115(0.097V_y + V_y^2); \quad (\alpha = 0^0) \tag{9}$$

obtained by the interpolation using data from figure 8a, and from figure 8b

$$\sigma_y = 0.0133(0.089V_x + V_x^2); \quad (\alpha = 90^0) \tag{10}$$

For the both forms, the coefficients of V represent ≈≤10% of V^2 coefficients, so that, the empirical square forms (1) and (7) can be considered as practically satisfied.

In the case of an experimental wind speed measurement it is necessary to be known the characteristic $V_{x,y}$ as function of the stress $\sigma_{x,y}$ by the transformation

$$\sigma_{x,y} = b_{y,x}V_{y,x} + a_{y,x}V_{y,x}^2 \quad to \Rightarrow \quad V_{y,x} = \frac{-b_{y,x} + \sqrt{b_{y,x}^2 + 4\sigma_{x,y}a_{y,x}}}{2a_{y,x}} \tag{11}$$

this characteristic being plotted in figure 8c.

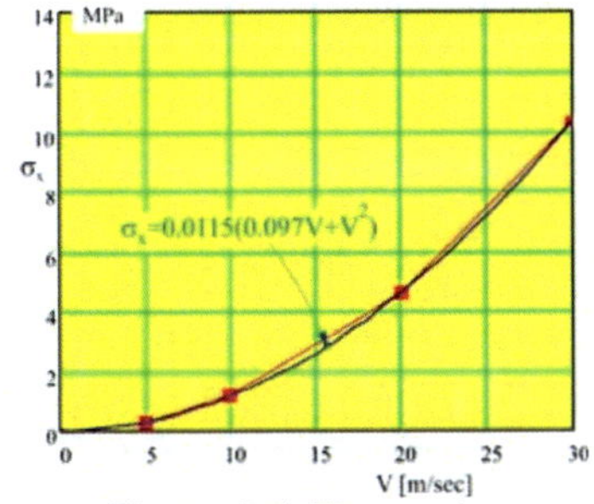

Figure 8a) Flowmeter characteristic, stress σ_x, - speed of direction $\alpha=0^0$

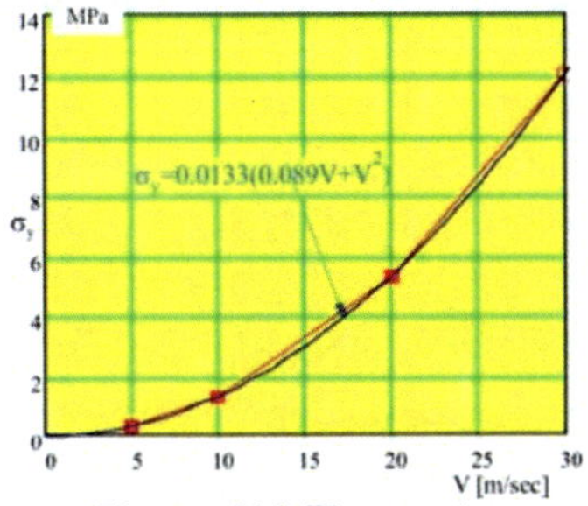

Figure 8b) Flowmeter characteristic, stress σ_y, - speed of direction $\alpha=90^0$

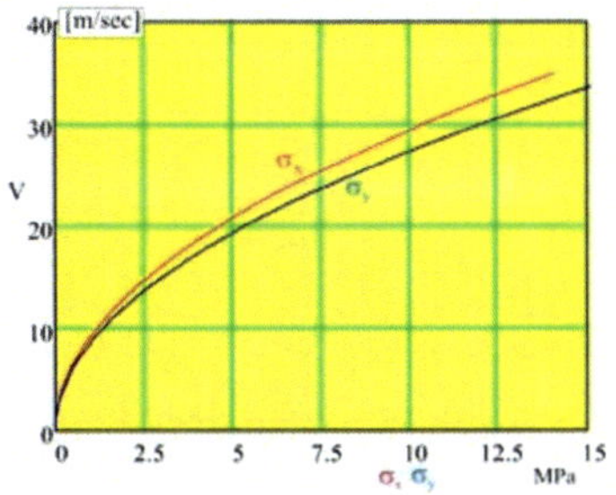

Figure 8c) Flowmeter characteristics, speed $V_{x,y}$ - stress σ_{xy} -

2.2. Flowmeter simulation response to a variable wind speed

Figure 9a presents a simulated sequence of time history wind speed V(t) composed from two component, the constant one of value Vc=30 m/sec and second one, a random variable of root mean square values V_{rms}, and frequency spectrum plotted in figure 9b.

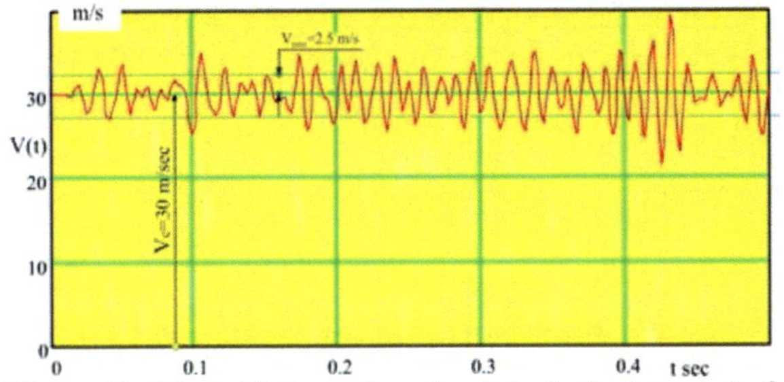

Figure 9a Time history of a simulated wind speed of two components: a constant one of value Vc=30 m/s and a random component of value V_{rms}=2,5 m/s (figure 9b plot frequency spectrum)

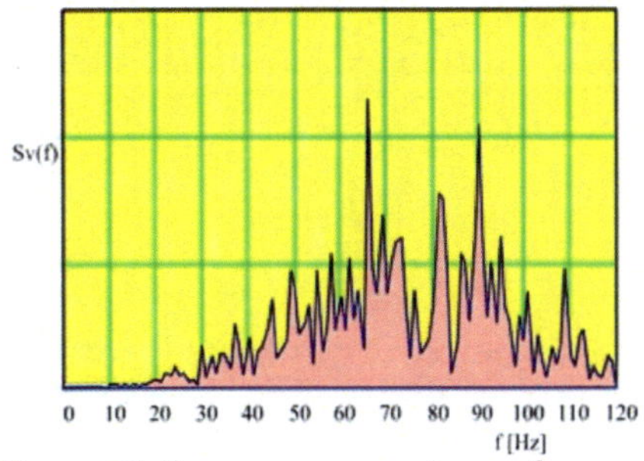

Figure 9b Frequency spectrum of random wind component (figure 9a)

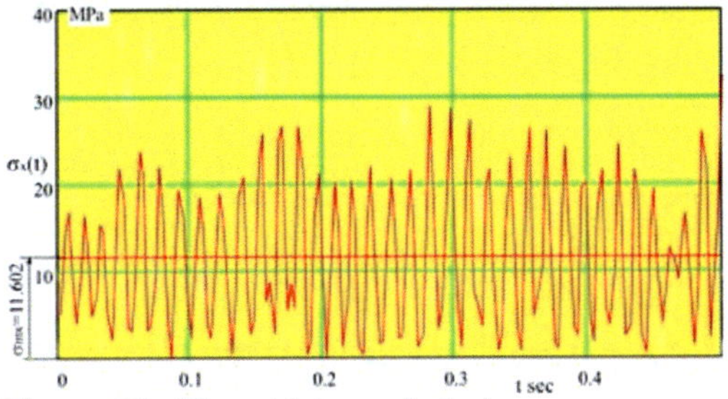

Figure 10a Time history of strain gage montage signal $\sigma_x(t)$

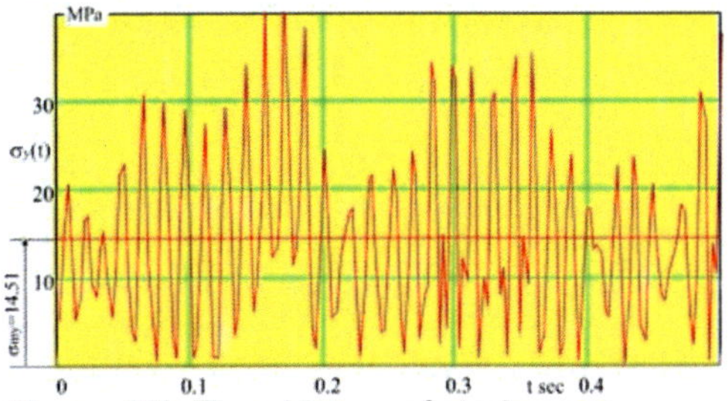

Figure 10b Time history of strain gage montage signal $\sigma_y(t)$

The direction of the wind blow, (direction of vector $\vec{V}$) was programmed at value $\alpha=45^0$. For these input data results, by simulation, the bending stresses σ_x and σ_y, which are considered as recorded, by the two strain gages montages. The time histories of the two signals $\sigma_x(t)$ and $\sigma_y(t)$ are plotted in figures 10a and 10b.
Taking in consideration, the characteristics equations (9-10) and transformation equation (11), in figure 11a is plotted the time history of resulting, by simulation, wind speed component $V_y(t)$, of the mean value V_{my}=20.66 m/sec and of the mean square root V_{yrms}=8.93 m/sec.

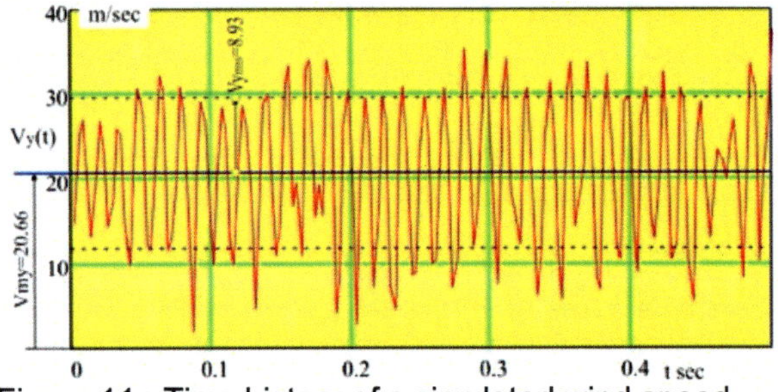

Figure 11a Time history of a simulated wind speed component $V_y(t)$

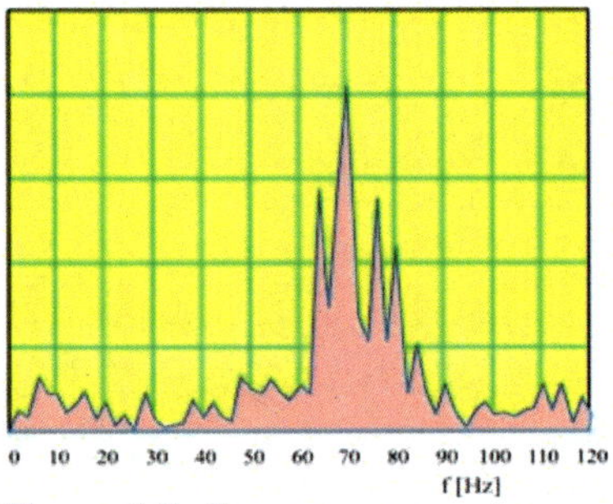

Figure 11b Frequency spectrum of wind speed component $V_y(t)$

The same plots are given, in figures 12a and 12b, for the simulated $V_x(t)$ component, and, finally, being plotted, in figures 13a and 13b the wind speed V(t)

$$V(t)=\sqrt{V_x(t)^2+V_y(t)^2} \tag{12}$$

considered as be experimental recorded.

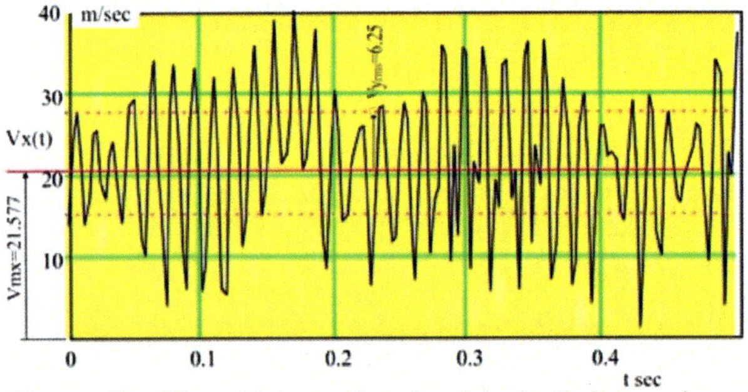

Figure 12a Time history of a simulated wind speed component $V_x(t)$

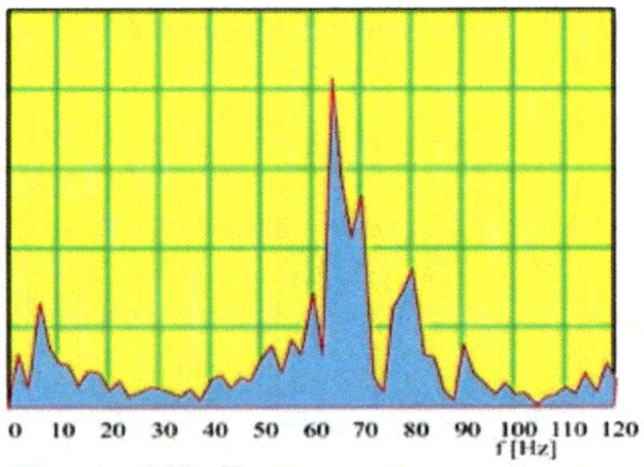

Figure 12b Frequency spectrum of wind speed component $V_x(t)$

Now, comparing the figures, 9a and 9b with 13a and 13b, which should be the same, it is find out very small deviation between the speed mean values, V_c=30 m/sec of simulation input data, and V_m=30.501 m/sec of output simulation. Also, there are large deviation in the variable components V_{rms}=2,5 m/sec, of input data, and V_{rms}=7.636 m/sec of output simulation, normally, because the frequency band of excitation (figure 9b) is located in spectral band of the first two natural frequencies of 75 and 82 Hz, determined by applying algorithm [2].
The instant angle direction $\alpha(t)$, of the blowing simulated wind is determined by equation

$$\alpha(t)=\frac{180}{\pi}tan^{-1}\left(\frac{V_y(t)}{V_x(t)}\right) \tag{13}$$

of which time history is plotted in figure 14a.
The mean value α_m during 0.5 seconds is 46.2 Deg, close to input simulation constant of the value of 45 Deg. The large instant deviation $\alpha(t)$- α_m occurs due to dynamic resonant effects, as above motivation. This effect can be diminishing applying low pass filtering process in frequency domain of wind blowing, 0 to 30 Hz (figure 14b) [3]
The filtering may be applied direct to output signals $\sigma_x(t)$ and $\sigma_y(t)$, perturbation components being eliminate in this phase of the signal process

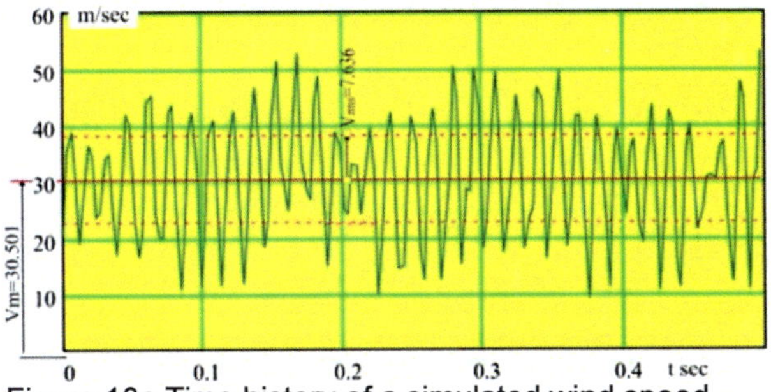

Figure 13a Time history of a simulated wind speed V(t)

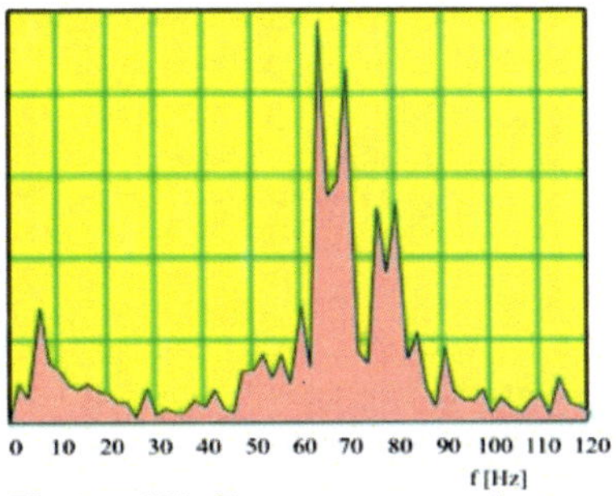

Figure 13b Frequency spectrum of wind speed V(t)

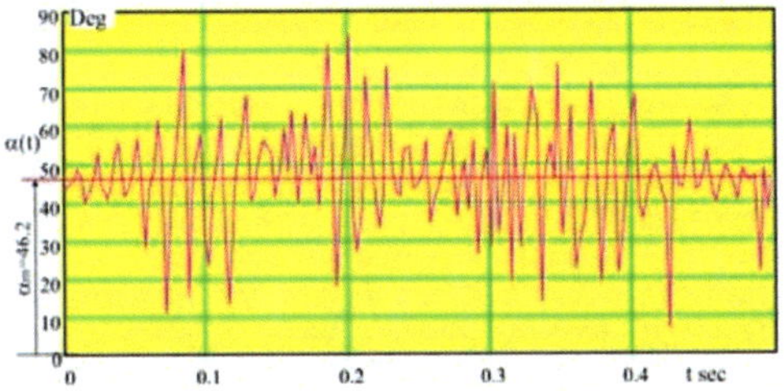

Figure 14a Illustration of instant angle direction α(t), of the blowing simulated wind

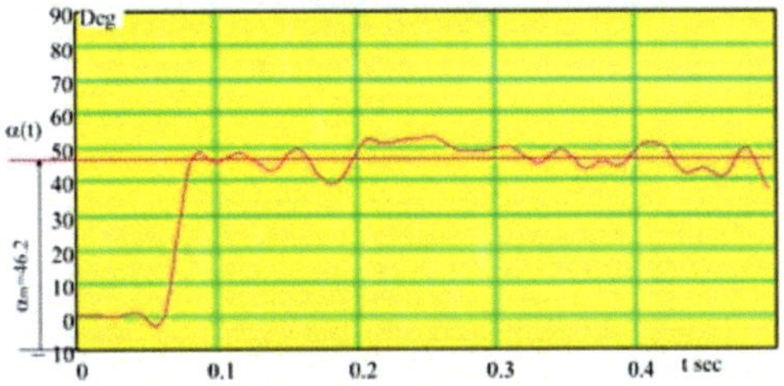

Figure 14b Time history of angle α(t) filtered by a low band pass filter 0 to 30 Hz

3. EXPERIMENTAL TESTS AND CONCLUSIONS

3.1. Wind parameters measurement on a building roof

One of the flowmeter application was to record the wind parameters on a building roof were is placed a parabolic dish parabolic antenna, on a tall metallic pillar (figure 15). The main aim of this setup was to determine wind load induced on the antenna structure. In parallel to load are recorded the wind speed components, using the flowmeter above analyzed. In figure 16 is plotted the time history of a 280 seconds sequence, including a strong gust.

Figure 15 Illustration of the flowmeter placed on a metallic pillar on a building roof.

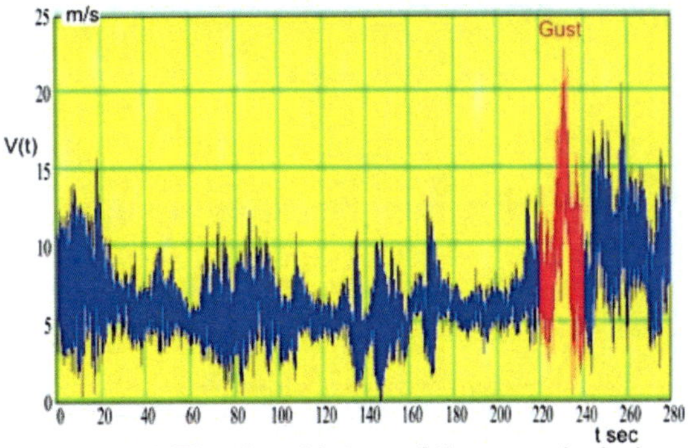

Figure 16 The time history of the speed wind V(t) recorded during of 280 seconds by flowmeter

The gust extension in figure 17 is a turbulent flow from two speed components

$$u_x(t) = \overline{u_x(t)} + \tilde{u}_x(t) \qquad (14)$$

$\overline{u_x(t)}$ being the meaning component, and $\tilde{u}_x(t)$ the variable component, implied in the Reynolds equation

$$\frac{\partial \overline{u_i}}{\partial t} + \overline{u_1}\frac{\partial \overline{u_i}}{\partial x_1} + \overline{u_2}\frac{\partial \overline{u_i}}{\partial x_2} + \overline{u_3}\frac{\partial \overline{u_i}}{\partial x_3} = F_i - \frac{1}{\rho}\frac{\partial p}{\partial x_i} + \nu\left(\frac{\partial^2 \overline{u_i}}{\partial x_1^2} + \frac{\partial^2 \overline{u_i}}{\partial x_2^2} + \frac{\partial^2 \overline{u_i}}{\partial x_3^2}\right) + A_i; \quad (i = 1,2,3) \qquad (15)$$

governing the turbulent flow.

The two components $\overline{u_x(t)}$ and $\tilde{u}_x(t)$, plotted in figure 17, are separated from recorded signal $u_x(t)$ using a moving block mediation of band T (figure 17),[3].
The gust sequence recorded by the flowmeter, of average speed ≈10 m/sec, is multiplied, for simulation to a strong gust flow, three times more, resulting, for 3.2 seconds of processing the displacement x(t) (figure 18) of the upper point P on the flowmeter (figure 15)

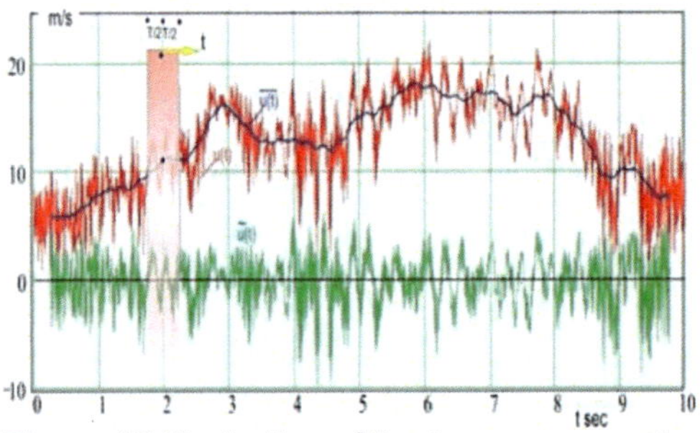

Figure 17 Illustration of the two components. $\overline{u_x(t)}$ $\tilde{u}_x(t)$ separation from $u_x(t)$

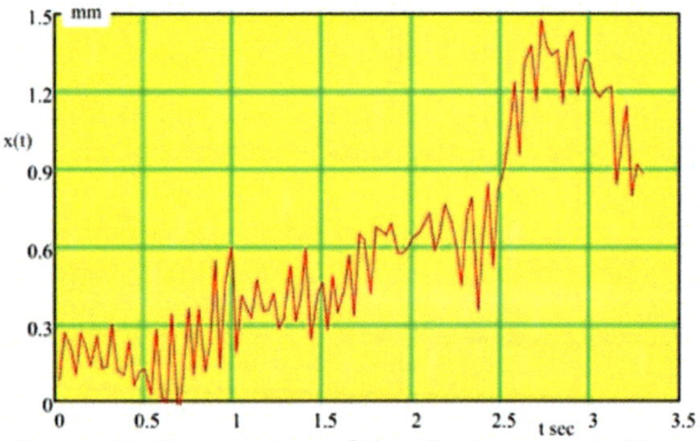

Figure 18 Time history of the displacement x(t) in point P (figure 15) during gust (figure 17)

3.2. Testing antenna and flowmeter in the open air carried at different transport speed

In view to test the antenna and flowmeter to loads induced by air strong flow, high of V=10 m/s, the antenna and flowmeter were placed on a car roof (figure 19), so, it was possible, in this situation of car power, to reach the maximum speed of the car at 70 km/h (≈20 m/sec).

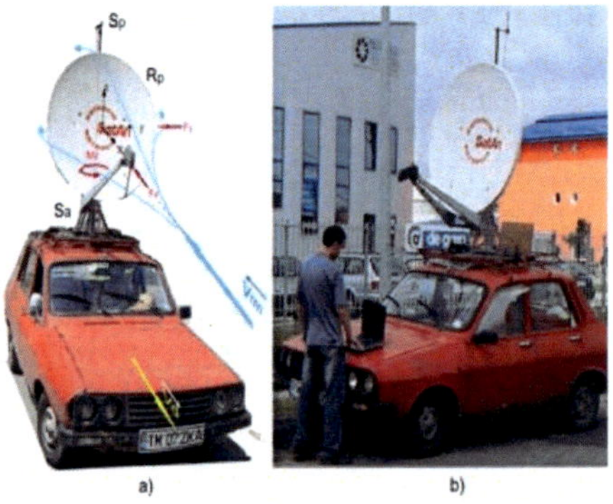

Figure 19 Illustration of the flowmeter placed with parabolic antenna on a car roof

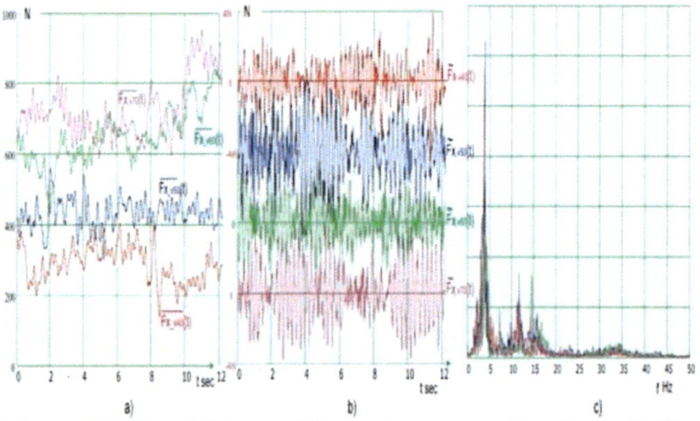

Figure 20 Time history of the antenna reaction force $F_X(t)$, induced by the air flow, for four different velocity of the car 40, 50, 60 and 70 km/h.

In the figure 20 are presented the plotted, mean $\overline{F_x(t)}$ and $\tilde{F}_x(t)$ components, of the antenna horizontal reaction force $F_x(t)$, against relative air stream. In spite of no car constant speed the records were useful to be analyzed, having the air stream measured by flowmeter [1]

REFERENCES

[1] Simoiu D, "Theoretical and Experimental Studies on Mechanical Structures Loads Induced by Wind. Application on Parabolic Dish Antenna". PhD Thesis, University "Politehnica" of Timisoara 2008, ISBN 978-973-625-645-5. Politehnica Press.

[2] Cioara T, Gh., "On a Curve Fitting Algorithm for Modal Parameter Estimation ", Proc.6th Int. Modal Analysis Conference, Orlando, USA p.717-723, 1988

[3] Cioara T, Gh, "Vibration measurement", Lectures, electronic edition, (in Romanian)